T0329605

Interconnected Modern Multi-Energy Networks
and Intelligent Transportation Systems

Interconnected Modern Multi-Energy Networks and Intelligent Transportation Systems

Towards a Green Economy and Sustainable Development

Edited by

Mohammadreza Daneshvar
University of Tabriz
Iran

Behnam Mohammadi-Ivatloo
LUT University, Finland
University of Tabriz, Iran

Amjad Anvari-Moghaddam
Aalborg University
Denmark

Reza Razzaghi
Monash University
Australia

IEEE Press Series on Power and Energy Systems
Ganesh Kumar Venayagamoorthy, Series Editor

IEEE PRESS

WILEY

Published by John Wiley & Sons, Inc., Hoboken, New Jersey.
Published simultaneously in Canada.

For general information on our other products and services or for technical support, please contact our Customer Care Department within the United States at (800) 762-2974, outside the United States at (317) 572-3993 or fax (317) 572-4002.

Wiley also publishes its books in a variety of electronic formats. Some content that appears in print may not be available in electronic formats. For more information about Wiley products, visit our web site at www.wiley.com.

Library of Congress Cataloging-in-Publication Data

Names: Daneshvar, Mohammadreza, editor. | Mohammadi-Ivatloo, Behnam, editor. | Anvari-Moghaddam, Amjad, editor. | Razzaghi, Reza, author.
Title: Interconnected modern multi-energy networks and intelligent transportation systems : towards a green economy and sustainable development / Mohammadreza Daneshvar, Behnam Mohammadi-Ivatloo, Amjad Anvari-Moghaddam, Reza Razzaghi.
Description: Hoboken, New Jersey : Wiley, [2024] | Includes index.
Identifiers: LCCN 2023049029 (print) | LCCN 2023049030 (ebook) | ISBN 9781394188758 (hardback) | ISBN 9781394188765 (adobe pdf) | ISBN 9781394188772 (epub)
Subjects: LCSH: Intelligent transportation systems. | Renewable energy sources. | Sustainable development.
Classification: LCC TE228.3 .D346 2024 (print) | LCC TE228.3 (ebook) | DDC 625.7/94–dc23/eng/20240116
LC record available at https://lccn.loc.gov/2023049029
LC ebook record available at https://lccn.loc.gov/2023049030

Cover Design: Wiley
Cover Image: © TK

Set in 9.5/12.5pt STIXTwoText by Straive, Pondicherry, India

Contents

List of Contributors

Saman Ahmadi
School of Engineering
STEM College, RMIT University
Melbourne, Australia

Syed Muhammad Nawazish Ali
School of Engineering
STEM College, RMIT University
Melbourne, Australia

Ali Moradi Amani
School of Engineering
STEM College, RMIT University
Melbourne, Australia

Amjad Anvari-Moghaddam
Department of Energy (AAU Energy)
Aalborg University
Aalborg, Denmark

M. Imran Azim
Department of Electrical and
Computer Systems Engineering
Monash University
Melbourne, Victoria, Australia

Mariem Ahmed Baba
Engineering for Smart and Sustainable
Systems Research Center
Mohammadia School of Engineers
Mohammed V University in Rabat
Rabat, Morocco

Miraj Ahmed Bhuiyan
Faculty Member School of Economics
Guangdong University of Finance and
Economics
Guangzhou, China

Mohamed Cherkaoui
Engineering for Smart and Sustainable
Systems Research Center
Mohammadia School of Engineers
Mohammed V University in Rabat
Rabat, Morocco

Christos Chronis
Department of Informatics and
Telematics, Harokopio
University of Athens
Athens, Greece

Mohammadreza Daneshvar
Department of Electrical and
Computer Engineering
University of Tabriz
Tabriz, Iran

Rahman Dashti
Clinical-Laboratory Center of Power
System & Protection
Faculty of Intelligent Systems
Engineering and Data Science
Persian Gulf University
Bushehr, Iran

M. Edwin
Department of Mechanical
Engineering, University College of
Engineering, Nagercoil
Anna University Constituent College
Nagercoil, Tamilnadu, India

Georgios Ellinas
KIOS Research and Innovation
Center of Excellence
Department of Electrical and
Computer Engineering
University of Cyprus
Nicosia, Cyprus

M. C. Eniyan
Department of Mechanical
Engineering, University College
of Engineering
Nagercoil
Anna University Constituent College
Nagercoil, Tamilnadu, India

Reza Gharibi
Clinical-Laboratory Center of
Power System & Protection
Faculty of Intelligent Systems
Engineering and Data Science
Persian Gulf University
Bushehr, Iran

Hessam Golmohamadi
Department of Computer Science
Aalborg University
Aalborg, Denmark

Ankita Jain
Prestige Institute of Global
Management
Indore, Madhya Pradesh
India

Mahdi Jalili
School of Engineering
STEM College, RMIT University
Melbourne, Australia

Vikas Khare
School of Technology Management
and Engineering, NMIMS
Indore, Madhya Pradesh, India

Mohsen Khorasany
Department of Electrical and
Computer Systems Engineering
Monash University
Melbourne, Victoria
Australia

Yigit Cagatay Kuyu
R&D department, Karsan Otomotiv
Sanayi ve Tic., Bursa, Turkey

Michalis Mavrovouniotis
KIOS Research and Innovation
Center of Excellence
Department of Electrical and
Computer Engineering
University of Cyprus
Nicosia, Cyprus

G. Antony Miraculas
Department of Mechanical
Engineering, St. Xavier's Catholic
College of Engineering
Chunkankadai
Nagercoil, Tamilnadu, India

Behnam Mohammadi-Ivatloo
LUT University, Finland
University of Tabriz, Iran

M. Saranya Nair
School of Electronics Engineering,
Vellore Institute of Technology
Chennai, Tamilnadu, India

Mohamed Naoui
Research Unit of Energy Processes
Environment and Electrical Systems
National Engineering School of Gabes
University of Gabés
Gabés, Tunisia

Farzad H. Panahi
Department of Electronics and
Communication Engineering
University of Kurdistan
Sanandaj, Iran

Fereidoun H. Panahi
Department of Electronics and
Communication Engineering
University of Kurdistan
Sanandaj, Iran

Tania Panayiotou
KIOS Research and Innovation
Center of Excellence
Department of Electrical and
Computer Engineering
University of Cyprus
Nicosia, Cyprus

Reza Razzaghi
Department of Electrical and
Computer Systems Engineering
Monash University
Melbourne, Victoria
Australia

Samaneh Sadat Sajjadi
School of Engineering
STEM College, RMIT University
Melbourne, Australia

S. Joseph Sekhar
Department of Engineering
University of Technology and
Applied Sciences-Shinas
Al-Aqar, Oman

Konstantinos Tserpes
Department of Informatics
and Telematics
Harokopio University of Athens
Athens, Greece

Behrooz Vahidi
Department of Electrical Engineering
Amirkabir University of Technology
(Tehran Polytechnic)
Tehran, Iran

Iraklis Varlamis
Department of Informatics and
Telematics
Harokopio University of Athens
Athens, Greece

About the Editors

 **Mohammadreza Daneshvar**, PhD, is an Assistant Professor at the Faculty of Electrical and Computer Engineering, University of Tabriz, Tabriz, Iran. Prior to that he was a postdoctoral research fellow in the field of modern multi-energy networks at the Smart Energy Systems Lab of the University of Tabriz. He obtained his MSc and PhD in Electrical Power Engineering from the University of Tabriz, Tabriz, Iran, all with honors. He has (co)authored more than 50 technical journal and conference articles, 6 books, and 26 book chapters in the field. Dr. Daneshvar is a member of the Early Career Editorial Board of the Sustainable Cities and Society Journal and also serves as a guest editor for the *Sustainable Cities and Society*, and *Sustainable Energy Technologies and Assessments* journals. Moreover, he serves as an active reviewer with more than 45 top journals of the IEEE, Elsevier, Springer, Wiley, Taylor & Francis, and IOS Press, and was ranked among the top 1% of reviewers in Engineering and Cross-Field based on Publons global reviewer database. His research interests include Smart Grids, Transactive Energy, Energy Management, Renewable Energy Sources, Integrated Energy Systems, Grid Modernization, Electrical Energy Storage Systems, Microgrids, Energy Hubs, Machine Learning and Deep Learning, and Optimization Techniques.

Dr. Behnam Mohammadi-Ivatloo, currently holds the position of Professor specializing in sector coupling within power systems at LUT University in Finland. He began his academic journey at the University of Tabriz in 2012 as an Assistant Professor and was later elevated to the rank of Professor in 2019. He also has research experience at Aalborg University, Aalborg, Denmark, and the Institute for Sustainable Energy, Environment and Economy, University of Calgary, Canada. He obtained his MSc and PhD in electrical engineering from the Sharif University of Technology, Tehran, Iran. He has a mix of high-level experience in research, teaching, administration, and voluntary jobs at the national and international levels. He was PI or CO-PI in 20 externally funded research projects. He has been a Senior Member of IEEE since 2017 and a Member of the Governing Board of Iran Energy Association since 2013, where he was elected as President in 2019. His main areas of interest are integrated energy systems, renewable energies, energy storage systems, microgrid systems, and smart grids.

Amjad Anvari-Moghaddam (S'10 -M'14 -SM'17) received his PhD (Hons.) in Power Systems Engineering in 2015 from the University of Tehran, Tehran, Iran. Currently, he is an Associate Professor and Leader of the Intelligent Energy Systems and Flexible Markets (iGRIDS) Research Group at the Department of Energy (AAU Energy), Aalborg University where he is also acting as the Vice-Leader of Power Electronic Control, Reliability and System Optimization (PESYS) and the coordinator of Integrated Energy Systems Laboratory (IES-Lab). His research interests include planning, control, and operation management of microgrids, renewable/hybrid power systems, and integrated energy systems with appropriate market mechanisms. He has (co)authored more than 300 technical articles, 7 books, and 17 book chapters in the field. Dr. Anvari-Moghaddam is the Editor-in-Chief of *Academia Green Energy* journal and serves as an Associate Editor of several leading journals such as the *IEEE Transactions on Power Systems*, *IEEE Systems Journal*, *IEEE Open Access Journal of Power and Energy*, and *IEEE Power Engineering Letters*. He is the Chair of IEEE Denmark, a Member of IEC SC/8B- Working Group (WG3 & WG6) as well as Technical Committee Member of several IEEE PES/IES/PELS and CIGRE WGs. He was the recipient of 2020 and 2023 DUO–India and SPARC Fellowship Awards, DANIDA Research Fellowship grant from the Ministry of Foreign Affairs of Denmark in 2018 and 2021, IEEE-CS Outstanding Leadership Award 2018 (Halifax, Nova Scotia, Canada), and the 2017 IEEE-CS Outstanding Service Award (Exeter-UK).

Reza Razzaghi (Senior Member, IEEE) received his PhD in electrical engineering from the Swiss Federal Institute of Technology of Lausanne (EPFL), Lausanne, Switzerland in 2016. In 2017, he joined Monash University, Melbourne, Australia, where he is currently a Senior Lecturer with the Department of Electrical and Computer Systems Engineering and an Australian Research Council DECRA fellow. His research interests include distributed energy resources, power system protection, dynamics, and transients. He has been the recipient of the 2019 Best Paper Award from the IEEE Transactions on EMC and the 2013 Basil Papadias Best Paper Award from the IEEE PowerTech Conference.

Preface

Modern grids aim at enabling modern functionalities (e.g., self-healing systems, smart transportation systems, sustainable energy networks, and multi-dimensional community of intelligence agents) and integrating smart technologies in the body of multi-energy networks (MENs) aiming to supply more reliable and efficient energy. In this context, transportation networks have recently experienced significant growth in terms of vehicle systems, physical infrastructures, public transit system management, autonomous vehicles, traveler information systems, and traffic management, to name a few. In this area, emerging new intelligent and hybrid energy systems, along with the substantial development in energy conversion technologies, as well as increasing demand for secure, comfortable, and reliable vehicle systems declare the fact that modern grids need to include intelligent interconnected transportation systems. Herein, diverse and intelligent transportation devices play a pivotal role in meeting the growing demand for vehicle systems that can be operated in a collaborative manner to make the modern structure of the energy network more efficient, reliable, resilient, and stable. However, how intelligent transportation systems (ITSs) can be interconnectedly operated for maintaining the sustainability of cleaner modern MENs is a key question that needs to be addressed in deep detail. As reliable transportation services are critical for future modern energy grids, a great need is felt for sustainable intelligent interconnected transportation systems to support the system in realizing modern energy services goals under the cleaner and modern structure of energy networks. As a pioneering book that presents the fundamental technologies and solutions for real-world problems in the context of intelligent interconnected transportation systems, it covers a conceptual introduction to modern transportation systems, highlights potential technologies and vehicle systems in this area, and discusses requirements for coordinated exploitation of modern transportation systems. Moreover, the current book presents innovative ways for interconnecting ITSs to ensure the sustainability of modern MENs with a high/full share of renewable energy sources.

The current book consists of 15 chapters. Chapter 1 aims to inspect the necessity for modernizing the coupled structure of ITSs and MENs. Moreover, it clarifies the different applications of ITSs as well as the coupled structure of ITSs and MENs. Chapter 2 provides a review of the development of green transportation (GT). It also states the concept of GT, current transportation issues such as traffic congestion and greenhouse gas emissions, and the relationship between various GT components. Chapter 3 presents an overview of techno-economic-environmental approaches and the benefits of GT systems to emphasize their applicability in today's world. Further, green transports are assessed from both economic and environmental points of view in this chapter. Chapter 4 emphasizes on the necessity of sustainable transportation by highlighting the catastrophic effects of greenhouse gas emissions from conventional transportation on climate change and public health while discussing some potential approaches for mitigating these emissions. It also presents some key modes of sustainable transportation along with their benefits and existing cases around the globe as well as elaborates the sustainable transportation in modern urban advancement. Chapter 5 examines various aspects of multi-energy technologies in developing GT for global sustainability. It also investigates novel technologies to promote GT systems as well as describes limitations and challenges related to present travel needs that are impeding GT adoption. Chapter 6 surveys the flexibility potentials of power-To-X (P2X) plants including the power-to-hydrogen, -methane, -heat, -mobility, and -chemical systems. It also investigates the flexibility opportunities of electric vehicles and hydrogen fuel cell fleets along with the flexibility opportunities of electricity demands in residential, industrial, agricultural, and commercial sectors. Chapter 7 aims to review original research works about modeling, management, and intelligent controls of the multi-energy system (MES) integrating EV routing and charging. It also clarifies unavoidable interdependences between the energy and transport infrastructure in the MES. Chapter 8 provides a systematic review to cover key fundamentals that make vehicles autonomous and their applications in intelligent transportation networks. Chapter 9 focuses on the brushless motor's storage technologies and control systems in an electric vehicle. It also discusses the different types of batteries used in electric vehicles, generally made from lead acid, nickel, lithium metal, silver, and sodium-sulfur. Chapter 10 presents a review of the different aspects of EVs, including the design and control features of EVs, in a comprehensive way. It also shows the need for several technological developments to grow the EV market and create a pollution-free environment. Chapter 11 investigates the interconnection between energy networks and ITSs by presenting the management of electric vehicle charging in parking structures. It also discusses the main related challenges along with various optimization targets and technologies. Chapter 12 highlights the role of multi-energy management schemes in the sustainable development of interconnected ITS. It also examines the sustainability of interconnected ITS in the view of energy management and describes the challenges in its implementation. Chapter 13 presents a blockchain-based financial and economic analysis of green vehicles as a path toward the ITS. It also provides financial and economic analysis as well as an analysis of the speed and range of EVs by using statistical software. Chapter 14 proposes a consistent and cost-aware energy procurement framework for an unmanned aerial vehicle powered concurrently by laser beams emitted by locally deployed laser beam directors and local renewable energy sources. Finally, Chapter 15 explores the problem of the personalization of the autonomous driving experience, leveraging the existing advanced driving assistance systems through a combination of reinforcement learning algorithms and federated learning techniques.

As any research achievement may not be free of gaps, the Editors kindly welcome any suggestions and comments from the respectful readers for improving the quality of this work. The interested readers can share their valuable comments with the Editors via m.r.daneshvar95@gmail.com.

1

The Necessity for Modernizing the Coupled Structure of Intelligent Transportation Systems and Multi-Energy Networks

Mohammadreza Daneshvar¹, Amjad Anvari-Moghaddam², and Reza Razzaghi³

¹ Department of Electrical and Computer Engineering, University of Tabriz, Tabriz, Iran
² Department of Energy (AAU Energy), Aalborg University, Aalborg, Denmark
³ Department of Electrical and Computer Systems Engineering, Monash University, Melbourne, Victoria, Australia

1.1 Introduction

In recent years, the growing need for all carriers of energy has driven energy networks to be reconstructed in a way to effectively match energy supply and demand [1]. In this transformation, grid modernization is introduced to pave the realization of required changes that make the energy structure cleaner, more efficient, affordable, sustainable, flexible, stable, and secure than the previous paradigm [2]. One of the prominent features of modernized energy networks is the adoption of renewable energy sources (RESs) in energy generation premises [3]. This evolution is intended to facilitate the decarbonization plans for an energy transition toward a carbon-free and green energy structure. However, such a development was not only accompanied by economic and environmental benefits, but also critical challenges have emerged, especially concerning uncertain outputs of RESs [4]. Multi-energy systems along with other energy storage, management, and energy trading technologies are proposed to address such challenges in the modern energy grid [5]. However, effectively responding to such concerns requires more than just using the mentioned solutions. Indeed, dynamic energy balancing is more difficult from the network viewpoint when it is equipped with 100% RESs given the stochastic nature of energy production. In such a circumstance, interconnecting different sectors of energy grids can make an appropriate multi-energy coupling that supports the whole structure of the system in continuous energy serving. One of these important sectors is transportation. The transportation

Interconnected Modern Multi-Energy Networks and Intelligent Transportation Systems: Towards a Green Economy and Sustainable Development, First Edition. Edited by Mohammadreza Daneshvar, Behnam Mohammadi-Ivatloo, Amjad Anvari-Moghaddam, and Reza Razzaghi.
© 2024 The Institute of Electrical and Electronics Engineers, Inc.
Published 2024 by John Wiley & Sons, Inc.

network encompasses different parts such as power lines, air routes, railways, and road networks that possess diverse energy-consumed systems like electric vehicles (EVs), buses, boats, etc. [6]. By including various energy-dependent devices, the transportation network plays a vital role in energy interactions across the grid. As future energy networks are planned to be eco-friendly infrastructure with fully clean energy production [7], the transportation sector is also not exempt from this green economy movement. Hence, exploiting carbon-free energy systems is necessary for realizing a green transportation network. In this regard, recent advances in the technology of multi-carrier energy devices offer great opportunities for generating, storing, and converting different carriers of energy to each other [8]. Therefore, the operation of coupled intelligent transportation networks and multi-carrier energy units can procure appropriate conditions for implementing net-zero emission plans and realizing a green transportation network. This issue highlights the necessity for developing the application of multi-energy systems to their usage in the transportation network. This chapter aims to further clarify various dimensions of this necessity.

1.2 Applications of Intelligent Transportation Systems

Recent years have witnessed a considerable rise in energy demand along with the rapid development in the technologies of smart devices for smart grids. The deployment of intelligent systems is not only limited to the residential sector, but also the transportation premise has benefited in controlling a variety of energy interactions. Such evolutions have driven the system to face huge volumes of data generated by smart devices across the grid. This data is taken into account as valuable information for the decision-making process of the transportation network. How the mentioned information needs to be used for managing various processes as well as completing different duties is the main reason for creating transportation management systems (TMSs). The TMS is from the area of transportation that sets up for improving flexibility, load optimization, and effective route planning [9]. The utilization of machine learning (ML) techniques along with artificial intelligence (AI) has made TMSs more intelligent, enabling them to achieve accurate performance. The emergence of intelligent devices in recent decades has resulted in the development of diverse information systems for planning, mapping, routing, and logistics. The exploitation of such systems has significantly increased the capabilities of data processing leading to the appearance of intelligent transportation systems (ITSs) [10]. Indeed, ITS is a system with the ability to take appropriate decisions for different states of the transportation network using the data from vehicles with smart devices. The received data from the sensors and intelligent devices are monitored by ITSs to extract useful information that can be effectively used by businesses and governments for taking purposeful decisions. The used feedback mechanism enables ITSs to continuously improve the performance of different systems across the transportation network. ITSs consist of a set of subsystems to achieve such improvements that are indicated in Figure 1.1 [11].

As one of the ITSs' subsystems, intelligent public transportation systems are for controlling the public transportation network, maintaining the performance of the transportation structure, and providing up-to-date information regarding the network operation and trips for the decision-makers and passengers [12]. As a context-aware solution to effectively control and manage the traffic challenges of the transportation network, the intelligent traffic control and management system has been developed that uses real-time data coming from predictive analytics as well as connected road infrastructure [13]. The intelligent parking management system relies on satisfying the requirements of the Internet of things (IoT) device management, vehicle management, and user information management for

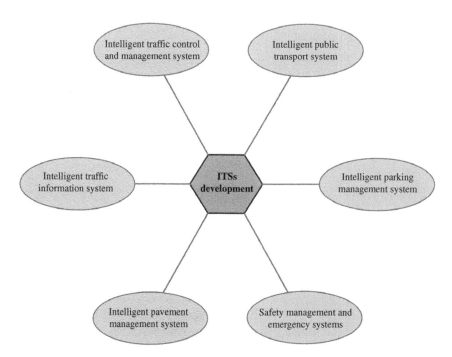

Figure 1.1 Different subsystems for the development of ITSs. *Source:* Adapted from [11].

managing vehicle parking [14]. The intelligent traffic information system is responsible for effectively monitoring traffic by using IoT to create interoperability among heterogeneous interconnected devices to avoid vehicle traffic congestion [15]. The application of IoT is not just limited to the mentioned ITSs' subsystems, but its integration with AI also procures appropriate platforms for safety management and emergency conditions. Given the significant role of pavement maintenance in megacities, the intelligent pavement management system pursues the key goal of scheduling road reviews as well as managing complaints [16]. Indeed, such systems can be most efficient in improving management capability, driving economic growth, and supporting the sustainability of the transportation network when accurate data is available from sensors. In this regard, different applications of ITSs can be classified into four main classes that are illustrated in Figure 1.2 [17].

All presented applications for ITSs in Figure 1.2 are based on using the collected data from vehicles aiming to improve the transportation process, facilitate public transportation services, and increase driver safety. Thus, ITSs not only ease the decision-making process for government authorities, but they can also manage and control planning for the transportation network in an appropriate manner, resulting in efficient road management, better driver experience, and a proper degree of passenger comfort [9].

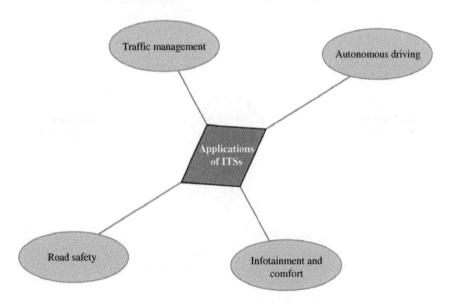

Figure 1.2 Main applications of ITSs. *Source:* Adapted from [17].

1.3 Coupled Structure of ITSs and Multi-Energy Networks

The described applications of ITSs highlight their undeniable place in the future energy network infrastructure. The rise in transport demand has substantially increased for different types of transportation services in recent years. In this respect, vehicles with traditional fossil fuels were mainly responsible for satisfying the transport demand in the transportation network. However, considerable advancements in the technology of clean energy systems have changed the mentioned trend in the usage of fossil fuel-based devices in the transportation sector. The environmental problems caused by traditional systems have driven the transportation sector to operate carbon-free devices in both personal vehicles and public transport services. Moreover, recent endeavors have been focused on significantly declining the carbon emissions in the transportation premise and actualizing green transportation more than ever before. However, rapid developments in energy technologies do not accept such a reduction degree and define ambitious goals in constructing zero-emission transportation networks. Therefore, the modernization of the transportation network is accompanied by realizing 100% RES goals in the grid modernization process. However, uncertainties in clean energy production by RESs have procured difficult conditions for making such decarbonization plans reliably implementable in the practice. Indeed, most clean energy production systems rely on climate-dependent systems whose outputs vary simultaneously with climate changes. Hence, incorporating RESs in the transportation sector donates the challenge of their uncontrollable energy generation, which is already a great concern for energy networks. In this circumstance, although ITSs can pave the successful adoption of diverse technologies for providing an acceptable degree of safety and services in the transportation network, reliable exploitation of the transportation sector still remains a considerable challenge when planning to realize zero-emission goals through full usage of RESs. Herein, given the great achievements of deploying multi-energy systems in boosting the ability of energy grids in coping with the intermittences of RESs, their utilization in the transportation network can open new doors for this sector to act most flexibly in making green transportation. Indeed, multi-energy networks offer a set of multi-energy interactions using different technologies such as power-to-X to enable renewable-dominant structures for supplying energy. The aforementioned benefits are a key portion of opportunities that the transportation grid with RESs requires to keep its sustainability in the green economy. Therefore, the considerable advantages of multi-energy systems in easing the adoption of a high level of RESs highlight the necessity of interconnected exploitation of the transportation network with multi-energy grids. Indeed, coupling them is an essential step for facilitating the implementation of decarbonization plans as well as realizing green transportation infrastructure.

1.4 Summary

Recent evolutions in technology development along with the rapid growth in multi-energy demand have affected different sectors across the globe. The transportation network has also not been spared from these transformations and has witnessed widespread changes in different layers of transportation management. One of the affected parts is ITSs which are developed in a way to improve synergies while promoting interoperability between various participants of the transportation network. In this respect, ITSs cannot satisfy the modernization expectations of transportation alone. On the other hand, multi-energy systems render attractive privileges for utilizing RESs and efficiently constructing green transportation. Thus, there is a great necessity for coupling ITSs and multi-energy networks to easily reach the grid modernization goals. This chapter highlights this necessity by mostly focusing on the grid modernization viewpoint. Moreover, it also clarified the different applications of ITSs in the transportation network.

References

1 Daneshvar, M., Mohammadi-Ivatloo, B., and Zare, K. (2023). An innovative transactive energy architecture for community microgrids in modern multi-carrier energy networks: a Chicago case study. *Scientific Reports* 13 (1): 1529.

2 Daneshvar, M., Mohammadi-Ivatloo, B., and Zare, K. (2022). A novel transactive energy trading model for modernizing energy hubs in the coupled heat and electricity network. *Journal of Cleaner Production* 344: 131024.

3 Daneshvar, M., Mohammadi-Ivatloo, B., and Zare, K. (2022). A fair risk-averse stochastic transactive energy model for 100% renewable multi-microgrids in the modern power and gas incorporated network. *IEEE Transactions on Smart Grid* 14: 1933–1945.

4 Daneshvar, M., Mohammadi-ivatloo, B., Asadi, S., and Galvani, S. (2020). Short term optimal hydro-thermal scheduling of the transmission system equipped with pumped storage in the competitive environment. *Majlesi Journal of Electrical Engineering* 14 (1): 77–84.

5 Huang, W., Zhang, X., Li, K. et al. (2021). Resilience oriented planning of urban multi-energy systems with generalized energy storage sources. *IEEE Transactions on Power Systems* 37 (4): 2906–2918.

6 Shao, C., Li, K., Qian, T. et al. (2022). Generalized user equilibrium for coordination of coupled power-transportation network. *IEEE Transactions on Smart Grid* 14: 2140–2151.

7 Daneshvar, M., Mohammadi-Ivatloo, B., Zare, K., and Anvari-Moghaddam, A. (2022). Transactive energy strategy for energy trading of proactive distribution company with renewable systems: a robust/stochastic hybrid technique. *e-Prime Advances in Electrical Engineering, Electronics and Energy* 2: 100028.

8 Kazemi, B., Kavousi-Fard, A., Dabbaghjamanesh, M., and Karimi, M. (2022). IoT-enabled operation of multi energy hubs considering electric vehicles and demand response. *IEEE Transactions on Intelligent Transportation Systems* 24: 2668–2676.

9 Iyer, L.S. (2021). AI enabled applications towards intelligent transportation. *Transportation Engineering* 5: 100083.

10 Zapata Cortes, J.A., Arango Serna, M.D., and Andres Gomez, R. (2013). Information systems applied to transport improvement. *Dyna* 80 (180): 77–86.

11 Agarwal, P.K., Gurjar, J., Agarwal, A.K., and Birla, R. (2015). Application of artificial intelligence for development of intelligent transport system in smart cities. *Journal of Traffic and Transportation Engineering* 1 (1): 20–30.

12 Elkosantini, S. and Darmoul, S. (2013). Intelligent public transportation systems: A review of architectures and enabling technologies. *2013 International Conference on Advanced Logistics and Transport*, Sousse, Tunisia, 233–238. IEEE.

13 (March 31, 2023). Intelligent traffic management systems: a lowdown of software & hardware components. https://intellias.com/intelligent-traffic-management/ (accessed 31 March 2023).

14 Lyu, L. and Fan, N. (2022). Research and design of intelligent parking management system based on UML technology. *2022 IEEE Conference on Telecommunications, Optics and Computer Science (TOCS)*, Dalian, China, 919–922. IEEE.

15 Al-Sakran, H.O. (2015). Intelligent traffic information system based on integration of Internet of Things and Agent technology. *International Journal of Advanced Computer Science and Applications (IJACSA)* 6 (2): 37–43.

16 Moradi, M. and Assaf, G.J. (2023). Designing and building an intelligent pavement management system for urban road networks. *Sustainability* 15 (2): 1157.

17 Ben Hamida, E., Noura, H., and Znaidi, W. (2015). Security of cooperative intelligent transport systems: Standards, threats analysis and cryptographic countermeasures. *Electronics* 4 (3): 380–423.

2

Green Transportation Systems

Reza Gharibi[1], Behrooz Vahidi[2], and Rahman Dashti[1]

[1] Clinical-Laboratory Center of Power System & Protection, Faculty of Intelligent Systems Engineering and Data Science, Persian Gulf University, Bushehr, Iran
[2] Department of Electrical Engineering, Amirkabir University of Technology (Tehran Polytechnic), Tehran, Iran

2.1 Introduction

2.1.1 Motivation and Problem Description

The most significant issue facing contemporary society is population growth. The demand to consume various resources grows even more as the population grows. For stability and forward movement, the world has to guarantee dependable and stable access to its basic requirements. At first glance, food, energy, and water are the three basic needs that are thought to be necessary for a society. The environment must be destroyed in order to fulfill these systemic needs.

The society and organizations that make policies give these three demands a lot of consideration. One of society's fundamental needs is social welfare, although it receives less attention. One of the main pillars of social welfare is transportation. Unreasonable population expansion, inadequate planning for the construction of roads, and an unchecked increase in the number of automobiles have all caused issues for human society [1].

The development of technology opened up a new area for the planning and redesign of conventional systems [2]. Redesign and special look at the changes in traditional systems to save energy consumption were applied and implemented in large networks such as electricity. In the power system, the coordination between different energy carriers led to significant economic benefits [3]. The transportation system made a move toward redesign to enhance its conditions as one of the most significant energy consumers and generators of greenhouse gases. Along

Interconnected Modern Multi-Energy Networks and Intelligent Transportation Systems: Towards a Green Economy and Sustainable Development, First Edition. Edited by Mohammadreza Daneshvar, Behnam Mohammadi-Ivatloo, Amjad Anvari-Moghaddam, and Reza Razzaghi.

with energy and greenhouse gas emissions issues, the traditional transportation system struggles with car congestion due to its inadequate infrastructure. Green transportation (GT) was proposed as a solution to these issues and as the cornerstone of the new transportation system, notwithstanding these issues and the fundamental requirement to rethink the traditional transportation system [4]. The following objectives and problems have been addressed in this chapter, which is centered on the GT system:

- Discussing the development and expansion of the transportation system.
- Population growth and the issues it brings are studied statistically.
- Vehicle traffic as a consequence of the community's growing population and people's time wastage.
- Pollution of the environment and greenhouse gas emissions from the transportation system.
- Effects and advantages of GT versus traditional transportation systems.
- Issues that prevent the growth of GT.
- Effect of new technologies on GT's acceleration.
- Intelligent transportation systems and methods for connection between various components.

2.1.2 Literature Review

One of the biggest problems brought on by the use of traditional transport systems is global warming. Without changing the current system's structure, every attempt to reduce transportation density will have a negative impact on both the general satisfaction of society and the transportation industry [5]. In Ref. [6], it is investigated how increasing the usage of fossil fuels puts both the global environment and global society at irreversible harm. It is recommended that the global economy shift toward carbon reduction so that the rise in global temperature does not exceed 2°. There are serious challenges in reducing carbon with the goal of lowering temperatures. The use of renewable energy sources is one of these difficulties. In order to overcome these obstacles, there must be broad international cooperation. As a result, the United Nations member states decided to work together in this direction [7].

The Green Climate Fund was formed in 2010 to help finance this objective. The big businesses then started to reduce their carbon emissions. Businesses are encouraged to invest in this area, and transportation companies strive to act in a less carbon-intensive manner, thanks to the assistance of the international community and the adoption of legislation [8]. The ultimate purpose of the transportation system is to make it easier for people to go about their normal, everyday lives without wasting time. Another major objective of better transportation is to move commodities more efficiently while having less of a negative impact on the environment [9]. In light of these circumstances, the world community has set sustainability in transportation as a goal. This means finding ways to meet societal requirements while also protecting the environment and future generations' access to resources [10]. Several governments have modified their policies to solve this problem and are working to achieve sustainable transportation [11]. Despite government attempts to promote sustainable transportation, the use of conventional and fossil fuels and the resulting greenhouse gas emissions have created an unstable environment [12]. One of the key causes of declining societal health and environmental destruction is the emission of greenhouse gases [13]. The use of clean fuels and the electrification of cars rather than the use of fossil fuels in the transportation sector can work to address these concerns and problems [14, 15]. The use of sustainable transport was recognized, and suggestions were made specifically for public transportation [16]. There are still some unknowns despite all the recommendations and alternative solutions to the transportation issues. Researchers recognized the urgent need to develop a GT system after assessing the severity of the system's issues. A system of GT is one in which all modes of transportation cause the least amount of damage to the environment and to people's health [4]. These concepts and topics are also covered in this chapter:

1) Investigated the development of transportation from the invention of the wheel to modern green and smart transportation.
2) Problems brought on by population growth, such as the escalation of traffic and the contamination of the environment.
3) Agreements and treaties established in relation to reducing greenhouse gas emissions.
4) Challenges that need to be overcome to implement GT.
5) Technology advancements that are practical as well as new are speeding up the development of greener transportation.

2.1.3 Chapter Organization

The remainder of this chapter first reviews and presents the history of the transportation system from the age of the invention of the wheel to the present, and then lists the issues and difficulties that the current transportation faces as a result of population increase. After reviewing the international accords that have been put forth to address the issues of GT and global warming, the GT system is described and its difficulties are looked at. Multiple energy networks and GT's effects are investigated. The methods for implementing GT are offered after the definition of the GT system. The newest technology utilized by the GT system is discussed and presented at the end.

2.2 History of Transportation

Studying the history and current state of transportation systems is crucial for researching and evaluating various vehicle types. The welfare of society is considerably increased by transportation systems. The movement of people and products both vertically and horizontally may be referred to as transportation. When it comes to vertical transportation, equipment is utilized for lifting tasks like removing objects from atop structures, metal from mines, and other things. The Shadow apparatus, which was used in Mesopotamia to lift water, is one of the historical examples of a vertical transportation system [17–20].

The invention of the first wheel in 3500 BC can be regarded as the start of horizontal transportation [21]. Around 2000 BC, small chariots developed in Syria or northern Mesopotamia and were the first horizontal wheeled means of transportation, which swiftly swept throughout the Middle East [22]. Although it was primarily utilized for military purposes, this kind of chariot was also used for travel and hunting. Around 1550 BC, chariot use spread throughout Greece as a result of trade and migration. Chariot use led to the development of roadways, which were primarily tied to the building of bridges across rivers and streams [22]. Horses were utilized to pull chariots throughout this time period, and they were regarded as the most crucial component of the transportation system [3].

Up until the year 1400 AD, when Europeans invented the horse-drawn wagon, chariots were still in use [23]. James Watt's invention of the steam engine in 1769 sparked a massive revolution in both industry and daily life [24]. Boats and steamships were among the first modes of transportation to make significant strides using Watt steam engines in 1783 [25]. Locomotives were the first land vehicles to benefit from steam engines. The locomotive was unveiled in 1801, and its "Salamanca" commercial counterpart was conceived and constructed in 1812 [25]. A more advanced version of this locomotive with eight wagons was shown two years later. Thirty tons of coal could be moved by this locomotive at a speed of 4 miles per hour [25]. The first air transport method was the hot air balloon, which was invented and developed in 1783 [26].

The invention of the bicycle in the 1790s marked the beginning of the personal vehicle age [25]. In 1886, the first motorized vehicle was created, and people started using them [27]. A century or so later, in 1980, a more advanced model of the automobile without a driver was invented [28]. Parallel to this progress, motorized airplanes first flew in 1903; electric subways began operating in 1890; cruise ships appeared in 2002; and lastly, from 2010 onward, electric automobiles have significantly increased horizontal transportation and impacted people's lives [29]. Figure 2.1 depicts the development and growth of transportation based on the history as described.

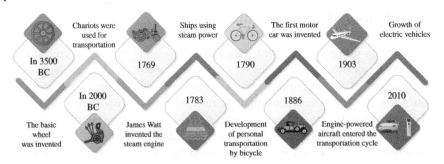

Figure 2.1 Development of the transportation system across time.

2.3 Transportation Expansion Issues

There are several issues with the traditional transportation system. These issues are caused by an increase in the number of vehicles, a lack of infrastructure, and facilities within the system, as well as factors relating to the local environment and society. The spread of urbanization and the rise in population are just two problems that have a negative impact on the traditional transportation system. The transportation system has also imposed problems on the environment, such as the rise in traffic and the release of greenhouse gases. Below, these concerns and issues are discussed in more detail.

2.3.1 Urbanization's Growth

Around the world, urban areas are more populous than rural ones. In 2018, 55% of people worldwide lived in urban settings, according to United Nations figures. Urbanization is expected to increase from 30% of the global population in 1950 to 68% of the global population by 2050 [30]. According to figures from 2018, North America now has the greatest rate of urbanization, with 82% of the population living in cities. Europe has 74%, Oceania has 68%, and Latin America and the Caribbean have 81%. Asia has now reached a 50% urbanization rate [30].

While only 43% of Africans live in cities, the continent remains predominantly rural. Since 1950, the world's rural population has slowly increased, and it is anticipated that it will soon hit an all-time high before starting to drop. The number of rural people worldwide is predicted to decline from its present level of about 3.4 billion to about 3.1 billion in 2050. Nearly 90% of the world's rural population lives in Africa and Asia. With 893 million people, India has the most people living in rural areas, followed by China with 578 million. Since 1950, the global urban population has increased significantly, from 751 million to 4.2 billion people in 2018. This is in contrast to the rate at which the rural population is growing. Asia now has 54% of the world's urban population while being less urbanized than other continents. With 13% apiece, Europe and Africa are next [30].

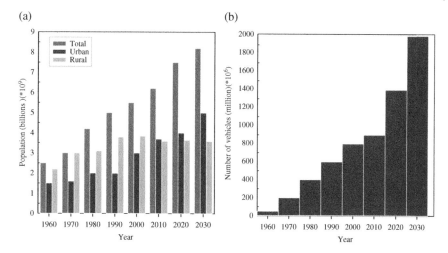

Figure 2.2 (a) Population increase in urban and rural areas and (b) the increasing number of vehicles worldwide.

The economic structure has changed as urbanization has increased. According to the data provided in Ref. [30], the gross domestic product (GDP) of agriculture has declined from 27% to 10%, and the GDP of industry has climbed from 32% to 43% as a result of the rise in urbanization demand [31]. The total number of registered automobiles has climbed from 200 million in 1960 to 1430 million worldwide in 2018 due to the growing population and the rising desire of the younger generation for a personal vehicle [32]. Three hundred sixty-three million commercial cars and 1.06 billion private vehicles make up the 1.42 billion vehicles now in use globally. Figure 2.2 depicts the rise in both population and vehicles [30].

The three key problems of social, economic, and environmental satisfaction are intimately tied to urban expansion and the rise in cars. Despite population expansion, well-managed urbanization can still offer a number of advantages if long-term population patterns are taken into account. Planning and management can simultaneously reduce environmental damage and other negative impacts brought on by an increase in city dwellers [30, 33]. The idea of GT is one approach to doing this.

2.3.2 Traffic Growth

One of the most fundamental necessities of humans and industry in modern life, with the development of industry and technology, is quick mobility. As a result, members of society employ a variety of modes of transportation, including cars, trains, and bicycles. However, due to their practicality and ease, personal automobiles continue to be the most popular form of transportation out of all these options. As a result, given the assumption that the population will continue to rise, more private and public cars will be present in major cities. The issue is that as the number of cars increases, there has not been a matching growth in the creation and extension of the transportation infrastructure. Traffic congestion will consequently become a serious problem.

It is true that more traffic and congestion may result from industrial expansion or other factors, and while they may help many people economically, they will also cause more issues for society as a whole. Public transportation will suffer from overcrowding, and the depreciation of public vehicles will grow. The society is shifting toward the usage of personal automobiles, which is increasing traffic in metropolitan areas. Increasing the number of personal vehicles on the road would undoubtedly increase the dangers for motorcyclists, cyclists, and pedestrians. Additionally, people without access to a car will have a harder time taking advantage of urban living's advantages and will have a lesser quality of life, a situation that will only get worse as the population ages. There are additional disadvantages to traffic in every society. One of these disadvantages is that every member of society wastes time, which is one of the barriers to the progress of society.

Figure 2.3 displays the amount of lost community hours for various cities throughout the world over the course of a year. It is vital to remember that these

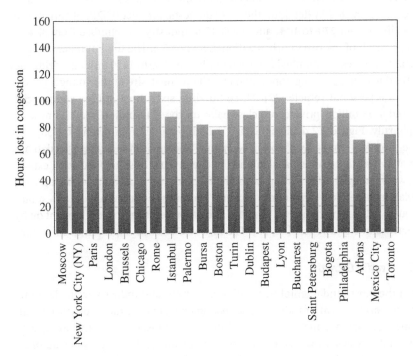

Figure 2.3 Average amount of time lost by people in some cities worldwide due to traffic.

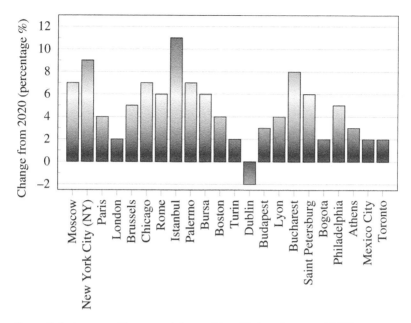

Figure 2.4 In some cities around the world, traffic has risen as of 2020.

hours vary depending on each city's population and geography. Figure 2.4 depicts the percentage growth in traffic from 2020 in several places throughout the world [34].

2.3.3 Environmental Issues

The transportation services industry played a bigger role in people's daily lives as a result of population growth. Additionally, the transportation industry serves as an essential and undeniable infrastructure for the growth of any society's economy and industry, transporting millions of tons of cargo and a large number of passengers every day. It was anticipated that advancements in technology and the growth of the virtual world would lessen the need for physical transmission; however, despite the virtual world's expansion, the emergence of new opportunities, and society's encouragement to make more transfers, not only was the global dependence on this industry not reduced, but it also experienced rapid growth. Despite the necessity of transportation for the global community, the use of fossil fuels in the system has made it a threat to the environment and the health of society. Emissions of CO_2, CO, volatile organic compounds, nitrogen, and sulfur oxides, as well

Table 2.1 Statistics on the emissions of greenhouse gases in some global cities.

City	PM_{10}	$PM_{2.5}$	SO_2	NO_2	$GHH(t_{co2,co,...})$
Johannesburg	35	76	22	12	9.92
Cape Town	65	48	4	13	7.6
Sao Paulo	59	134	1	15	1.4
Mexico City	131	190	34	35	5.53
London	25	60	9	31	9.6
Delhi	240	243	14	35	1.5
Kolkata	138	197	33	9	1.1
Mumbai	155	117	19	41	1.2
Shanghai	60	102	17	49	12.9
Beijing	86	149	22	34	10.8
Seoul	55	107	7	71	4.1
Sydney	17	66	8	20	20.3

as fine particles, are primarily to blame for this threat (heavy metals, micro- and nanoplastics, etc.) [35].

Table 2.1 lists the emissions of CO_2, SOx, suspended particles PM_{10}, and $PM_{2.5}$ in various cities around the world [1]. There are many causes of PM_{10} and $PM_{2.5}$ emissions, including improper fuel combustion and tire and brake wear. In addition to the importance of transportation for the world community, the transportation sector and its related services have grown to be one of the biggest consumers of petroleum products globally, contributing significantly to the production of airborne pollutants. Approximately, 16% of all greenhouse gas emissions are attributed to the transportation system. 23% of CO_2 emissions from the combustion of fossil fuels globally are related to the transportation sector. From 1970 to 2007, transportation-related CO_2 emissions rose by 45%; by 2030, they are predicted to rise by about 40% [36].

For example, Figure 2.5 depicts the amount of CO_2 gas produced in the United States in 2020, which is equal to 5981 million tons, and 27% of the total CO_2 production is accounted for by the transportation system [37].

Despite all the advancements in air, sea, and rail transportation, road transportation is much more popular due to its lower cost, high flexibility, and quick response time. Road transportation accounts for 30–40% of all transportation-related CO_2 emissions, so finding and implementing a solution to this issue in society is only natural. The "greening" of road transportation has been suggested as a solution to this issue [37].

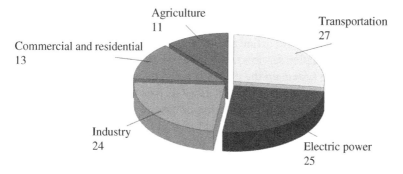

Figure 2.5 Different sectors in the United States that produce greenhouse gases.

2.4 Definition of Green Transportation

In order to address the transportation-related difficulties, the UN established a high-level advisory group in 2016 to offer guidance and address these problems. At the first World Summit on Sustainable Transport in 2016, this group's initial report was released [30]. Only 50% of metropolitan areas have trouble-free access to urban transportation, according to a report released by the UN in 2020 [38]. Informal transportation networks, which frequently lack safety measures, have grown in the remaining 50% of regions.

The worldwide community's travel habits have significantly changed as a result of the COVID-19 pandemic outbreak, which has led to an increase in the usage of personal transportation. Despite these modifications to the transportation system and the effect these services have on societal problems related to the economy, society, and the environment (Figure 2.6), technological and economic advancement should

Figure 2.6 Various impacts of the transportation system on society.

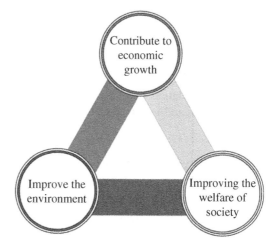

be taken into account to enhance this significant and influential system in people's lives. Due to these factors, scientists and researchers recognized the urgency of the situation and made an effort to define and present a GT system.

The idea of "transportation services that have a less detrimental impact than the current transportation system on the health of individuals in society and the surrounding environment" can be used to describe GT. As a vast field of technology, GT encompasses things like the effective use of electric vehicle technology, the more effective use of fossil and conventional fuels, and the use of novel fuels like hydrogen and biogas for private and public transportation [1].

2.5 Advantages of Green Transportation

The primary objective of GT is to plan transportation-related activities so that there is the least amount of money spent and the least amount of environmental harm. Before the idea of GT and the disregard for the environment, objectives were solely determined by economic factors. Environmental factors are taken into account alongside economic costs due to growing environmental concerns. Different benefits are produced in comparison to the conventional transportation system by implementing the GT concept [1, 39].

One of GT's most significant advantages is its ability to reduce the number of cars in urban regions. The danger of traffic accidents and collisions between cars can be reduced by fewer cars on the road and the use of new technologies. The sustainability of current energy sources is improved by using new energies as well as by planning and optimizing the transportation network. Reducing greenhouse gas emissions, including CO_2, is one of GT's main objectives. GT drastically decreases the use of automobiles while increasing the use of new technologies, renewable energy sources, and other forms of transportation. These acts contribute to reducing greenhouse gas emissions [39, 40].

Traffic is decreased as a consequence of the transit system being optimized as a result of GT, which decreases community members' wasted time and raises their satisfaction. People in the community and all sectors engaged in the transportation system can reduce costs by reducing traffic, enhancing vehicle fuel efficiency, optimizing transportation, and utilizing technology [39]. Figure 2.7 displays some of the advantages and results of GT.

2.6 International Agreements

The global community decided to take action and work toward removing these gases from the environment due to the environmental issues caused by the emission of greenhouse gases. This decision was made with an eye toward the future

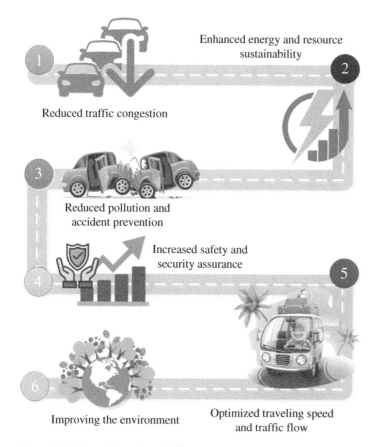

Enhanced energy and resource sustainability

Reduced traffic congestion

Reduced pollution and accident prevention

Increased safety and security assurance

Improving the environment

Optimized traveling speed and traffic flow

Figure 2.7 Several benefits of GT.

and taking into account the growth of these emissions. The European Union made a decision and published the European Commission's vision in a document known as a "White Paper" in 2011, focusing on the coordination of transportation policies as one of the most significant factors in the production of greenhouse gases [1].

In this document, it is stated that as transportation increased and fossil fuels were used more frequently, oil became more scarce. As a result, the price of oil would increase, which would have an impact on the economy and lower social satisfaction. Additionally, it was decided to keep temperature variations under 2° because of the environment and because it was stated in this document. It is necessary to reduce greenhouse gas emissions by 80–90% from 1990 levels until 2050 in order to meet this objective [1].

In these situations, oil fuel still accounts for 96% of EU transportation. In accordance with this document, the use of conventionally fueled vehicles will decrease by half by 2030 and end entirely in the major urban areas by 2050. Additionally, the

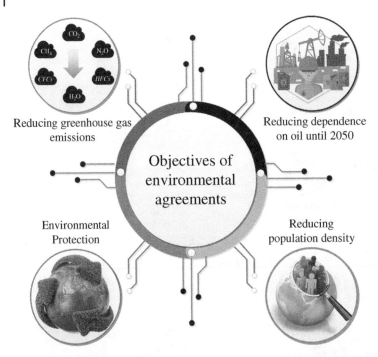

Figure 2.8 The goals of global agreements to protect the environment. *Source:* Mopic / Adobestock.

Paris Agreement was signed in 2016 to limit global temperature rise to 2° and protect the earth's climate [41]. Figure 2.8 illustrates these agreements' general objectives.

2.7 Challenges to GT

Due to its close ties to the personal and social lives of society's members and the institutionalization of this need in their lives since ancient times, the transportation system encounters significant resistance to changes. Society refuses change for two primary reasons. The first is the society's habit to an extremely antiquated and institutionalized system in the specifics of their lives, and the second is the economic issues and laws. These two factors together account for the majority of the resistance from the people of the society. It is expensive to switch from one system to another at first, and doing so necessitates amending the relevant laws and regulations, which will meet resistance from the general populace.

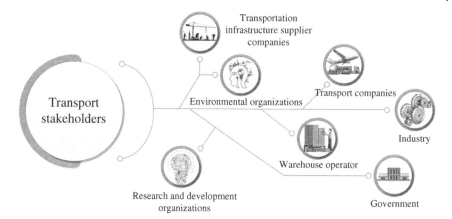

Figure 2.9 Profitable parts of the transport network. *Source:* Transport companies - FrankBoston / Adobe Stock; Industry - Alex Mit / Adobe Stock.

In general, when taking into account the challenges, all the stakeholders in the transportation system should be taken into account.

The companies that provide the infrastructure for the transportation system, environmental and environmental organizations, transportation companies, businesses, research organizations, and most importantly, society's citizens, are all examples of stakeholders. These stakeholders' actions and responses to the adjustments and difficulties they encounter vary from one another. The challenges and obstacles are broken down into five categories and each category is expressed separately in order to determine the best course of action for GT while also satisfying and controlling all stakeholders (Figure 2.9).

2.7.1 Institutional Challenges

Politicians and legislators have always faced a fundamental problem when their views diverged on a particular issue. This issue is a fundamental one for managers of smaller organizations as well as for politicians in order to achieve the same objectives within the organization. Due to its size and the presence of numerous stakeholders, the transportation system is struggling with the issue of legislation. The existence of geographic and political borders, as well as various legislative bodies in various nations, is the cause of this issue. For these reasons, creating coordination to bear the costs of creating a GT system can become a big challenge.

The societal adoption of new technologies and fuels is another significant issue that needs to be addressed through legislation. Society must understand how new technology affects the economy and personal comfort in order to be motivated to use it. The information on these technologies' effects on the economy and social life is scarce because of how recent they are. Customers and society are therefore not persuaded to purchase a vehicle based on the use of novel and environmentally friendly fuels.

2.7.2 Regulatory Challenges and Barriers

Numerous changes in the social and economic structure of the country have been brought about by population growth and urbanization. The expansion of the transportation system and its significant impact on people's lives are the results of these changes. The planning of governments and the management of countries is based on the idea that the development of each populated area should be balanced; however, as the population grows and moves to urban areas, there are more jobs available in the service sector. Urban areas become more densely populated as a result. This results in various issues, lowers social welfare, and obstructs city managers' ability to make sound policies. There is still competition among the conventional vehicles using traditional fuels in the transportation sector, despite the introduction of new vehicles using cleaner fuel and also as a result of investors' reluctance to invest in the new generation of vehicles. Applying a higher tax cannot be a good way to encourage society to use these fuel-efficient vehicles because it deviates from the intended goal and harms domestic transportation competition [42].

With the implementation of current policies, there is also the concern of a loss of national competitiveness because of the possibility for public unrest brought on by the capacity overload of the transportation and urban infrastructure due to population and vehicle growth. Despite these problems, policymakers must find a way to interact and communicate with the public because policies cannot be implemented without the support of society and because achieving the goals of GT requires widespread acceptance by society and the public.

2.7.3 Technology-related Barriers

The reduction of greenhouse gas emissions is one of the essential pillars of the execution of the idea of GT. The employment of technology is one of the most practical and efficient methods out of all the tools available for this aim. Technology can be used to manage automobiles completely and comprehensively while simultaneously lowering the degree of greenhouse gas emissions [43].

Even with all the positive aspects and advantages of the new system's use of technology, there are drawbacks. The majority of these issues originate from how challenging it is to integrate and adapt the conventional system to new technologies. A major problem is adapting to and having access to technology in many countries and communities. Because less developed countries have less access to technology than wealthy nations do, it will take time and significant advancement to bring the idea of GT to the level that is desired.

If traditional fuels are used, the concept of GT will encounter several issues due to the growth in people and the ensuing increase in automobiles. Under these circumstances, the use of fuels should shift toward high-quality fuels, and efforts should be made to improve the fuel's quality and environmental compatibility by implementing new technology. However, it is not an easy process to replace new fuels and automobiles that are compatible with these fuels. The large number of traditional cars using fossil fuels is one of the issues that lie ahead. The development of associated fuel station technology, as well as vehicle maintenance technology, is necessary to replace modern technology with sustainable fuel as a replacement for many traditional automobiles. The general satisfaction of society is yet another problem with technological advancement. The development of technology must be followed by the observation of its effects on people's lives throughout time in order for it to result in their satisfaction.

In light of this, one of the challenges facing emerging eco-friendly transportation technology is public satisfaction. To show how technology improves people's lives and to hasten the adoption of new technologies, policymakers should offer solutions.

2.7.4 Financial Barriers

Depending on the country, different infrastructure is required to promote GT. As a result, each country should use a distinct approach when building the infrastructure required for GT. In developed countries, the necessary infrastructure is in good condition and does not require significant investment to get them ready for the introduction of GT. Numerous infrastructures in less developed countries are in poor shape, and it costs a lot of money to expand and improve them. Following are some of the infrastructures that need to be modified and ready for use with GT:

- Road
- Transport companies
- Vehicle control technology
- Green fuel station
- Data communication centers between vehicles
- GT-based management and planning agencies for urban transportation

GT has a variety of objectives, some of which are global in scope, and it will succeed in achieving those objectives if the international community supports it. Reducing greenhouse gas emissions and protecting the environment are two of these objectives. These objectives necessitate a global strategy for infrastructure provision and development in less-developed nations. Large investment firms should be encouraged to invest in these nations by these policies [43].

2.7.5 General Admission

When purchasing and using vehicles, the majority of society's members consider factors like fuel costs, vehicle dependability, car prices, and vehicle safety. Less attention is paid to environmental issues like greenhouse gas emissions. The traditional mindset of people is also not very conducive when it comes to public transportation, which has a special place in the concept of GT.

According to the ingrained social stereotype, public transportation is primarily used by schoolchildren, the elderly, persons of low socioeconomic status, etc., and is therefore less trustworthy and independent than a personal vehicle [42]. For these reasons, simply replacing the existing system with a new one without any policy or planning is not recommended. Instead, the public should be won over by enhancing the public transportation system's level of quality, dependability, and organization. Additionally, because they worry about society's acceptance, investors and transport businesses in the investment sector are less interested in reducing carbon emissions. By creating regulations and favorable conditions, policymakers should encourage these enterprises to transform and invest [44] (Figure 2.10).

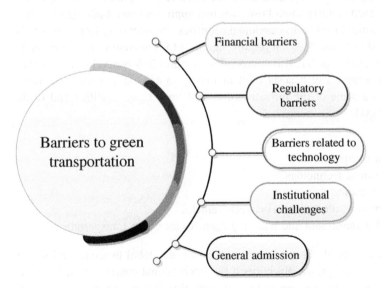

Figure 2.10 GT's challenges and barriers.

2.8 Green Transportation's Effects on Multi-Energy Networks

According to forecasts, the expansion of multi-carrier and hybrid networks will result in a favorable and profitable increase in natural gas consumption in the coming years. Multi-carrier networks have started to grow significantly in recent years [45]. In contrast to independent energy systems, multi-carrier systems exhibit integrated and interacting behaviors among various energy carriers. As an illustration, CHPs, which burn natural gas to generate both heat and electricity, increase the reliance on both thermal and electric energy [46]. In multi-carrier systems, several energy carrier quantities are combined with each other in order to supply various customers while pursuing various optimization goals. In comparison to single-carrier and independent systems, one of the most significant benefits of multi-carrier systems is the decrease in greenhouse gas emissions [45].

The GT system and its components are regarded as one of the largest consumers of multi-carrier networks due to the expansion of the consumption of various energy carriers by new vehicles, such as electricity, natural gas, hydrogen, etc. The transport system's continuity and interconnection to multi-carrier networks can be looked at from two distinct viewpoints. The dependency between multi-carrier energy network optimization and transport network optimization is the first viewpoint. The performance and optimization of multi-carrier energy networks can be improved by the coordination and optimization of various GT system components, which use energy supplied by the multi-carrier energy network [47].

The second viewpoint focuses on the transfer of energy between various transportation system components and multiple energy networks. Several storage mechanisms are used in modern vehicles. Modern vehicles use a variety of storage systems such as batteries and hydrogen storage devices. These vehicles can be employed as resources to optimize multi-carrier networks when necessary, which will lead to their stability and enhancement [48]. A multi-energy network and its connection to vehicles using various fuels are depicted in Figure 2.11 as an example.

2.9 Implementation Strategies for the Green Transportation System

The need for more infrastructure, such as roads and automobiles, as a result of population growth is one of the most significant issues with traditional transportation. This point of view has its own economic and social issues, and each of the

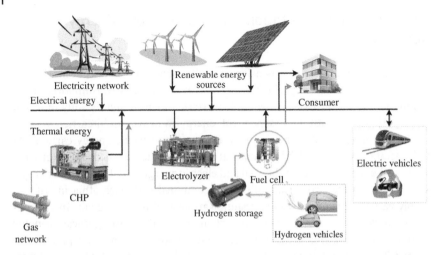

Figure 2.11 Examples of various types of vehicles and multi-carrier energy networks. *Source:* Renewable energy – pngimg.com; Gas networks – Cambeezy/cleanpng.

old infrastructures can only be expanded to a certain point in terms of size. Due to environmental issues, this old perspective and system cannot be expanded upon or maintained.

In 2004 (GIZ), a novel strategy concentrating on the demand side of transportation for large works was outlined and suggested as a means of overcoming the traditional transportation system. The following three fundamental ideas form the basis of this strategy [49]:

- Avoid or reduce
- Change
- Improvement

The first tenet of this strategy, known as "Avoid" asserts that, to the greatest extent possible, urban planning should be done such that the efficacy of transportation is maximized, mobility is decreased, and the basic requirements of the community are met regionally.

"Change" focuses on enhancing transportation by attempting to increase the community's use of nonmotorized modes of transportation like walking and bicycling. Public transportation should thus take precedence in society, followed by the usage of personal vehicles only under rare conditions.

Fuel economy and the optimization of transportation system components are the main topics of the third "Improvement" principle. By emphasizing the need to utilize fewer fossil fuels, this principle also paves the way for the creation of clean fuels like electricity and hydrogen. This stage also includes attempting to build the infrastructure for these fuels, including their stations and the vehicles that consume them.

City management can use and put into practice these concepts on the transportation system in one of two ways: through an attraction policy or a fine. Different transportation sectors execute these policies in different ways. For instance, in the public sector, the more regular, reliable, and comfortable they are, the more likely it is that people will choose them as their preferred mode of transportation. This will then make control and monitoring easier for city managers [31].

The presented approach is executed and applied differently in various countries. The amount of CO_2 gas is what causes the variation in implementation. The objective is to minimize gases, especially CO_2, in underdeveloped countries where the standard of vehicles is low and the level of greenhouse gas emissions is high. The objective shifts to lowering the concentration and eventually removing CO_2 in developed countries where infrastructure and vehicles have attained a good standard level [50] (Figure 2.12).

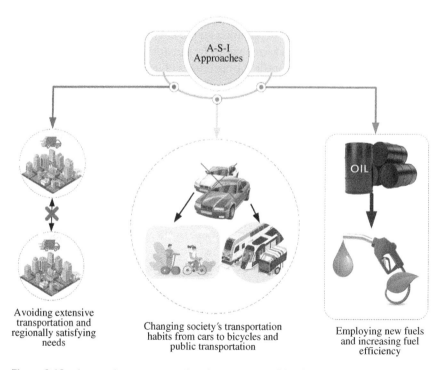

Figure 2.12 Approaches to overcoming the encountered barriers include avoid-shift-improve (A-S-I). *Source:* Employing new fuels and increasing fuel efficiency - AKS / Adobe Stock.

2.9.1 Actions Performed to Promote Green Transportation

The National Environmental Policy Act, passed in 1970, is one of the steps made to enhance the environmental behavior of transportation [51]. The Clean Air Law was then approved and put into effect. The law requiring vehicle adaptation to greenhouse gas emission laws was developed at the same time and put into effect in 1990. Every vehicle must abide by the specifications of these laws so that its greenhouse gas emissions do not go above a predetermined threshold. These regulations stipulate that cars must possess a pollutant level certificate, which must be updated and renewed every three months. This has two extremely practical benefits: The first is that the overall level of CO_2 and other greenhouse gas emissions is managed. The second benefit is that it raises public awareness of environmental contamination and the significance of environmental issues in society.

2.10 New Technologies for Green Transportation

As the idea of GT has grown and environmental laws and treaties have been adopted to reduce the impact of pollution on the environment, the rise in greenhouse gas emissions has been accompanied by penalties and legal pressure. These fines decrease revenue and impede the operations of transportation businesses. In this regard, in order to sustain both their revenue and client pleasure, transportation businesses are compelled to alter how they utilize current facilities. The technology significantly lowers greenhouse gas emissions, providing transportation businesses with a way to comply with and uphold environmental laws and regulations. Additionally, companies' productivity and profits have increased, and client satisfaction levels have risen as well [13].

2.10.1 Energy Technology

In society, it is typical to drive conventional cars with internal combustion engines. Internal combustion engines propel the vehicle by converting the chemical energy found in fossil fuels into mechanical energy through combustion. After internal combustion engines, transportation technology moved toward the use of batteries and fuel cells. However, using batteries is not always preferable to using internal combustion engines. When using batteries, characteristics like storage capacity, lifespan, usability, safety, and cost-effectiveness should be taken into account. Lithium-ion batteries are one type of battery that can support these features [52].

One of the effective ways to increase energy use is to use hybrid technologies. Reducing the losses brought on by combustion engines' internal friction is another way to improve. Around 15% of losses are brought on by friction, which can be greatly reduced by using new lubricants. These lubricants are the focus of ongoing laboratory and research work because they significantly improve energy efficiency. Greenhouse gas emissions are decreased by lowering losses and raising energy efficiency.

Another advancement and step toward environmentally friendly transportation is fuel technology. The role of fuel in internal combustion engines is crucial. The use of hybrid fuel technology can significantly increase vehicle efficiency. Gaseous fuels like hydrogen and natural gas, alcohol-based fuels like methanol and ethanol, as well as vegetable fuels and waste, alternative fuels that are environmentally friendly, are among the fuels that can be used as alternative or hybrid fuels in vehicles. By lowering greenhouse gas emissions, these fuels can be a very good replacement for traditional fossil fuels without affecting the performance of combustion engines and newer models of cars [1] (Figure 2.13).

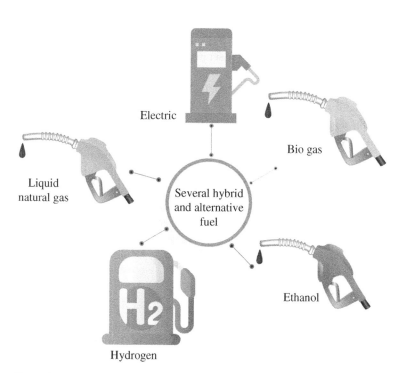

Figure 2.13 Some examples of fuels that can be used in green vehicles, including alternative and hybrid fuels.

2.10.2 Environmentally Friendly Technologies

In addition to the fuel issue, other features of cars that fit the concept of GT also contribute to lowering the emissions. Among these features are the use of environmentally friendly raw materials and their safe recycling, the reduction of the vehicle's weight to lower energy consumption, and the advancement of the vehicle's batteries to maximize its performance. In addition to improving green recycling, using environmentally friendly materials in the manufacture of cars lowers greenhouse gas emissions in auto factories.

Following the use of sustainable fuels to reduce emissions, the consumption of these fuels should be decreased in order to make them more economically viable and encourage society to adopt these technologies. The weight of the car affects the amount of fuel used. According to a general rule, you can reduce fuel consumption by 7% by losing 10% of the weight of the car [53].

2.10.2.1 Greener Tires

In the discussion of GT, tires need special consideration because they are one of the most crucial car parts and are closely related to environmental issues. The GT's tires have two distinct functions. First, they ought to have a feature that makes cars more dependable and increases passenger safety. Second, because of wear on the road and factory conditions, tires must be produced in a way that maximizes environmental compatibility. Vulcanization is a process used to make rubber tires durable and useful. Rubber is vulcanized while being exposed to sulfur. This approach can be abandoned to lessen carbon emissions.

Growing live moss inside tires is one of the suggested solutions for the future of tires. These mosses will work by soaking up carbon dioxide from vehicles and water from the roads. Additionally, these mosses will help to maintain tire pressure. The expression of this solution promises to persuade researchers and businesses to change tires in order to improve the environment, though it should be noted that this is just a proposed solution and is very far from being a finished product [54].

2.10.2.2 Reusing Energy

Beyond cutting down on fuel use and switching to sustainable fuels, the GT system takes a holistic approach to energy. One of the key components of GT is the reuse of consumed energy, which stores a portion of the kinetic energy produced by fuel consumption for use during the next acceleration of the vehicle. Various storage devices, including batteries and flywheels, are used to perform this recovery.

2.11 Intelligent Transportation System

The expression of the GT concept is heavily reliant on both environmental factors and technical advancement. The traditional transportation system has changed significantly in recent years as a result of technological advancements such as the development of self-driving and intelligent vehicles and new sensors. It became possible to apply smart programs with the development of technology and its penetration into transportation systems, as well as other new systems like robotic systems. The following are examples of elements of the transportation system that will develop and become more intelligent as technology advances:

- Road management and vehicle volume control
- Car information management
- Passenger information
- Traffic management
- Management of emergency vehicles

The integration of information across the many transportation sectors could offer a number of advantages. Its first and most significant benefit is a reduction in the pollution produced by vehicle movement. One of the most significant objectives of GT is pollution reduction, which may be accomplished by making it more intelligent. Other benefits of positive changes include a decrease in the number of cars and traffic, a decrease in accidents, and an increase in the dependability and safety of inhabitants [55].

Numerous concepts and improvements were developed in them as a result of the smart platform's integration into the transportation system and the growth of electric vehicles. The following can be included as some of the ITS-related fields and developments:

- Creating smart roads: The primary motivation for expressing and putting this notion into practice is to be aware of the number of vehicles on the road. It can be quite helpful to use this data for central planning and optimization, as well as to advise drivers about the status of the route and help them choose a route.
- Satellite traffic control: With this technology, information can be exchanged between ground stations and satellites to intelligently monitor and regulate traffic on the roads.
- Self-driving vehicles: With the development of electric vehicles, as well as advancements in software and robotics, self-driving vehicles can play a significant part in environmentally friendly transportation. If there is road infrastructure, these vehicles can use the internal computer to make the safest and most cost-effective choice.

- Vehicles with biometric technology: Compared to conventional security measures, using biometric technologies to lock and secure the vehicle is far more effective. You can utilize fingerprints or eye scans with this technology.
- Remote tracking: Using location-based technologies, the location of the vehicle can be traced. Insurance and social service expenses can also be intelligently calculated based on the distance traveled.
- Monitoring the driver's and passengers' health actively: The expansion of in-car sensors and the presence of a communication network connecting the vehicle to management centers allow for real-time monitoring of passenger health and the intelligent dispatch of emergency assistance in the event of an emergency.

2.11.1 Vehicle Communication in Intelligent Transportation

Communication between components is a key topic when it comes to intelligent transportation. The Internet of Things is one of the platforms suggested and offered for communication in the transportation industry. New channels of communication with the outside world and with other cars are made available via the Internet of Things. This communication has improved the efficiency of automobiles and environmentally friendly transportation. Here are three examples of intelligent transportation vehicle communication [2] (Figure 2.14).

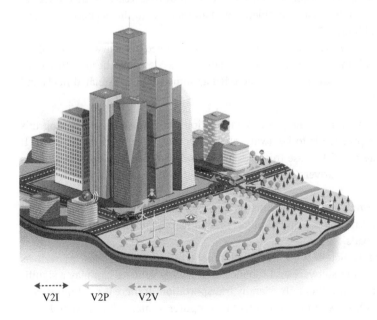

V2I   V2P   V2V

Figure 2.14 Examples of communication in GT.

- Vehicles communicating with infrastructure (V2I): This kind of communication serves as a coordinator. Transportation management carries out its duties for traffic control and environmental monitoring in this type of communication by gathering information about the surroundings as well as overall system information. In these communications, information about location, traffic alerts, accident warnings, and safety warnings is sent.
- Vehicle-to-vehicle (V2V) communication: Every vehicle, including cars, motorbikes, and new vehicles, can communicate with one another and exchange information through this link. This promotes the safety of transportation by transferring the sensor data from the vehicles to one another and determining details like the distance between them and their blind spots.
- Pedestrian-to-vehicle (V2P) communication: In this regard, electronic and smart gadgets that are linked to communication platforms are used by pedestrians to interact with moving vehicles. The benefit of this method of communication is that the pedestrian can update himself on the circumstances and facts surrounding him and make the best choice for his course of action and individual strategy. For instance, he can decide on the best course of action while selecting the mode of transportation, such as public or private, or his own route.

2.12 Conclusion

The necessity for the current transportation system to transform into a sustainable system is brought on by the world's population's continued development as well as the rise in urban population density. The idea of GT was developed in response to this demand and is one of the suggestions for the sustainability of this industry. In order to further explore this concept, the chapter discussed the evolution of the transportation system from the discovery of the wheel in 3500 BC to the development of the automobile and personal electronic devices. After describing the history of the transportation system, problems that have arisen owing to the conventional nature of the current system were discussed. Issues like traffic levels and the number of hours that each person wastes in traffic in some places across the world were statistically studied. One of the issues is the transportation system's emission of greenhouse gases, which was statistically investigated. After stating the existing problems, the concept of GT was defined and established as an idea and solution to overcome the problems, and the advantages of this system and the green concept were stated and examined. The advantages of this system and the green concept were discussed after stating the current issues and defining and establishing the concept of GT as an idea and a means of resolving them.

In order to express the idea of GT and address environmental threats, the international community's cooperation is necessary. The treaties that were approved and put into effect to lessen environmental threats were repeatedly mentioned after introducing the concept of GT. The Paris Agreement and the European Commission's White Paper were two of the agreements that were stated and investigated. The complexity of the system and its significant impact on society's day-to-day activities make the implementation of GT challenging. The following subject in this section's continuation deals with the difficulties and challenges associated with implementing GT. We can list organizational problems, financial problems, regulatory problems, the difficulty of societal acceptance, and technology-related improvements as some of the issues that were looked at and discussed. It was discussed and explained after stating the issues with the A-S-I solution that is suggested for the step-by-step implementation of this concept. The next issue that was looked at was the steps that have been taken so far to establish GT around the world. One of the key pillars in the development of the idea of transportation is the availability of technology. Fuel and energy technology, which was investigated and covered in the remainder of this part, is one of the crucial technologies that must be employed to realize the objectives of GT and make progress every day. The infrastructures needed to achieve intelligent transportation were among the topics covered in the intelligentization segment. Intelligent transportation heavily relies on the ability of the system's various components to communicate with one another in order to accomplish its desired goals. Three different V2V, V2I, and V2P communication types were studied and addressed at the conclusion of this chapter. The subjects discussed make it clear that the current transportation system needs to be changed in order to be environmentally friendly and prevent damage to the environment, and that GT is one of the best solutions. The main topics and reviewed results addressed in this chapter are summarized in the following list:

- History of transportation: Due to investigations, transportation began with the development of the wheel in 3500 BC, advanced with the development of steam engines, and today, with the introduction of electric cars, has transitioned into GT.
- Issues of transportation expansion: By 2018, there were 4.2 billion people living in urban areas, which led to an increase in traffic and the number of vehicles there. As a result, these vehicles generate greenhouse gases.
- GT advantages: It was discovered that the concept of GT has several advantages, like lowering greenhouse gas emissions, reducing traffic, increasing security, etc.
- International agreements: As global warming and environmental pollution increased, the world community came up with regulations and agreements to address the issue. One of these was the European Union's white paper.

- GT implementation challenges: There are a number of challenges and impediments that must be overcome, including technological, financial, institutional, and regulatory issues.
- A-S-I is one of the methods for achieving the concept of GT, which entails three components: avoidance, change, and improvement. It has been thoroughly examined and explained.
- New technologies: A review and discussion of new technologies, such as clean fuels, and new vehicle ideas, such as green tires, was conducted.
- Intelligent transportation system: The examination of intelligent transportation and the areas in which it is being used, as well as various communication techniques in intelligent transportation, were examined and debated.

References

1 Shah, K.J., Pan, S.-Y., Lee, I. et al. (2021). Green transportation for sustainability: review of current barriers, strategies, and innovative technologies. *Journal of Cleaner Production* 326: 129392.

2 Khan, W.U., Ihsan, A., Nguyen, T.N. et al. (2022). NOMA-enabled backscatter communications for green transportation in automotive-Industry 5.0. *IEEE Transactions on Industrial Informatics* 18: 7862–7874.

3 Gharibi, R. and Vahidi, B. (2022). Coordinated planning of thermal and electrical networks. In: *Coordinated Operation and Planning of Modern Heat and Electricity Incorporated Networks* (ed. M. Daneshvar, B. Mohammadi-Ivatloo, and K. Zare), 449–479. Wiley.

4 Björklund, M. (2011). Influence from the business environment on environmental purchasing – drivers and hinders of purchasing green transportation services. *Journal of Purchasing and Supply Management* 17 (1): 11–22.

5 Stern, N. (2013). The structure of economic modeling of the potential impacts of climate change: grafting gross underestimation of risk onto already narrow science models. *Journal of Economic Literature* 51 (3): 838–859.

6 Elie, L., Granier, C., and Rigot, S. (2021). The different types of renewable energy finance: a bibliometric analysis. *Energy Economics* 93: 104997.

7 Zhang, D., Zhang, Z., and Managi, S. (2019). A bibliometric analysis on green finance: current status, development, and future directions. *Finance Research Letters* 29: 425–430.

8 Yang, H., Shao, E., Gong, Y., and Guan, X. (2021). Decision-making for green supply chain considering fairness concern based on trade credit. *IEEE Access* 9: 67684–67695.

9 Sun, D., Zeng, S., Lin, H. et al. (2019). Can transportation infrastructure pave a green way? A city-level examination in China. *Journal of Cleaner Production* 226: 669–678.

10 Brundtland, G.H. and Khalid, M. (1987). *Our Common Future*. Oxford, GB: Oxford University Press.

11 Whitelegg, J. and Haq, G. (2003). *The Earthscan Reader on World Transport Policy and Practice*. London: Earthscan.

12 Van Fan, Y., Perry, S., Klemeš, J.J., and Lee, C.T. (2018). A review on air emissions assessment: Transportation. *Journal of Cleaner Production* 194: 673–684.

13 Van Fan, Y. and Klemeš, J.J. (2019). Emission pinch analysis for regional transportation planning: stagewise approach. *2019 4th International Conference on Smart and Sustainable Technologies (SpliTech)*, Split, Croatia (18-21 June 2019), 1–6. IEEE.

14 Gaede, J. and Meadowcroft, J. (2016). A question of authenticity: status quo bias and the International Energy Agency's World Energy Outlook. *Journal of Environmental Policy & Planning* 18 (5): 608–627.

15 Raymand, F., Ahmadi, P., and Mashayekhi, S. (2021). Evaluating a light duty vehicle fleet against climate change mitigation targets under different scenarios up to 2050 on a national level. *Energy Policy* 149: 111942.

16 Santos, G. and Nikolaev, N. (2021). Mobility as a service and public transport: a rapid literature review and the case of Moovit. *Sustainability* 13 (7): 3666.

17 Sacks, K.S. (2014). *Diodorus Siculus and the First Century*. Princeton University Press.

18 Rossi, C. (2014). *Some Examples of the Hellenistic Surprising Knowledge: Its Possible Origin from the East and Its Influence on Later Arab and European Engineers*, 61–84. Bologna: Patron.

19 Rossi, C., Russo, F., and Russo, F. (2009). *Ancient Engineers & Inventions*. Springer.

20 Feraudi, B. (1958). Meccanica tecnica: Dinamica delle macchine. Apparecchi di sollevamento e trasporto. 9. Hoepli.

21 Gambino, M. (2009). A salute to the wheel. smithsonianmag.com/science-nature/a-salute-to-the-wheel-31805121/ (accessed 29 May 2021).

22 Rossi, C., Chondros, T.G., Milidonis, K.F. et al. (2016). Ancient road transport devices: developments from the Bronze Age to the Roman Empire. *Frontiers of Mechanical Engineering* 11: 12–25.

23 www.world4.eu (2020). Coaches and carriages in 16th and 17th century. https://world4.eu/coaches-carriages (accessed 5 January 2021).

24 www.britannica.com (2019). Who-was-James-Watt. https://www.britannica.com/question/Who-was-James-Watt (accessed 15 January 2023).

25 Nguyen, T.C. (2020). The history of transportation. https://www.thoughtco.com/history-of-transportation-4067885 (accessed 29 May 2021).

26 B. T. B. B. M. a. Library (2017). The first gas balloon flight. http://www.bbml.org.uk/first-gas-balloon-flight/ (accessed 30 November 2019).

27 www.timetoast.com (2013). The history of cars. http://timetoast.com/timelines/ the-history-of-cars (accessed 5 April 2021).

28 Reilly, M. (2016). The self-driving van was born. https://www.technologyreview. com/2016/11/08/107226/in-the-1980s-the-self-driving-van-was-born (accessed 29 May 2021).

29 www.en.wikipedia.org (2002). Timeline of transportation technology. https://en. wikipedia.org/wiki/Timeline_of_transportation_technology (accessed 29 May 2021).

30 UN (United Nation) (2018). World urbanization prospects 2018. Department of Economic and Social Affairs Population Division. http://population.un.org/wup/ Publications/Files/WUP2018-Highlights.pdf (accessed 19 January 2021).

31 Roser, M. (2018). Future population growth. https://ourworldindata.org/future-population-growth (accessed 13 March 2018).

32 Davis, S.C. and Boundy, R.G. (2018). Transportation energy data book: edition 38.2. Energy and Transportation Science Division, Oak Ridge National Laboratory. https:// tedb.ornl.gov/wp-content/uploads/2022/03/TEDB_Ed_40.pdf (accessed 18 January 2021).

33 U. U. N. E. a. S. C. f. A. a. Pacific (2012). Low carbon green growth roadmap for Asia and the Pacific. Turning resource constraints and the climate crisis into economic growth opportunities. www.unescap.org/sites/default/files/Full-report.pdf (accessed 19 November 2018).

34 inrix.com (2022). Traffic growth. https://inrix.com/scorecard/ (accessed 5 December 2022).

35 Van Fan, Y., Klemes, J.J., Perry, S., and Lee, C.T. (2018). An emissions analysis for environmentally sustainable freight transportation modes: distance and capacity. *Chemical Engineering Transactions* 70: 505–510.

36 Leipzig, I. (2010). Reducing transport greenhouse gas emissions: trends & data. Background for the 2010 International Transport Forum, Berlin.

37 www.epa.gov (2022). Greenhouse-gas-emissions. https://www.epa.gov/ ghgemissions/sources-greenhouse-gas-emissions (accessed 15 December 2022).

38 W. W. C. Report (2020). The value of sustainable urbanization. https://unhabitat. org/sites/default/files/2020/10/wcr_2020_report.pdf (accessed 30 May 2021).

39 Saroha, R. (2014). Green logistics & its significance in modern day systems. *International Review of Applied Engineering Research* 4 (1): 89–92.

40 Larina, I.V., Larin, A.N., Kiriliuk, O., and Ingaldi, M. (2021). Green logistics-modern transportation process technology. *Production Engineering Archives* 27 (3): 184–190.

41 N. N. R. D. Council (2017). The Paris agreement on climate change. www.nrdc.org/ sites/default/files/paris-agreement-climate-change-2017-ib.pdf (accessed 15 November 2018).

42 Browne, D., Caulfield, B., and O'Mahony, M. (2011). Barriers to sustainable transport in Ireland, CCRP report, Environmental Protection Agency. http://www. tara.tcd.ie/handle/2262/93272 (accessed 23 January 2021).

43 Li, J., Pan, S.-Y., Kim, H. et al. (2015). Building green supply chains in eco-industrial parks towards a green economy: barriers and strategies. *Journal of Environmental Management* 162: 158–170.

44 Geels, F.W. (2002). Technological transitions as evolutionary reconfiguration processes: a multi-level perspective and a case-study. *Research Policy* 31 (8–9): 1257–1274.

45 Mancarella, P. (2014). MES (multi-energy systems): An overview of concepts and evaluation models. *Energy* 65: 1–17.

46 Zhang, X., Shahidehpour, M., Alabdulwahab, A., and Abusorrah, A. (2015). Optimal expansion planning of energy hub with multiple energy infrastructures. *IEEE Transactions on Smart Grid* 6 (5): 2302–2311.

47 Wu, Y., Wu, Y., Guerrero, J.M., and Vasquez, J.C. (2021). A comprehensive overview of framework for developing sustainable energy internet: from things-based energy network to services-based management system. *Renewable and Sustainable Energy Reviews* 150: 111409.

48 Shi, R., Li, S., Zhang, P., and Lee, K.Y. (2020). Integration of renewable energy sources and electric vehicles in V2G network with adjustable robust optimization. *Renewable Energy* 153: 1067–1080.

49 GIZ (The Deutsche Gesellschaft für Internationale Zusammenarbeit) (2018). Urban transport and energy efficiency. https://www.giz.de/en/downloads_els/Roadmap_Sumseec.pdf (accessed 11 November 2018).

50 EUEuropa (European Commission-Europa) (2011). Future transport fuels report. https:/ec.europa.eu/transport/sites/transport/files/themes/urban/cts/doc/2011-01-25-future-transport-fuels-report.pdf (accessed 12 May 2018).

51 Caldwell, L.K. (February 1999). *The National Environmental Policy Act: An Agenda for the Future*. Indiana University Press www.iupress.indiana.edu/product_info.php?products_id=19877 (accessed 15 August 2020).

52 Axsen, J., Burke, A., and Kurani, K.S. (2008). Batteries for plug-in hybrid electric vehicles (PHEVs): Goals and the state of technology circa 2008.

53 Lambert, F. (2018) BMW i3 gets a 100 kWh battery pack for 435 miles of range as a proof-of-concept by Lion smart. http://electrek.co/2018/09/07/bmw-i3-100-kwh-battery-pack-lion-smart/ (accessed 6 December 2020).

54 www.greenjournal.co.uk (2019). Green-car-technologies-of-the-future. https://www.greenjournal.co.uk/2019/08/green-car-technologies-of-the-future (accessed 15 December 2022).

55 Maimaris, A. and Papageorgiou, G. (2016). A review of intelligent transportation systems from a communications technology perspective. *2016 IEEE 19th International Conference on Intelligent Transportation Systems (ITSC)*, Rio de Janeiro, Brazil (1–4 November 2016), 54–59. IEEE.

3

Techno-Economic-Environmental Assessment of Green Transportation Systems

M. Imran Azim, Mohsen Khorasany, and Reza Razzaghi

Department of Electrical and Computer Systems Engineering, Monash University, Melbourne, Victoria, Australia

3.1 Introduction

Transportation system sustainability is one of the pressing concerns for modern urbanization in major cities across the world. Recent statistics published by the International Organization of Motor Vehicle Manufacturers demonstrate that nearly 1.06 billion personal/public vehicles and 363 million commercial vehicles are currently in use worldwide [1]. More than 90% of these vehicles entirely rely on fuels for the transmission system. Such huge fuel utilization actually contributes to around 60% of the total global fuel consumption [2] and labels vehicles as preeminent contributors to air pollutant emissions of carbon dioxide, nitrogen and sulfur oxides, volatile organic compounds (e.g., cyanide dioxide formaldehyde, and benzene), and other fine particulate matters (e.g., carbon black, micro- and nanoplastics, and heavy metals) [3]. For instance, in 2016, close to 8 gigatons of carbon dioxide were emitted by the vehicles globally, and the figure was around 71% higher than that of 1991. In fact, a further increase is likely to be observed in the upcoming years as the number of vehicles is expected to grow [4].

Given this context, governing bodies aim to plan for more energy- and climate-efficient transportation [5]. This has led to the introduction of green transportation systems aspiring to provide modern technology-facilitated vehicle services with fewer detrimental impacts on the immediate environment and human health and higher energy ratings contrasting with existing means of transport [6]. The general objectives of the policies undertaken to familiarize green transportation are to reduce air pollution and fuel dependency by at least 60% and

Interconnected Modern Multi-Energy Networks and Intelligent Transportation Systems: Towards a Green Economy and Sustainable Development, First Edition. Edited by Mohammadreza Daneshvar, Behnam Mohammadi-Ivatloo, Amjad Anvari-Moghaddam, and Reza Razzaghi.
© 2024 The Institute of Electrical and Electronics Engineers, Inc.
Published 2024 by John Wiley & Sons, Inc.

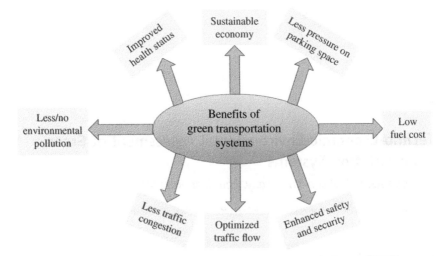

Figure 3.1 An overview of benefits offered by green transportation systems.

80%, respectively, by 2050 and to introduce green freight corridors efficiently limiting congestion [7]. Green transportation systems can contribute toward optimized traffic flow and traveling speed, enhanced security and safety, and offer extensive benefits to our economic budgets, environment, and health [8]. Some noteworthy beneficial aspects are captured in Figure 3.1 and demonstrated as follows [2]:

- *Cutting down fuel costs:* Embracing green transportation systems can save us a significant amount of costs related to fuel purchases at fuel stations.
- *Building a sustainable economy:* Manufacturing green transportation systems can create a substantial number of jobs in the transportation sector. Thus, socioeconomic disparities can be minimized, and a sustainable economy can be built.
- *Decreasing pressure on parking spaces:* Using green transportation systems with multiple occupant features can facilitate overpopulated workplaces and corporations to getting relief from parking problems.
- *Decreasing traffic congestion:* Adopting green transportation systems with multiple occupant features can also assist in reducing traffic congestion.
- *Less environmental pollution:* Shifting to green transportation systems can help us lower or completely get rid of toxic gas emissions into our atmosphere, resulting in fewer to no environmental pollutants.
- *Improving human health status:* Adapting green transportation systems can also facilitate us to avoiding cancer, bronchitis, type 2 diabetes, high blood pressure, and cardiovascular and respiratory diseases caused by toxic gas emissions and improve our health status.

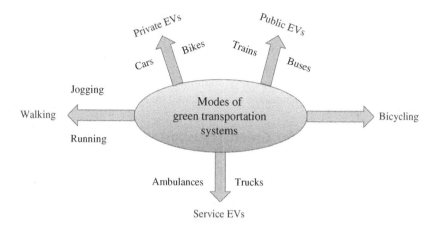

Figure 3.2 An overview of popular modes of green transportation systems.

Figure 3.2 depicts examples of some popular modes of green transportation systems, such as walking to destinations, bicycles, and electric vehicles (EVs) [2]. Walking (could be jogging or running) to work, institutions, shopping, and so on is not only free from any greenhouse gas emissions but also a good form of body exercise. Bicycling is also a sort of workout (along with being free from any greenhouse gas emissions) but faster than walking. On top of that, the purchasing and maintenance costs of bicycles are much lower than those of other vehicles. EVs are primarily powered by multi-carrier energy technologies, such as solar, geothermal, hydroelectric, battery, and so forth. Hence, less or no toxic emissions are caused by these vehicles. At present, three types of EVs are commonly used worldwide including private, service, and public EVs. Hybrid cars and electric bikes (motorcycles and scooters) are notable examples of private EVs. Examples of some service and freight EVs include hybrid ambulances, trucks, and lorries. On the other hand, some prominent examples of public EVs are hybrid trains, buses, and boats. Hydrogen fuel cell-driven vehicles are another type of green transport, which run on hydrogen and emit only water vapors, but their wide adoption has not been as popular as EVs due to significant production costs [9].

However, a number of barriers are also associated with the adoption of green transportation systems, such as regulatory barriers in terms of policy-making, institutional barriers (e.g., political issues to materialize undertaken policies, outdated infrastructure – road and freight networks, and taxation related to vehicle purchase and registration), and public support to prioritize environmental sustainability [10]. As such, it is important to articulate how available technologies can be employed effectively to attain the inherent benefits of green transportation systems. To this end, this chapter focuses on an overview of the techno-economic-environmental approaches and benefits of green transportation systems to help the readers understand their applicability in today's world. To do so, the following contributions are made:

- An overview of modern technologies, that can be utilized for green transportation systems, is demonstrated.
- Further, both the financial viability and environmental-friendly operation of green transportation systems are assessed based on some real-world data.

3.2 Technologies for Green Transportation Systems

Several modern technologies are currently available to develop and operate green transportation systems. They include eco-friendly, energy-efficient, energy reusing, intelligent system, integrated management, and distributed ledger technologies. The benefits, statistical facts, and current applicability of these technologies are outlined in this section. Also, a succinct overview is summarized in Table 3.1.

Table 3.1 Overview of features of different technologies for green transportation systems.

Technologies	Overview of features	References
Eco-friendly and energy efficient	• To develop, optimize, and operate transportation in eco-friendly and energy-efficient ways. • To manage acceleration and deceleration of vehicles by reusing energy.	[8, 11–17]
Intelligent system	• To incorporate telecommunication, sensor system, positioning system, computing hardware, data processing, planning processing, and virtual operation features to improve the operational comfort and safety of modern transportation.	[8, 18–23]
Integrated system	• To integrate patterns of transport growth and systematic use of available lands, emission rates of toxic gases, advanced traffic management, and adequate infrastructure management. • To achieve five major goals, including baseline review, target setting, political commitment, implementation and monitoring, and evaluation and reporting.	[8, 24]
Distributed ledger	• To contract, execute, and store transactions and billings in a decentralized way with confidentiality, privacy, and security. • To allow transport owners to track and monitor their charging and discharging transactions in an automated way. • To incorporate digital currency, designated as a token, to introduce an electronic billing system in the transportation sector.	[25–32]

3.2.1 Eco-Friendly and Energy-Efficient Technologies

An eco-friendly technology can optimize vehicle performance by utilizing eco-friendly materials for production and less rolling resistance, scaling down vehicle weight, and advancing driving attributes, for instance, transmission, braking, acceleration, and cornering [11]. The use of eco-friendly materials in vehicle construction products can result in fewer greenhouse gas emissions. Also, the reduction in a vehicle's weight can diminish fuel consumption. It is reported in Ref. [8] that approximately 5–7% of fuel usage can be decreased by a 10% minimization in vehicle weight. The BMW i3 – with an underfloor battery pack and an optional range-extending fuel engine to power the car – is a glaring example in which reinforced carbon fiber is applied in vehicle construction instead of steel to lower the vehicle weight by 30–50% contrasting with steel. The weight of the battery pack is also counterbalanced by the light body of the BMW i3 [12].

A solar impulse technology is a type of eco-friendly technology, in which solar energy is converted into electrical energy by photovoltaic (PV) cells to drive vehicles. This technology (designated as solar impulse-2's lightweight construction) left its mark on the world in 2015 by building the first-ever solar-driven round-the-world flight [13]. Besides, the first-ever self-charging solar-powered cargo bike, with a 545 W peak capacity, was launched in 2021. The cargo bike gathers about 50% of its power on average from solar energy, but it goes up to 100% during sunny days enabling the cargo bike to operate as an EV [14].

An energy-efficient technology-based green transportation system is predominantly driven by batteries (with optimized scheduling, longevity, safety, and cost), hydrogen gas, and hybrid combustion systems. According to an analytical analysis published in Ref. [15], lithium-ion batteries are found to be more supportive in providing optimized operation efficiently as they can be charged and discharged with 15% more efficiency compared to traditional lead acid batteries. Thus, vehicles with lithium-ion batteries can be operated with greater energy efficiency. On the other hand, hydrogen gas can generate energy around 30% more efficiently in comparison with conventional combustion-governed power plants and can lead to the development of energy-efficient vehicles [33]. Vehicles with hybrid combustion systems use biodiesel instead of existing diesel to cut back on toxic gas emissions without compromising the combustion engine performance of the vehicles [16].

An energy reusing technology focuses on more air usage in place of fuel in green transportation systems. It also incorporates kinetic energy recovery technology to recover and reuse the induced energy – originated from the break of an EV. The induced energy can be stored in batteries and flywheels for later utilization under acceleration. This technical approach has already been used in both racing and passenger cars [8]. An energy control system is also presented in Ref. [17] to manage the acceleration and deceleration of vehicles by adopting this technology.

3.2.2 Intelligent System Technologies

Intelligent system technologies have paved the way to reshape modern transportation by incorporating new technological features into it, such as telecommunication, sensor system, positioning system, computing hardware, data processing, planning processing, and virtual operation [8]. Some noteworthy uses of intelligent system technologies in modern transportation are listed below [18]:

- *Smart roads and parking:* The number of vehicle tyres moving through the ground can be sensed by such roads. In addition, drivers can receive instant notifications with regard to traffic patterns and avoid congested roads subsequently. Smart technologies can also manage parking availability dynamically and contribute toward decreasing the time and fuel consumption required to search for a parking slot.
- *Electronic toll collection (ETC):* The ETC system can facilitate the direct electronic debit from the registered vehicle owners without stopping them at a tollbooth, and thus contribute to reducing delays, especially during high traffic. The ETC system has already been deployed in many parts of the world. In particular, Taiwan has implemented this technology on all of its motorways [19].
- *Automated vehicles:* A calibrated road technology integrated with sophisticated cameras is utilized in these vehicles to drive automatically, i.e., no drivers are required. However, its safety on complex roads has yet to be examined.
- *Biometric Access:* The vehicle owners can unlock and start vehicles with their fingers and eyeballs – which are read and scanned flawlessly by the biometric technology. Due to some operational advantages, fingerprint readers are more widely accepted contrasting with retina scanners [20].
- *Remote shutdown:* This technology has primarily been introduced to prevent theft. In fact, stolen vehicles can be closed down remotely with the help of this technology. Remote shutdown is customarily implemented on the smartphones of the vehicle owners. Because of its protection, the global market is growing, which is expected to reach US\$723.65 million by 2026 [21].
- *Head-up display:* This active window display can help the vehicle driver navigate by providing vibrant images in regard to the distance between nearby vehicles.
- *Health monitoring:* The heartbeats of the drivers can be tracked either by seat belt sensors or steering wheel sensors by dint of this technology, and hence the active health monitoring of the vehicle driver can be possible. The idea has been proposed by Ford company and they expect to bring it to reality in upcoming years [22].
- *Real-time information and comprehensive tracking:* Real-time information in regard to traffic conditions and congestion, road closures, and rerouting can be provided by modern technologies to organize travel in the shortest possible time. In fact, the travel distance of a vehicle can be tracked using comprehensive

tracking technology. Some state governments and insurance companies are intending to deploy the comprehensive vehicle tracking system [23].

- *Road charging and congestion pricing:* Intelligent technologies can be employed to implement dynamic road pricing schemes to charge vehicles for the utilization of congested roads with the purpose of evading traffic congestion and improving road sustainability.

- *Safety of heavy freight vehicles:* Radar-driven active acceleration and braking system, vehicle-to-vehicle and vehicle-to-infrastructure communication, and Peloton technologies (i.e., automated mechanisms to solve challenges of fuel use and crashes) can enable heavy freight vehicles to achieve energy saving, reduced delivery time and accidents, and operational safety.

3.2.3 Integrated Management Technologies

Multifarious facets of travel needs are required to be taken into account while addressing transportation and urbanization issues, which mainly include pattern of transport growth and systematic use of available lands, emission rates of toxic gases, advanced traffic management, and adequate infrastructure management. All these features are interconnected and cannot be solved in an isolated fashion. This necessitates the use of integrated management technologies in transportation [24].

In general, these technologies have five major goals, including baseline review, target setting, political commitment, implementation and monitoring, and evaluation and reporting. These goals are achieved by implementing technologies in three different stages. In the first stage, substantial objectives are determined. The second stage defines the indicators and task forces. The deliverable items are finalized in the third stage [8]. A brief description of primary targets along with subsequent tasks and deliverable items is encapsulated as follows:

- *Baseline review:* It is the first and foremost goal of integrated management technologies in the transportation system. It essentially reviews all the related data and maps of emerging trends and issues, political priorities, organizational setup and responsibilities, and consultation with stakeholders. Time saving and minimized congestion are two ultimate deliverable items of the baseline review.

- *Target setting:* The second objective is to organize a favorable organizational structure to drive the system. Significant priorities, issues, and indicators are identified, and potential action plans are laid out to reach the responsibility and time frame targets. Travel and vehicle efficiency are considered deliverable items in the target setting.

- *Political commitment:* It deals with government strategic programs and vision, public information strategy and sensitization, and partnership commitments with private businesses as well as national and regional authorities. The deliverable items are increased travel and vehicle efficiency.

- *Implementation and monitoring:* They aim at performing strategic actions appropriately, arranging internal review of the actions undertaken regularly, refining any action if necessary, monitoring results, and providing information to all engaged stakeholders. Two paramount deliverable items are increased public transportation and a lower health risk.
- *Evaluation and reporting:* This is the last goal of integrated management technologies, which focuses on evaluating the process and outcome and reporting on actions. It also involves internal auditing in accordance with monitoring results. Moreover, certification and public information are also taken care of. Increased private investment is the principal deliverable item.

3.2.4 Distributed Ledger Technologies

Distributed ledger technologies denote the technological protocols that facilitate the protection and well functioning of a digital and encrypted database, including contemporary access, validation, and immutable records, without a direct intervention of a central authority [25]. Blockchain is the most popular form of distributed ledger technology at present. It essentially integrates a set of chronologically arranged blocks that are irreversible and organized by consensus-based protocols and exercises an automatic hash algorithm to validate and store transactions in these blocks [26]. Once any transaction is validated and recorded, the immutable feature of blockchain does not permit changes [27].

There are four different categories of blockchain, namely public blockchain, private blockchain, consortium blockchain, and hybrid blockchain. A public blockchain allows everyone to participate in the blockchain network without any permission in a decentralized manner, i.e., a permissionless blockchain. Examples are Bitcoin, Ethereum, and Litecoin. Private blockchains are entirely maintained by a single authority and participation is subject to permission. Hyperledger and Solana are examples of private blockchain. In consortium blockchain, permission is needed to join, but unlike private blockchain, it is controlled by a group of authorities to improve security. A single authority controls hybrid blockchains like private blockchains, but some processes are executed without any permission [28].

Blockchain can find its application in green transportation systems to guarantee traceability and transparency in the supply chain and manage resources efficaciously. The technology can also handle the charging and discharging of EVs in a decentralized fashion. Blockchain-enabled coordination of EVs is driven by multilateral computerized embedded agreements, commonly known as smart contracts [29]. EV owners can also participate in blockchain-empowered peer-to-peer (P2P) mechanisms to charge and discharge their vehicles based on individual preferences [30]. Once coordinated/P2P charging and discharging patterns, rules, and conditions are mutually agreed upon, the smart contracts are written on the blockchain and cannot be removed or altered. Furthermore, a blockchain is rugged against a single point of failure and network members can be invalidated if it is required for sturdy functioning [31].

Blockchain can be adopted to securely store various data sets of green transportation systems, which include data on fuel usage, carbon emission, running and maintenance costs. This can assist in conducting better analysis, whenever necessary, to report sustainability features of green transportations. Also, the privacy and security of blockchain-driven transportation systems can be ensured by employing challenge-response protocols, digital signatures, entity authentication, symmetric and asymmetric encryption, pseudonyms, non-likability, and asymmetric cryptographic private and public keys in its platform. Mixing, zero-knowledge proofs, and ring signatures are typical examples of such blockchain-based methods [32].

In addition, blockchain can allow EV owners to track and monitor their charging and discharging transactions in an automated way. Transactions' data and history are always preserved in the blockchain database and are also available to the energy retailers (if they have blockchain accounts) for periodic billing. Digital currency, designated as a token, can also be incorporated to settle payments with confidentiality. In other words, blockchain has the potential to introduce an electronic billing system in the transportation sector [31].

3.3 Economic Implications of Green Transportation Systems

Adequate infrastructure, efficient operational technologies, and coordinated network systems of green transportation can play a vital role in economic growth as basic goods and services can easily be accessed swiftly at cheaper prices while environmental sustainability is guaranteed [34]. A statistic states that environmental pollutants (considered in the long-range energy alternative planning system model) like nitrogen oxide, nitrous oxide, sulfur oxide, sulfur dioxide, carbon dioxide, carbon monoxide, methane, and non-methane volatile organic compounds incur US\$0.33/kg, US\$0.91/kg, US\$1.16/kg, US\$0.0019/kg, US\$0.018/kg, US\$0.049/kg, and US\$1.35/kg, respectively [35]. These costs can be avoided completely or mitigated remarkably by the adoption of sustainable transportation.

The financial benefits of green transportation can be analyzed from two points of view in general: cost savings to vehicle owners and employment creation in the job market [2].

3.3.1 Cost Saving

The latest research undertaken by the World Resources Institute reports that the world needs US$2 trillion in capital investment on low-carbon transportations, and it is projected that US$300 billion could be saved annually compared to fossil-fueled transportations. This emphasizes the necessity of investing in green transportation when cost-effective transportation is one of the targets of modern society [36].

The overall running and maintenance costs of sustainable transportation are also substantially lower than those of internal combustion engine-driven vehicles, although the capital prices of batteries – which are the heart of EVs – are still comparatively expensive. Proterra, an EV manufacturer, demonstrates that an electric bus can save approximately €49,000 in running and maintenance costs over the period of five years in comparison with a standard passenger-carrying bus. Another EV manufacturer, called New Flyer, reports that a typical electric bus can cut down around €393,550 and €123,000 in fuel and maintenance expenses, respectively, over its lifetime [37].

The investment in electric cars can also bring significant cost savings in the long run. According to Australia's Electric Vehicle Council (EVC), the average cost of powering petrol-driven vehicles ranges between AU$1.50/l and AU$2/l. On the contrary, the equivalent EV cost would be close to AU$0.33/l, resulting in an impressive fuel cost saving. The motoring body also claims that the annual running cost savings could be around AU$1600 on average (70% cheaper) if fuel-powered cars are replaced by electric cars [38].

Further, electric cars do not experience rigorous maintenance like their fuel-based counterparts. In fact, electric cars are free from some well-known services required for internal combustion engines, such as fuel changes and prolonging brake life through regenerative braking, resulting in lower maintenance costs. An analysis of Consumer Reports finds that electric car owners need to spend half as much on maintenance and repair costs on average, contrasting with petrol-driven vehicle owners. In other words, electric car owners can save 50% more in maintenance and repair costs [39].

On top of it, governmental incentives are also available to EV owners to save money on vehicle charges that include registration fees and stamp duties. For instance, in the Australian Capital Territory (ACT), the registration fees of electric car owners are waived for two years while stamp duties are completely exempted.

Moreover, zero-interest loans, from AU$2000 to AU$15,000, are also offered to EV owners to promote the usage of zero-emission vehicles [40].

3.3.2 Job Creation

A wide range of employment opportunities can be created by investing in green transportation systems. These jobs could be of three types: direct jobs, indirect jobs, and induced jobs. Direct jobs can be generated from the construction, manufacturing, and maintenance activities related to green transport, which include the assembly and repair of various EVs, the construction of bicycle lanes and parking, pavements, the coordination of modern technologies, and wholesale and retail market settlements. The production of materials and parts for EVs and bicycles, initialization of green transport system layout, and development of required technologies are typical examples of some indirect jobs. On the other hand, green transportations aim at providing cost savings to owners (e.g., private EVs) and users (e.g., public EVs). These savings are more likely to be spent in the overall economy, which can also contribute to creating induced jobs in society [41].

Some cities in the world are putting significant efforts to promote the bicycle job market. One of them is the city of Portland, Oregon, USA. It launched its Bicycle Master Plan back in 1996 to encourage its dwellers to cycle and enlarge the biking network. At present, nearly 6–8% of its commuters are bicycle commuters. The sizes of its bikeways, bike lanes, and off-street bike paths are 314 mi, 202 mi, and 76 mi, respectively [42].

The implication of advocating the use of bicycles has been outstanding on the local job market of Portland. The job opportunities in different sectors, such as entertainment and recreation, manufacturing, retail, and wholesale trades, in the city from 2015 to 2025 are depicted in Table 3.2. As is observed, total employment opportunities are projected to increase in 2025 compared to 2015 and 2020. Jobs related to bicycle manufacturing and retail trade may diminish slightly in

Table 3.2 Job opportunities in the cycling sector in Portland, Oregon, USA.

Employment sector	2015 (Actual)	2020 (Actual)	2025 (Projected)
Entertainment and recreation	6985	7668	8152
Manufacturing	27,195	27,118	26,391
Retail trade	34,515	34,139	33,855
Wholesale trade	20,529	21,810	22,810
Total	**89,224**	**90,735**	**91,208**

Source: Adapted from [43].

Table 3.3   Opportunities projected to be provided by green transportation globally in 2030.

Transport measure	Incremental investment (US$ trillion)	Returns annually (US$ billion)	Supported jobs (million)
Passenger transport	8.61	320.42	3.6
Freight transport	0.59	79.85	0.1
Total	**9.20**	**400.27**	**3.7**

Source: Adapted from [44].

2025. However, both wholesale trade and the entertainment and recreation sectors are expected to create greater numbers of jobs in the city, resulting in an overall escalation in the job market [43].

Nevertheless, EVs also have the potential to bring dramatic changes to job sectors globally. A case study published in [44] suggests that US$8.61 and US$0.59 trillion of incremental investment in passenger and freight EVs would facilitate the world receiving US$320.42 and US$79.85 billion in annual returns, respectively, in 2030. Besides, 3.6 and 0.1 million jobs, respectively, would be created in passenger and freight transportations in the same year. In total, a US$7.20 trillion investment in EVs across the world would see US$7.20 billion in annual returns and creation of 3.7 million jobs, as illustrated in Table 3.3.

3.4   Environmental Implications of Green Transportation Systems

Environmental assessment mainly addresses the issues caused by transportation in the environment as part of the decision-making process in modern societies. The characteristics of transportation are considered for the base year and horizon year (i.e., the expected growth year) while analyzing the impacts on the environment [45]. The pivotal environmental concerns of transportation are toxic gas emissions and threat to human health [37]. Thus, green transportation aims at scaling down environmental pollution (for better air quality and noise level) and improving human health status.

3.4.1   Lowering Emission of Pollutants

Complying with the Paris Agreement, many countries in the world have set their targets to lower acidifying gas emissions and ambient pollution (e.g., carbon dioxide, nitrogen oxide, and suspended particulate matter) from various sectors that include transportation. For instance, Australia's 17% greenhouse gas emissions come from transportation, figuring at 96 megatons per year. While the transportation sector in Australia is experiencing unprecedented growth in recent times, the country is committed to achieving zero net emissions by the end of 2050 [46].

According to Australian National Greenhouse Accounts (NGA) factors [47], a standard diesel-powered bus consumes nearly 0.3-l fuel/km traveled and emits 0.815232 kg of carbon/km traveled. The average carbon emission intensity becomes 4.402253 kg/km for a fleet. The adoption of EVs can contribute greatly to reducing carbon emission intensity toward zero and cut down on significant expenditures on carbon prices. The analysis has found that the abated carbon emission of a bus and a fleet would be 652 and 35,218 tons, respectively, over the period of eight years if conventional vehicles were replaced by EVs. In return, around AU$19,566 and AU$1,056,541 could be saved, respectively, as a result of carbon offsets.

The United States Environmental Protection Agency (EPA) reports that hybrid vehicles held at share of 9% of all produced vehicles in 2021, and this figure is expected to grow in the upcoming years. In 2021, the carbon emission rate of all vehicles in the USA plummeted to 347 g/mi on average, which was 25% lower than that of the model year 2004. The real-world average fuel economy (i.e., the relationship between traveled distance and fuel consumption) increased by 32% or 6.1 mpg. For this study, vehicles were divided into five classes, which include SUV, sedan/wagon, minivan/van, truck SUV, and pickup truck. Among them, sedans and wagons were labeled as the vehicles with the lowest carbon emissions and highest fuel economy [48].

Vehicle manufacturers are also contributing substantially to achieving compliance with the greenhouse gas emission standards by employing several cutting-edge technologies. While Tesla did not produce any carbon due to their all-electric fleets as displayed in Table 3.4, other manufacturers generated some sorts of carbon gases ranging from 309 g/mi (the lowest achieved by Subaru) to 385 g/mi (the highest achieved by Ford) in 2021.

Compared to 2016, Kia and Toyota ensured the largest reduction in carbon emissions, by 28 g/mi each, in 2021. The fuel economy of Kia and Toyota increased by 2.5 mpg and 2.1 mpg, respectively, over this period. Honda, Nissan, and BMW also had an abatement in carbon emissions during the course of five years, figuring at 3 g/mi, 7 g/mi, and 10 g/mi, respectively. Meanwhile, their fuel economy exhibited an upsurge at 0.7 mpg, 0.8 mpg, and 2.1 mpg, respectively. Mercedes, on the contrary, showed unchanged variations in both carbon emissions and fuel economy between model years 2016 and 2021 [48]. The detailed comparisons among vehicle manufacturers are demonstrated in Table 3.4.

Table 3.4 Carbon emissions and fuel economy comparison for some vehicle manufacturers.

Vehicle manufacturer	Carbon emissions (g/mi) in 2016	Carbon emissions (g/mi) in 2021	Fuel economy (mpg) in 2016	Fuel economy (mpg) in 2021
Tesla	0	0	96.8 (mpge)	123.9 (mpge)
Honda	315	312	28.2	28.5
Subaru	317	309	28.1	28.8
Nissan	318	311	27.8	28.6
Kia	338	310	26.2	28.7
Toyota	355	327	25.0	27.1
BMW	349	339	25.0	27.1
Mercedes	376	376	23.6	23.6
Ford	389	385	22.8	22.9

Source: Adapted from [48].

3.4.2 Improving Human Health Status

Toxic emissions from fuel-driven vehicles are responsible for numerous human health issues, including cancer, bronchitis, type 2 diabetes, high blood pressure, and cardiovascular and respiratory diseases, for example. In Australia, between 900 and 2000 early deaths are caused every year by the air pollution coming from motor vehicles. Further, between 900 and 4500 bronchitis, cardiovascular, and respiratory disease cases are recorded, costing between AU$1.5 and AU$3.8 billion for health treatments [49]. Green transportations can protect and promote human health by dropping down health risks and diseases resulting from vehicular air pollutants, saving a lot of money otherwise required for health treatments [2].

A European research study has found that the life expectancy in urban areas is reduced by nine months on average due to ambient particulate matter pollution generated from vehicles. Each year, pollution also leads to 482,000 premature deaths in Europe. Moreover, the overall cost of these health concerns is valued at around US$1.2 trillion annually. Based on the study, people's health is jeopardized owing to motorized transportation with significant greenhouse gas emissions. Elderly, homeless, and poor people, along with children, are affected disproportionately, causing health inequalities at an alarming rate [50].

To this end, the World Health Organization (WHO) Europe introduced a health framework related to sustainable transportations in 2020 with the purpose of reducing health concerns by minimizing emissions of pollutants and strengthening the link between public health and transportation in European regions. The governing bodies are also working toward developing coherent policies to invest in environment-friendly transportation to improve human health status [50].

Another research study conducted by Chinese researchers in 2019, reported in Ref. [51], outlines that thousands of lives could be saved by the initiation of green transportations. In this analysis, they modeled a scenario in which 27% of the total transportations was assumed to be EVs, and a mix of renewable sources and coals were considered as primary power sources. They adopted an atmospheric chemistry transport layout to determine the concentrations of nitrogen dioxide, ozone, and particulate pollution. Their experiment showed positive air quality and better health benefits. In fact, the cuts in atmospheric pollution could prevent approximately 17,500 deaths by 2030.

Finally, a pilot project is going on in Accra, Ghana, to demonstrate how green transportations can protect and promote health [52]. In particular, to integrate health action in transport, the Urban Health Initiative (UHI) model is processed in the project in six steps as follows:

- *Step 1:* Existing transport policies mainly related to air pollution and human health are mapped considering the involvement of key stakeholders.
- *Step 2:* Policy-makers are directed to build competencies in evaluating health and financial effects of transportation policies.
- *Step 3:* Various tools, such as integrated sustainable transport and health assessment tool (iSThAT), integrated transport and health impact modeling tool (ITHIM), and health economic assessment tool (HEAT), are locally used to carry out the health and economic assessment of green transportations.
- *Step 4:* Alternative scenarios are also considered to estimate potential health and financial impacts.
- *Step 5:* Campaigns are arranged to intensify the need for green transportations.
- *Step 6:* Incentives are provided to act, monitor, and track green transportation initiatives.

3.5 Conclusion

An overview of the techno-economic-environmental approaches and benefits of green transportation systems has been carried out in this chapter to highlight their suitability in the present world. The chapter has begun by discussing various prominent advantages of green transportation systems along with their operational modes, and the following contributions have been made:

- Some modern technologies, such as eco-friendly, energy-efficient, energy reusing, intelligent systems, integrated systems, and distributed ledger technologies, have been introduced to develop, optimize, and operate green transportations in leading-edge, efficacious, protected, and comfortable manners.

- Furthermore, the application of green transportations, in providing financial returns to vehicle owners and opening employment opportunities to drive the social economy, has been analyzed by dint of up-to-date research and experimental data.
- Moreover, this chapter has also demonstrated the environmental evaluation of sustainable transport systems by considering numerous existing case studies related to transportation and the environment, and the analysis stresses the importance of adopting green transportation to reduce/completely remove hazardous vehicular emissions from the environment and human health concerns.

References

1 OICA (2021). OICA 2021 production statistics. https://www.oica.net/category/production-statistics/2021-statistics/ (accessed 16 November 2023).

2 Rinkesh. "Modes and benefits of green transportation". https://www.conserve-energy-future.com/modes-and-benefits-of-green-transportation.php (accessed 16 November 2023).

3 Van Fan, Y., Perry, S., Klemeš, J.J., and Lee, C.T. (2018). A review on air emissions assessment: transportation. *Journal of Cleaner Production* 194: 673–684.

4 Lu, M., Xie, R., Chen, P. et al. (2019). Green transportation and logistics performance: an improved composite index. *Sustainability* 11 (10): 2976.

5 Sun, D., Zeng, S., Lin, H. et al. (2019). Can transportation infrastructure pave a green way? A city-level examination in China. *Journal of Cleaner Production* 226: 669–678.

6 Bjorklund, M. (2011). "Influence from the business environment on environmental purchasing – drivers and hinders of purchasing green transportation services," *Journal of Purchasing and Supply Management*, vol. 17, no. 1, pp. 11–22, Mar. 2011.

7 European Commission (2011). Roadmap to a single European transport area: towards a competitive and resource efficient transport system. https://op.europa.eu/en/publication-detail/-/publication/bfaa7afd-7d56-4a8d-b44d-2d1630448855/language-en (accessed 16 November 2023).

8 Shah, K.J., Pan, S.-Y., Lee, I. et al. (2021). Green transportation for sustainability: Review of current barriers, strategies, and innovative technologies. *Journal of Cleaner Production* 326: 129392: 1–13.

9 Corby, S. (2021). "Hydrogen vs electric cars: what's the difference and which is better?" https://www.carsguide.com.au/ev/advice/hydrogen-vs-electric-cars-whats-the-difference-and-which-is-better-82898 (accessed 16 November 2023).

10 Rajak, S., Parthiban, P., and Dhanalakshmi, R. (2021). Analysing barriers of sustainable transportation systems in India using Grey-DEMATEL approach: a supply chain perspective. *International Journal of Sustainable Engineering* 14 (3): 419–432.

11 Van Fan, Y., Klemeš, J.J., Walmsley, T.G., and Perry, S. (2019). Minimising energy consumption and environmental burden of freight transport using a novel graphical decision-making tool. *Renewable and Sustainable Energy Reviews* 114: 109335.

12 Lambert, F. (2018). "BMW i3 gets a 100 kWh battery pack for 435 miles of range as a proof-of-concept by Lion Smart". https://electrek.co/2018/09/07/bmw-i3-100-kwh-battery-pack-lion-smart/ (accessed 16 November 2023).

13 Toppa, S. (2025). "The first solar-powered round-the-world flight has begun". https://time.com/3736858/fueless-flight-bertrand-piccard-solar-impulse-2-andre-borschberg/ (accessed 16 November 2023).

14 Bowden, A. (2021). "Dutch firm launches the SunRider – the world's first self-charging solar-powered e-cargo bike". https://ebiketips.road.cc/content/news/dutch-firm-launches-the-sunrider-the-worlds-first-self-charging-solar-powered-e-cargo (accessed 16 November 2023).

15 PowerTech Company. Lithium-ion battery. https://www.powertechsystems.eu/home/tech-corner/lithium-ion-battery-advantages/ (accessed 16 November 2023).

16 US Department of Energy. Biodiesel benefits and considerations. https://afdc.energy.gov/fuels/biodiesel_benefits.html (accessed 16 November 2023).

17 (2013). Green transportation: five innovations that are driving efficient vehicle technology. https://www.politico.eu/article/green-transportation-five-innovations-that-are-driving-efficient-vehicle-technology/ (accessed 16 November 2023).

18 Maimaris, A. and Papageorgiou, G. (2016). A review of intelligent transportation systems from a communications technology perspective. *Proceedings of the International Conference on Intelligent Transportation Systems*, Rio de Janeiro, Brazil, 54–59 (November 2016).

19 (2017). FETC innovation from highway toll to ITS Taiwan smart city. https://www.worldhighways.com/feature/fetc-innovation-highway-toll-its-taiwan-smart-city (accessed 16 November 2023).

20 Clark, M. "Iris recognition scanners vs fingerprint scanners: compare and contrast". https://www.bayometric.com/iris-recognition-scanners-vs-fingerprint-scanners/ (accessed 16 November 2023).

21 (2019). Remote vehicle shutdown market to reach USD 723.65 million by 2026 – reports and data. https://www.globenewswire.com/en/news-release/2019/12/17/1961557/0/en/Remote-Vehicle-Shutdown-Market-To-Reach-USD-723-65-Million-By-2026-Reports-And-Data.html (accessed 16 November 2023).

22 Subramaniyam, M., Singh, D., Park, S.J. et al. (2018). Recent developments on driver's health monitoring and comfort enhancement through IOT. *IOP Conference*

Series: Materials Science and Engineering, Kattankulathur, India (22–24 March 2018), vol. 402, 012064: 1–9.

23 Fortune Business Insights. Vehicle tracking device market size, share and industry analysis. https://www.fortunebusinessinsights.com/industry-reports/vehicle-tracking-device-market-101787 (accessed 16 November 2023).

24 Oracle. What is a transportation management system? https://www.oracle.com/scm/logistics/transportation-management/what-is-transportation-management-system/ (accessed 16 November 2023).

25 Nevil, S. (2023)."Distributed ledger technology (DLT): definition and how it works". https://www.investopedia.com/terms/d/distributed-ledger-technology-dlt.asp (accessed 16 November 2023).

26 Azim, M.I. (2022). Peer-to-peer energy trading in low-voltage distribution networks. PhD thesis. The University of Queensland, Australia.

27 Casino, F., Politou, E., Alepis, E., and Patsakis, C. (2019). Immutability and decentralized storage: an analysis of emerging threats. *IEEE Access* 8: 4737–4744.

28 Wegrzyn, K.E. and Wang, E. "Types of blockchain: public, private, or something in between". https://www.foley.com/en/insights/publications/2021/08/types-of-blockchain-public-private-between (accessed 16 November 2023).

29 Giancaspro, M. (2017). Is a 'smart contract' really a smart idea? Insights from a legal perspective. *Computer Law and Security Review* 33 (6): 825–835.

30 Azim, M.I., Tushar, W., Saha, T.K. et al. (2022). Peer-to-peer kilowatt and negawatt trading: a review of challenges and recent advances in distribution networks. *Renewable and Sustainable Energy Reviews* 169: 112908: 1–23.

31 Khan, S.N., Loukil, F., Ghedira-Guegan, C. et al. (2021). Blockchain smart contracts: applications, challenges, and future trends. *Peer-to-Peer Networking and Applications* 14 (5): 2901–2925.

32 Aitzhan, N.Z. and Svetinovic, D. (2018). "Security and privacy in decentralized energy trading through multi-signatures, blockchain and anonymous messaging streams," *IEEE Transactions on Dependable and Secure Computing*, vol. 15, no. 5, pp. 840–852, Sep. 2018.

33 Mutuku, S. "Advantages and disadvantages of hydrogen energy". https://www.conserve-energy-future.com/advantages_disadvantages_hydrogenenergy.php (accessed 16 November 2023).

34 DBSA. How green transport can be a catalyst for economic development and growth? https://www.dbsa.org/article/how-green-transport-can-be-catalyst-economic-development-and-growth (accessed 16 November 2023).

35 Shahid, M., Ullah, K., Imran, K. et al. (2022). Economic and environmental analysis of green transport penetration in Pakistan. *Energy Policy* 166: 113040: 1–10.

36 Yadav, N. and Levefre, B. (2016). "Sustainable transport investment could save $300 billion a year – within existing financial flows". https://www.wri.org/insights/

sustainable-transport-investment-could-save-300-billion-year-within-existing-financial (accessed 16 November 2023).

37 (2022). EV transition: why should you switch to electric buses? https://stratioautomotive.com/ev-transition-electric-buses/ (accessed 16 November 2023).

38 evHub. How much money do you save with an electric car? https://electricvehiclehub.com.au/information-centre/how-much-money-do-you-save-with-an-electric-car/ (accessed 16 November 2023).

39 (2020). Electric vehicle owners spending half as much on maintenance compared to gas-powered vehicle owners, finds new CR analysis. https://advocacy.consumerreports.org/press_release/electric-vehicle-owners-spending-half-as-much-on-maintenance-compared-to-gas-powered-vehicle-owners-finds-new-cr-analysis/ (accessed 16 November 2023).

40 ACT Government. Motor vehicle registration and renewal. https://www.accesscanberra.act.gov.au/s/article/motor-vehicle-registration-and-renewal-tab-zero-emissions-vehicle-registration (accessed 16 November 2023).

41 (2021). Job creation through green transport, 2021. https://www.greengrowthknowledge.org/case-studies/job-creation-through-green-transport (accessed 16 November 2023).

42 Walljasper, J. (2010). "Portland finds bike-friendly policies boost local economy". https://www.shareable.net/portland-finds-bike-friendly-policies-boost-local-economy/ (accessed 16 November 2023).

43 Ibsen, M. and Bump, T. (2015). "The economic impact of the bicycle industry in Portland". https://www.portland.gov/bps/documents/economic-impact-bicycle-industry-portland-2015/download (accessed 16 November 2023).

44 Urbantransitions. Climate emergency: urban opportunity. https://urbantransitions.global/wp-content/uploads/2019/09/Climate-Emergency-Urban-Opportunity-report.pdf (accessed 16 November 2023).

45 Dhingra, S.L., Rao, K.V.K., and Tom, V.M. (2003). "Environmental impact assessment for sustainable transport". https://www.emerald.com/insight/content/doi/10.1108/9781786359513-017/full/pdf?title=environmental-impact-assessment-for-sustainable-transport (accessed 16 November 2023).

46 Climate Council. Transport emissions: driving down car pollution in cities. https://www.climatecouncil.org.au/wp-content/uploads/2017/09/FactSheet-Transport.pdf (accessed 16 November 2023).

47 (2021). National greenhouse accounts factors: 2021. https://www.dcceew.gov.au/climate-change/publications/national-greenhouse-accounts-factors-2021 (accessed 16 November 2023).

48 Highlights of the automotive trends report. https://www.epa.gov/automotive-trends/highlights-automotive-trends-report (accessed 16 November 2023).

49 VIC Government. Sustainable transport. https://www.health.vic.gov.au/planning-infrastructure/sustainable-transport (accessed 16 November 2023).

50 Health 2020: transport and health. https://www.euro.who.int/data/assets/pdf file/ 0020/324641/Health-2020-Transport-and-health-en.pdf%3Fua%3D1 (accessed 5 February 2023).

51 Liang, X., Zhang, S., Wu, Y. et al. (2019). Air quality and health benefits from fleet electrification in China. *Nature Sustainability* 2 (10): 962–971.

52 World Health Organization (2021). Sustainable transport for health. https://apps. who.int/iris/rest/bitstreams/1349942/retrieve (accessed 16 November 2023).

4

Urban Integrated Sustainable Transportation Networks

Syed Muhammad Nawazish Ali, Saman Ahmadi, Ali Moradi Amani, and Mahdi Jalili

School of Engineering, STEM College, RMIT University, Melbourne, Australia

4.1 Introduction

One of the main sectors responsible for the worldwide climate crisis is transportation. Transportation is mainly known as the second main contributor to greenhouse gas (GHG) emission, after energy, around the world, with almost 27% pollution in the United States [1] and approximately 18% in Australia. In particular, freight transportation accounts for roughly 50% of all transport emissions and 7–8% of all CO_2 emissions [2]. Furthermore, according to [3], given predictions for worldwide economic and population expansion as well as the present average distance covered per freight unit [4], freight transport operations are expected to be more than twice by 2050 in comparison with 2015. The goals set as per the Paris Agreement at the EU level to be accomplished in 2025–2030 and by 2050 are markedly different from this scenario. Additionally, in September 2020, the European Commission (EC) proposed to increase the reduction target of GHG emissions for 2030, which includes both removals and emissions, to atleast 55% below the levels of 1990 in accordance with the European Green Deal. Yet, considering the present rate of GHG emissions, the most concerning fact is that warming is expected to approach 1.5 °C by 2030s [5].

Urbanization, on the other hand, is contributing to increased energy consumption and demand for larger transport networks. Hence, planning a low-emission, responsive, and well-designed transport system is crucial for achieving sustainability. However, there are numerous challenges in implementing low-emission and environmentally friendly transportation systems. From the economic

Interconnected Modern Multi-Energy Networks and Intelligent Transportation Systems: Towards a Green Economy and Sustainable Development, First Edition. Edited by Mohammadreza Daneshvar, Behnam Mohammadi-Ivatloo, Amjad Anvari-Moghaddam, and Reza Razzaghi.
© 2024 The Institute of Electrical and Electronics Engineers, Inc.
Published 2024 by John Wiley & Sons, Inc.

perspective, transforming the transport sector into a sustainable system requires significant investment in both infrastructure and technology. Safety and security risks associated with promoting alternative modes of transportation are also among the challenges needing serious attention.

Although achieving accessible, affordable, and efficient urban transport requires addressing various challenges in an integrated manner, there are several social, environmental, and economic benefits associated with the implementation of sustainable transport. Improved access to affordable transport services, increased physical activity, and reduced social inequalities are some instances of social benefits that can be achieved by sustainability. Environmental benefits include reduced local air pollutants and GHG emissions, improved air quality, and reduced noise pollution. From the economic point of view, benefits include cost savings for transport users and providers, reduced healthcare costs, and new job opportunities. A well-designed integrated transport system can help create more sustainable cities by reducing dependencies on personal vehicles as an economic objective. These significant benefits are the major motivation for promoting, implementing, and adopting the modes of sustainable transportation.

These modes are adopted in various countries based on the availability of facilities and government policies that encourage their usage. Some of these modes are quite healthy in terms of exercise and activeness such as walking and cycling although they have constraints regarding longer distances and extreme weather conditions. On the other hand, some modes are significantly beneficial for our environment and their adoption can help not only in cleaning the atmosphere from GHG emissions but also in the reduction of serious health issues pertaining to adverse effects of polluted environment. Examples of such sustainable modes are electric bikes, electric scooters, electric cars, electric buses, and electric trains. Some modes are relatively inexpensive and provide social benefits such as carpooling. A consistent adoption of these modes at a global scale can shift the current adverse climate change situation upside down. The implementation of these modes of sustainable transportation requires successful urban planning.

Urban planning is essential to guarantee that cities have the equitable and sustainable transportation infrastructure they require for urban growth. To promote healthy transportation systems, urban planners must promote walking, cycling, and other efficient modes of travel. Sustainable transport not only protects the environment and people's well-being, but it also provides better connectivity and drives productivity in modern smart cities. In other words, sustainable transport has an immense role to play in urban economic growth as it safeguards resources. Sustainable transport can reduce transportation costs, create job opportunities, and attract investments, businesses, tourists, and other economic drivers. Furthermore, sustainable transport infrastructures like bike lanes or pedestrian walkways and transit systems make it simpler for people and businesses to get goods and services, ultimately stimulating economic activity. In short, urban planning plays a significant role in the implementation of sustainable transportation infrastructure that requires a comprehensive integration of different infrastructures.

These infrastructures include mobility, energy, and society. Traditionally, these different sectors have had separate planning and operation processes without taking care of each other. For example, transportation planners have not been concerned about the electricity grid when they design city bus routes and schedules. However, with the emergence of electric buses, the transport network and electricity grid should be more coordinated. From the governance perspective, establishing sustainable transportation requires significant investment in upgrading the transport and energy infrastructure. It includes the cost of upgrading the electrical system, e.g., for transformers and substations, and modifications in the transport network, such as the construction of charging stations.

It has been shown that investing in major infrastructure is likely to be ineffective during transition to sustainable transport, unless accompanied by local actions and policies using small-scale investments. Thus, establishing an integrated governance framework for transition to sustainable transportation is key. Last, but not least, is the society engagement. A glance at trials shows that this item is even more important than technical problems in the success of transition to public transport. For example, investment in establishing a cycling track in an area where people have difficulties in basic infrastructures, such as water or electricity, may cause a negative pushback for the public.

This chapter encompasses various significant aspects of sustainable transportation networks that are shown in Figure 4.1. This chapter mainly contributes to providing a holistic comprehensive review of these aspects considering cutting-edge technologies and case studies at a global scale that is missing in the existing literature. It begins with highlighting the true motivation for transitioning toward sustainable transportation and proceeds by providing insights into the challenges and opportunities for its implementation. It further elaborates on various modes of sustainable transportation along with modern urban advancement and concludes with a detailed discussion on sustainable infrastructure. This holistic review not only adds considerable value to the existing literature but also provides a starting point for those researchers, who are newcomers in this field and want to have a proper structure of the problem. Moreover, it is quite beneficial for the policymakers and industrialists who are interested in this area. It provides an overview of key features enabling such critical integration, while putting together some of the recent advances in each studied aspect of sustainable transportation.

The structural breakdown of this chapter is as follows: Section 4.2 introduces the adverse effects of conventional transportation on the deterioration of environment and public health at a global scale, hence focusing on the necessity of sustainable transportation to overcome these significant issues. Section 4.3 highlights major challenges such as financial, planning, security challenges, etc., and opportunities

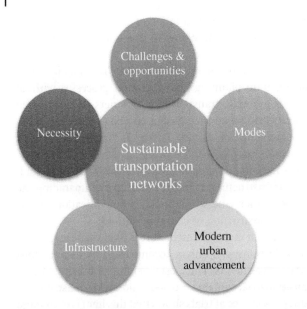

Figure 4.1 Significant aspects of sustainable transportation networks.

such as social, environmental, and economic benefits. Section 4.4 elaborates on the modes of sustainable transportation, which include electric vehicles (EVs), sustainable technology-based green trains, electric buses, electric bikes, bicycles, etc. Section 4.5 discusses how the potential of an urban area can improve the sustainability of the transportation network. Section 4.6 presents the infrastructure for sustainable transportation focusing on governance, interaction with electricity infrastructure, and transition to sustainable transportation. Section 4.7 gives the concluding remarks by covering the overall discussion.

4.2 Necessity of Sustainable Transportation

In view of the continuously increasing issues regarding climate change and public health caused by the conventional fuel-based transportation system, the implementation of sustainable transportation has become inevitable. The impact of these critical issues is discussed below.

4.2.1 Impact of Conventional Transportation on Climate Change

One of the sectors that contributes the most to pollution and emissions in urban areas is transportation [6]. Transportation-related emissions, particularly the amount of carbon dioxide (CO_2) emissions, are a major contributor to both climate change and environmental pollution [6]. Air pollutant emissions (APE) and GHG emissions are primarily two forms of emissions that have a negative impact on the environment [7]. CO_2 is one of the GHGs that is released into the atmosphere and has the ability to trap additional solar heat. The resultant rise in temperature is known as the "Greenhouse Effect," which causes unfavorable changes in climate. This hazardous GHG is released into the atmosphere in large quantities as a result of motor vehicle use [6].

On the contrary, APE is a mixture of numerous chemicals, including nitrogen oxide (NO_2), carbon monoxide (CO), and many other dangerous organic compounds, like benzene. It also consists of emissions of particulate matter. These pollutants are a by-product of vehicle emissions into the atmosphere. These pollutants have the potential to cause fatal conditions like heart disease, cancer, lung infection, and aggravated asthma [8].

The excellent performance of the transportation sector is crucial for our economic stability and social life, yet it continues to be one of the main causes of environmental damage in the modern world. Over 70% of pollution is caused by transportation, including cars, wagons, and public transportation in the EU [9]. Adverse climate shifts and severe weather conditions, which have reportedly happened globally throughout time, are mostly the result of the uncontrolled utilization of fossil fuels and excessive carbon emissions. The Sustainable Development Goals (SDGs) that the United Nations (UN) has established also witness to the need for mitigating ecological impact and promoting all preventive and precautionary measures [10].

4.2.2 Impact of Transportation-related Emissions on Public Health

Nowadays, roads are flooded with private vehicles as a result of which the transportation sector shares the responsibility for the increasing issues for public health such as smog, air acidification, road accidents, and carbon dioxide emissions. The road transportation industry, which also contributes significantly to elevated levels of other elements including hydrocarbons and carbon monoxide, is the main source of these air pollutants in the majority of international cities. These elevated levels play an important role in a number of respiratory and cardiovascular diseases. Transport-related pollutants have been directly linked to bronchitis, heart attacks, asthma, strokes, and other illnesses in numerous epidemiological studies [11]. Out of the total CO_2 emissions, the transport sector produced 24% [12].

The interaction of humans with the environment is greatly influenced by nonrenewable energy sources-based transportation systems. In 2010, CO_2 emissions of 7 gigatonnes were produced directly by the transportation sector. Compared to emissions from other sectors (i.e., agriculture, industry, power, commercial, or residential), GHG emissions of annual transportation are rising more quickly. In the upcoming years, it is anticipated that transportation demand will rise significantly as income levels rise and infrastructure around the globe expands. By 2050, annual emissions from the transportation industry are anticipated to double [13].

According to the Environmental Protection Agency (EPA) of the US, the transportation industry was the greatest producer of GHG emissions in the country in 2016, responsible for 28.5% of all national GHG emissions, which are related to the energy sector. As per the most recent Energy Information Administration (EIA) figures, from October 2015 to September 2016, the transportation sector in the US produced CO_2 emissions of 1893 million metric tonnes (MMt) which was more than that of the power sector i.e., 1803 MMt [14]. If increased renewable energy production reduces the amount of electricity produced using fossil fuels, which further reduces emissions from the power sector, this trend is expected to continue [15].

4.2.3 Role of Road Transportation in Carbon Emissions

In comparison with other forms of transportation including rail, marine, and aviation, road transportation accounts for the biggest percentage of energy consumption and CO_2 emissions. The GHG emissions from the transportation sector in the United States in 2016 were 22.9%, 18%, and 41.6% for freight trucks, light-duty trucks, and passenger cars, respectively [14]. Emissions by road transportation must be a primary priority of mitigation policies because they surged more than emissions generated by any other industry between 1990 and 2016. It is, therefore, possible to significantly reduce the environmental effects of the transportation industry as a whole by systematically developing and implementing innovative technology to limit the impacts of road transportation [15].

One of the primary areas where emissions are still increasing is transportation, which contributes to around one-fourth of worldwide GHG emissions [16–18]. Road transportation makes up over 50% of all GHG emissions related to the transportation sector, making it the single-largest emitter within the transportation industry. The world's efforts to minimize GHG emissions caused by the transportation sector are being impeded by the rapidly expanding demand for mobility and the ownership of private vehicles [5]. The mitigation of GHG emissions in transport to control climate change issues will be more difficult compared to other sectors [19, 20] due to society's continued dependence on fossil fuels.

4.2.4 Existing Global Energy Market

Technological advancements and growing population around the globe have resulted in the rapid increase in energy demand. Although the use of fossil fuels has threatening impact on human health [16, 17] as well as the atmosphere, they still maintain a dominating position in the energy sector. Burning of fossil fuels causes the emission of hazardous gases like nitrous oxide, carbon dioxide, and methane. It is also predicted that there will be a tremendous increase in such emissions due to the growth of industrial and urban sectors. If we will not shift toward green energy resources, the increasing levels of these detrimental gases will not only deteriorate our health and environment but also cause unexpected changes in our whole ecosystem [18]. Humans will experience the most adverse effects in the form of potential health problems and unavoidable climate catastrophes due to the excessive use of fossil fuels [19, 21].

4.2.5 Potential Approaches for Mitigating Emissions

Different countries have taken decisive steps to get rid of the hazards caused by transportation-related emissions. To reduce these issues, governments have begun reconsidering their energy policies and regulations. Several strategies have been proposed to either completely or partially reduce GHGs and the consequences they cause. Other potential approaches include increasing the effectiveness of existing technologies [20, 22], creating new, efficient technology that has less negative environmental effects [23, 24], and/or making a partial or complete switch to renewable energy sources [25, 26]. Indeed, it is the most potential approach to quickly phase out fossil fuels [27].

Decarbonatization is among the many steps that have been taken around the globe to address the pollution crisis caused by traffic congestion. In recent years, there has been an increase in the emissions reductions of the transportation sector through the use of EVs in urban areas. EVs are anticipated to have the potential to lessen both environmental pollution and the current energy problem [6]. Passenger automobiles account for about 12% of the total EU emissions of carbon dioxide, while light commercial vehicles or vans account for about 2.5% [28]. The EU has implemented a number of strict regulations to decrease emissions from transportation. One amongst them would be to establish goals for a percentage decrease in CO_2 emissions starting in 2021 [29]. For cars, it is 15% reduction in 2025 and 37.5% reduction in 2030. For vans, in 2025 and 2030, there will be a 15% and 31% reduction, respectively.

4.3 Challenges and Opportunities Associated with the Implementation of Sustainable Transportation

Achieving sustainable transport requires solution strategies that can pave the way for the development of low-emission and environment-friendly transport systems. However, there are various challenges in the underlying system that can hinder the successful implementation of sustainable transportation. This section summarizes some of the major challenges and benefits associated with the development of sustainable integrated transport as follows.

4.3.1 Growing Car Sector

The latest energy and carbon emission figures show that transport accounts for 24% of global CO_2 emissions (if only energy-related emissions are considered), with road transport responsible for almost 76% of the carbon emissions in the transport sector [30, 31], a considerable proportion that can potentially get larger with the continuously growing number of cars worldwide. According to the latest global figures, it is projected that the motorization rate will increase from 92 vehicles per 1000 people in 2020 to 173 vehicles per 1000 people by 2050 [32]. It is also estimated that the number of light-duty vehicles worldwide will continue to grow to over two billion vehicles by 2050, with around 70% of them still operating with fossil fuels. It is worth mentioning that the COVID-19 pandemic and its related restrictions have amplified the situation, with urban dwellers willing to use their private cars rather than traveling with others due to higher risks of infection [33, 34]. Consequently, the transport sector will inevitably need to deal with the increased use of conventional private vehicles.

4.3.2 Urban Growth

There is a direct relationship between urbanization, the gradual shift in residence of the human population from rural to urban areas, and energy used for transportation, mainly because the transport network becomes larger when urban areas spread out. The latest revision of World Urbanization Prospects [35] estimates that the world's population in urban areas will increase from 55% in 2018 to 68% by 2050. With the rapid urbanization trend around the world, accelerated by the COVID-19 pandemic and the emerging trend of flexible working, suburban areas have been growing faster, and we can expect increased transport demand in outer suburbs. Hence, transport providers may need to increase their supply to account for those shifted demands.

4.3.3 Transformation Cost

A fully sustainable transport is not achieved unless all its key elements are transformed in a way that they become part of an environmentally friendly ecosystem. Nonetheless, funding plays a key role in the fulfillment of a successful transformation. In many cases, this transition requires significant investment in both infrastructure and technology, with the associated costs later reflected in the relative price of services. According to the Global Infrastructure Outlook [36], it is estimated that $2 trillion in transport infrastructure investments would be needed every year until 2040 to upgrade the infrastructure [37]. Hence, when it comes to sustainable transport, it is crucial for the transport sector to budget for the transformation cost well before implementation, and to ensure their services are still affordable and accessible for all users when they are operational.

4.3.4 Planning Challenges

One of the main objectives of sustainable transport is improving the efficiency of transport in urban areas by either provisioning new means of transport, such as providing bike lanes, or the efficient use of resources and urban space, such as by designing integrated transport services. In addition, a responsive transport system may require structural changes over the long run to implement new technologies or to reflect changes in policies or demands. However, the lack of enough resources allocated to the development and implementation of an environmentally friendly transport system (such as land, money, equipment, etc.) is a critical challenge that can seriously limit the benefits of sustainability in such integrated systems, especially in areas that host large-scale transport demands. To this end, achieving sustainability in the transport sector requires careful planning to ensure a well-designed and responsive transport system via demand-oriented strategies such as push-and-pull and avoid-shift-improve (ASI) [38, 39], which offer a more comprehensive approach for an overall sustainable transport system design.

4.3.5 Safety Risks

With an increased focus on reducing carbon emissions and improving air quality, sustainable transport requires promoting cycling and walking as alternative modes of transportation. However, the lack of dedicated bike lanes and safe pedestrian crossings can lead to accidents and injuries. According to global reports, road traffic accidents are the top cause of death for individuals between the ages of 15 and 29, with pedestrians and cyclists accounting for approximately 35% of these fatalities [11, 40]. Furthermore, the emergence of low-noise transport technologies, such as EVs, has raised new safety concerns due to their quiet operation. Unlike conventional fuel-powered vehicles, the electrified driveline in modern vehicles produces significantly less noise, which can make them harder for pedestrians and cyclists to hear, particularly for those with impaired hearing or sight. Nonetheless, the installation of high-capacity infrastructure, such as fast charging equipment for EVs, may pose a significant danger to public safety, especially in areas with high population density. This includes the risk of electric shock caused by cable or plug damages, which can be exacerbated by severe weather conditions.

4.3.6 Security Challenges

Integration in transport systems often relies on utilizing data, software, and sensors to coordinate and optimize the operation of the key system components and, consequently, enhance the driving experience. Hence, one of the primary issues in smart transport platforms is the vulnerability of such integrated systems to cyberattacks. As integrated transportation systems become more prevalent, they rely on more complex networks of interconnected technologies, creating a larger attack surface for cybercriminals. In addition, implementing sustainable transport infrastructure, such as EV charging stations, involves substantial investment, making them vulnerable to copper theft and vandalism. Such criminal activities can result in the exposure of wiring, posing a significant risk of injury or even fatalities.

As discussed above, to achieve accessible, affordable, and efficient urban transport, various challenges need to be addressed in an integrated manner. In general, urban transport is intricately linked with urban life, and thus it is critically important to understand the benefits of sustainable transport to society as a whole. Recent studies show that increasing public awareness of traffic problems and perceived benefits of sustainability can play a significant role in producing more informed policy interventions and promoting sustainable transport [41, 42]. Besides key advantages such as safer transport services and less traffic congestion, as shown in Figure 4.2, there are various social, environmental, and

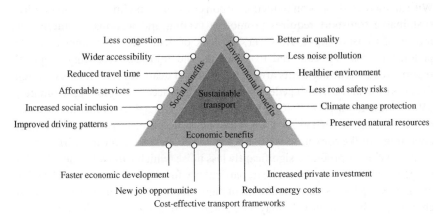

Figure 4.2 Schematic representing the social, environmental, and economic benefits of sustainable transport.

economic aspects in which the transport sector can reap the benefits of sustainability.

4.3.7 Social Benefits

Implementing a sustainable, integrated transport system can provide the society with improved access to affordable transport services and increased physical activity. It can also help to better recognize and address the specific needs of vulnerable groups and communities by reducing social inequalities regarding access to transport services. Sustainable modes such as public transport and biking infrastructure can help to increase social inclusion by providing affordable and accessible transportation for people with limited mobility, low-income households, and other marginalized groups. Meanwhile, reduced travel time in optimized demand-serving scenarios can potentially lead to time-efficient transport modes, saving the transport user's time by shortening overall trip duration. Furthermore, a sustainable transport system can make the use of new transport modes possible or even help change traditional transport modes to more sustainable ones. Some examples are building bike and walking paths in city areas to help society take advantage of reduced congestion, less parking pressure, and lower health risks, or changing conventional bus services with bus rapid transit (BRT) as a cost- and time-efficient transport mean with dedicated bus lanes.

4.3.8 Environmental Benefits

One of the key features of sustainable transport is the reduced number of motorized/conventional vehicles, known as the source of local air pollutants, combustion noise, and greenhouse effect. Sustainable transport options such as biking, walking, and using electrified transport services can improve the quality of life for individuals living in urban areas by preserving natural resources and reducing noise pollution. More importantly, by incorporating sustainability and integration into urban mobility, the transport sector can help reduce transport emission and contribute to a healthier environment through enabling active transport modes (e.g., walking or cycling). For example, a recent case study on a sustainable public transport project in Rzeszów, Poland, estimates that replacing 52 old buses with new low-emission buses contributes to a reduction of 201.9 tonnes in GHG emissions [33]. Besides actions taken to reduce greenhouse emissions, emerging technologies such as electric and fuel-cell vehicles can play critical roles in improving air quality. A transition away from fossil fuel-powered fleets to zero-carbon solutions, such as EVs, can not only alleviate climate change concerns but also generate co-benefits for wider use of renewable energy sources.

4.3.9 Economic Benefits

In view of the transport user's perspective, sustainable transport methods such as biking, walking, and public transport are affordable options that can effectively help save the increasing cost of fuel, parking, and maintenance associated with private cars. In addition, sustainable transport options such as biking and walking can promote physical activity, which can reduce healthcare costs associated with inactive lifestyles. With energy-efficient technologies implemented in the transport system, from high-efficient electrified fleets to optimized route-planning strategies, transport provides a significant reduction in the operation cost in terms of total energy consumption. Furthermore, socioeconomic factors such as changes in driving patterns due to less congested roads, less demand for car travel due to improved public transport accessibility, shifting conventional travel modes to more efficient ones due to effective urban planning, and presenting policies enforcing efficiency improvements can all help achieve a cost-effective integrated transport framework. The other benefit associated with the implementation of sustainable transport is the presence of entirely new transport solutions incorporating a range of new technologies. As part of transitioning to sustainable transport, new technologies are needed to accommodate the need for eco-friendly systems. The deployment of new technologies can positively impact economic and social aspects of sustainability, mainly because new job opportunities are created in such integrated systems. For instance, a recent study on Europe's 2050 carbon-neutral targets illustrates that more than 1.5 million new jobs could be expected as part of the gradual transition to renewable energy production in the transport and electricity sectors [43]. Nonetheless, a well-designed, integrated, transport system can help create more sustainable cities by reducing dependencies on vehicles as an economic objective (via digital transformation and the consequent service economy) [44, 45]. In other words, sustainable transport can foster economic development via improving the efficiency and accessibility of the overall transport system.

4.4 Modes of Sustainable Transportation

There are several modes of sustainable transportation. Their implementation on a global scale can effectively reduce GHG emissions and can contribute significantly to the improvement of existing environmental issues. These modes are elaborated below.

4.4.1 Walk

The simplest mode of sustainable transportation for local places is walking, which can not only save money but also keep us healthy. Indeed, it is an excellent form of exercise without following any particular schedule. Walking longer distances can be made comfortable with shoe inserts and good-quality jogger shoes. Move your body and benefit the environment by going for a walk with friends, family, or even by yourself [46]. It is preferable to walk to work, school, grocery stores, and other nearby destinations. Walking is free, it emits no GHGs, and it's also an excellent type of physical training for the body [47]. Walking benefits your heart's health, your quality of life, and the health of your city. Hence, if you make five trips of less than 2 km/week on foot, you cut 86 kg off your annual carbon footprint. By preferring walking over driving personal conveyance, one can contribute significantly to the reduction of air pollutants, which is quite essential for a clean environment and healthy lifestyle. A study demonstrates that the health advantages of walking in a typical metropolitan setting outweigh the drawbacks by walking only for 16 hours [48].

4.4.2 Bicycle

This is an obvious choice. Bicycling to work is a fantastic way to travel sustainably. Bicycling instead of using a car makes a significant contribution to reducing GHG emissions. Although walking is a fantastic environmentally friendly way of transportation, riding a bicycle has numerous advantages because it is quicker and provides its own sort of exercise. These days, bicycles can be purchased for incredibly affordable prices. Purchasing and maintaining a bicycle only costs a small portion of that of an automobile [47]. The use of bicycles to reach where you want to go has been around for centuries. You never have to be concerned about these forms of transportation having a negative influence on the environment by making loud noises or releasing hazardous emissions such as carbon monoxide into the atmosphere [46].

It should not be a surprise that walking and riding a bicycle are the most environmentally friendly modes of transportation. In addition to having zero carbon dioxide emissions at every level, they are also fun and healthy. Of course, it would not always be possible for you to ride a cycle everywhere you want to go, but if you have to choose between a 10-minute bus ride and a 20-minute bicycle ride, the latter would be better for the environment and for you. Numerous nations and towns are launching campaigns and initiatives to promote cycling. The goal of the USA is to build a coast-to-coast bicycle path. From the nation's capital to the west of Seattle (Pacific Ocean), the Great American Rail Trail travels 3700 miles through 12 states. A 62-mile cycling expressway will be opened in Germany just for cyclists. Ten western cities will be connected by this motorway, along with four colleges, in a move to promote more environmentally friendly transportation.

Cycling is indeed environmentally friendly, social, cathartic, calming, and a great way to see the world. They are also the best and most affordable method to see a new city [49]. The statistics for cycling are equally encouraging: if you pedal 5 miles to work 4 days a week, you avoid driving 3220 km. At least 750 kg of CO_2 emissions and 380 l of gasoline consumption can be saved. Cycling is frequent and the quickest form of transportation in cities, especially during rush hours. As a result, cyclists are less likely to be exposed to air pollution since they spend relatively less time in traffic. In this aspect, a report demonstrates that the health advantages of cycling in an urban setting only exceed the drawbacks after six hours of cycling [48].

4.4.3   Electric Bike/Scooter

Electric bikes are excellent eco-friendly means of transportation because they do not pollute the environment with hazardous emissions. It only takes a little pedaling to move an electric bike. Yet, there are many legal restrictions on how fast electric bikes can go. Some nations cap the top speed of electric bikes at 20 mph. Yet, in certain nations, riding an electric bike requires a specific registration, insurance, and license [47]. Supporters of low-emissions modes of transportation, like bicycles, now have several options. You might also wish to think about traveling by moped or scooter. Riders choose electric scooters over hoverboards mostly because they are safer and more compact than bikes. Those who are overweight, however, often worry that this mode of transportation would not be able to carry them. Yet, there are scooters available for riders who weigh more than 300 pounds, making it possible for heavier riders as well [46]. Electric motorcycles emit no pollution similar to other EVs. Usually, they run on batteries. If renewable energy resources are used to recharge the batteries, the pollution becomes negligible using this mode of sustainable transportation [47].

4.4.4   Carpooling

Although not a novel idea, carpooling is nevertheless beneficial and pertinent today. In short, if you're all traveling to the same place, you ride together. This significantly reduces the number of vehicles on the road by reducing traffic and harmful pollutants. Arizona is one state that has included this approach in its laws and regulations. There are currently designated lanes that are only used for driving in carpools. However, if you are caught driving in such lanes alone, you could be fined heavily. These sustainable transportation solutions are now being adopted by carshare applications like Lyft and Uber. You can enjoy a less expensive trip by sharing it with others [46]. These initiatives are useful in motivating people to cultivate greater environmental awareness. The flourishing global economy has contributed to the rise in automobile production. Although many people are thrilled about this achievement, pollution levels have dramatically risen. Carpooling, another name for multiple-occupant vehicles, helps to minimize pollution levels by reducing the number of cars on the road. Vehicles with several occupants are a very beneficial and environmentally friendly form of transportation. When frequently traveling in the same direction, groups of co-workers and friends can share one vehicle. Using one automobile to transport a number of passengers to their destination is much more affordable and environmentally sustainable than having multiple people drive their own vehicles in the same direction. This is undoubtedly an excellent method to save money and gasoline [47].

4.4.5 Electric Car

In Europe, cars produce 12% of CO_2 emissions out of which 73% of emissions are caused by newly sold cars. Majority of them are company cars. As a result, there has been a strong push for electric cars future that prevents the use of fossil fuels. Sales of EVs increased by more than three times in 2020, reaching 10.5% in Europe [49]. Over the past few years, the number of cars that do not need petrol to run has increased dramatically. Instead of rushing to gas stations, drivers now plug their electric cars into charging stations. Using domestic 110 V or 13A plugs at home or when traveling makes charging these cars pretty simple. Tesla, led by CEO Elon Musk, is one of the most well-known companies of electric cars to have gained popularity in the environmentally friendly community. EVs are available for a reasonable price starting at around $23,000. Overall, this increases the practicality of broad ownership and may result in lower pollution levels and improved air quality [46].

Although there are several excellent electric cars available, proper research is required to select the best one. Also, you need the compatible amenities, which you may purchase from an EV cable store. Despite the fact that power plants that generate energy may also release poisonous emissions, electric cars that are totally powered by electricity do not produce any harmful gases. Even so, renewable energy sources including hydroelectric, geothermal, wind turbines, and solar energy can be used to generate power. In comparison with gasoline-fueled conventional cars, electric cars with alternative fuels and cutting-edge vehicle technologies ease environmental pressure [47]. An electric automobile has a more environmentally friendly footprint than a thermal car if we compare the two. This is because clean energy-based electricity is required to operate an electric car. It also uses rechargeable batteries and an electric motor for propulsion. This indicates that the use of electrical energy to propel the car reduces CO_2 emissions from getting into the atmosphere [48].

4.4.6 Public Transportation

It is obvious that using public transportation has several advantages for the environment. Millions of people could avoid using their personal conveyance and take public transportation, which would significantly reduce the pollution that cities produce. The bus is the best means of transportation for cities, especially hybrid and battery electric buses. New technologies have made it possible to design and produce automobiles that emit as little pollution as possible. It is crucial that public transportation offers a comfortable environment as a part of its high-quality service, which can encourage more people to adopt it. Hence, it needs to be secure, efficient, and reliable [48]. Some of the most unique green buses on the market today are already available. The Mercedes-Benz Ciatro G BlueTec Hybrid Bus, which makes use of lithium-ion batteries and four-wheel hub motors, is a prime example. The largest battery in the world is thought to be the lithium-ion one. Large amounts of energy extracted from a generator (diesel) can be stored in the battery.

Trains are becoming more environmentally friendly thanks to hybrid electric locomotives and other cutting-edge green technologies as the majority of governments throughout the world are now more committed than ever to promoting sustainable transportation. Similar technologies used in hybrid electric cars are also employed in hybrid electric locomotives. The conductor rail, overhead power cables, and energy-storage technologies like batteries and fuel cells are all used by modern electric trains. These electric trains have the advantage of reaching maximum speeds of greater than 200 mph while retaining high safety levels [47]. Using public transportation has significant benefits to the environment. Usually, carpooling and public transportation operate on the same principles. It might be advantageous to occasionally use the bus or train instead of your car [46].

4.5 Sustainable Transportation in Modern Urban Advancement

As cities continue to grow and populations increase, the need for sustainable transportation becomes increasingly vital: the more a city grows, the more transportation it needs, and the more sustainable transport it needs to provide. Transportation plays a crucial role in urban advancement, and sustainability is essential for creating more livable, equitable, and environmentally friendly cities. This section explores the roles of sustainable transport in urban growth and how it can contribute to the overall well-being of urban residents.

4.5.1 Importance of Sustainable Transport in Urban Growth

Urban growth is an essential aspect of economic development, but it often comes at a significant cost to the environment and public health. Nonetheless, sustainable transport can help cities mitigate the negative impacts of conventional transport systems on the environment by reducing the reliance on private vehicles via promoting alternative, more efficient, modes of transportation. Examples are electric buses and bicycles that can help reduce congestion and GHG emissions. Sustainable transport also promotes social equity by providing affordable and accessible transportation options for all members of society, including those who cannot afford to own a car. Further, it helps improve urban planning to accommodate novel modes of transport, enable smart cities, and increase economic growth.

4.5.1.1 Urban Planning

Given the global drive toward more sustainable transport modes, an efficient distribution of urban space is paramount in creating feasible transport systems. Thus, urban planners have a vital role in guaranteeing that cities have sustainable and equitable transport infrastructures. Sustainable transport necessitates urban plans that encourage walking, cycling, and other eco-friendly transportation modes while decreasing reliance on private cars to minimize their negative environmental effects. Nevertheless, urban space transformation to promote sustainable transport modes has gained widespread attention in the literature, despite its time-consuming implementation process. Many studies have investigated the effects of transforming urban spaces in sustainable transport modes such as cycling and walking [50–52]. A successful example of integrating sustainable transport and urban planning is the city of Copenhagen in Denmark, where cycling and public transportation are given priority. Copenhagen's commitment to cycling infrastructure, including dedicated lanes, separated pathways, and secure bike parking, has resulted in a sustainable urban design that encourages over 60% of inner-city trips to be made by bicycle, outnumbering cars in the area and improving quality of life for its residents [53].

4.5.1.2 Smart Cities

Smart cities make use of technology and data to boost sustainability, economic growth (ecosystem stability), and living standards in urban areas [54]. The key to building smart cities is sustainable transport, which can be supported by a range of innovative and modern technologies. Figure 4.3 illustrates some of the key characteristics of sustainable transportation that are directly connected to the development and maintenance of smart cities. These features contribute to environmental stability as well as improving the quality of urban life. The development of sustainable transport infrastructure and technology, like EVs and optimized public transit, can address environmental concerns as well as provide greater connectivity, which allows people to travel more quickly and efficiently. To create intelligent solutions for urban areas, it is therefore necessary to establish secure infrastructure and integrate technology, which enables travelers to stay connected [55]. Sustainable transport can provide the residents of smart cities with improved access to

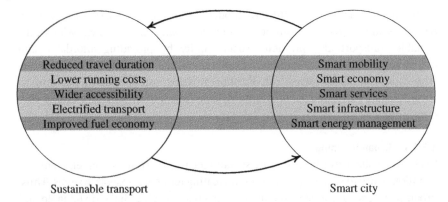

Reduced travel duration
Lower running costs
Wider accessibility
Electrified transport
Improved fuel economy

Smart mobility
Smart economy
Smart services
Smart infrastructure
Smart energy management

Sustainable transport   Smart city

Figure 4.3 Schematic of some of the key features connecting smart cities and sustainable transport.

essential services, such as healthcare, education, and employment, all contributing to the economic growth and development of the city. By creating a more equitable and accessible transport system, smart cities can promote sustainable and inclusive growth.

4.5.1.3 Economic Growth

Sustainable transport in modern cities can contribute to economic growth and productivity by reducing transportation costs and creating job opportunities. In addition, sustainable transport can attract new investors, businesses, and tourists. Copenhagen, Denmark, shows best practices for sustainable urban cycling and bicycle tourism. Approximately 7% of foreign visitors use bicycles while they are in the city, highlighting the success of this sustainable development [56]. Sustainable infrastructure such as bike lanes, pedestrian walkways, and electrified public transport systems can make it easier and more affordable for people to access goods and services. This can also increase economic activity. Additionally, there are many job opportunities that can be created by sustainable transport, such as in the construction, maintenance, and operation (e.g., food and drink industry) of transport systems. For example, there are over 35,000 new jobs created in the United Kingdom due to investments in bike infrastructure and cycle tourism, showing a remarkable cost-benefit ratio of 13 : 1 [57, 58].

4.5.1.4 Promoting Sustainable Transport

In order to ensure sustainable urban development and growth, sustainable transport must be promoted in urban areas. The following are strategies that cities can use to promote sustainable transport:

- Implementing sustainable urban planning: Urban planners can design cities in ways that encourage sustainable transport. This means creating mixed-use areas that reduce the need to travel long distances and promoting local amenities accessible via public transport, cycling, or walking.
- Investing in public transportation: Public transport is a vital element of sustainable transport. It is often cheaper and more efficient than private transportation and can help reduce congestion. Public transportation can be expanded and made more accessible to all.
- Encouraging active modes: Active transportation such as walking, or cycling, is more sustainable and healthier than traditional modes. Building bike lanes and walking pathways in cities can encourage active transportation.
- Promoting electric vehicles: EVs offer an environmentally friendly alternative to conventional fossil fuel-powered vehicles. Cities can promote EVs by providing charging infrastructure and offering incentives, such as tax rebates for EVs and free parking.

4.6 Infrastructure for Sustainable Transportation

In the context of sustainable development, sustainable transportation improves the quality, convenience, and livability of the city through constructing a comprehensive range of transportation infrastructure services to achieve the objectives of economic efficiency and sustainability in urban transportation. Several transportation facilities, such as pathways, cycling routes, trains, buses, and taxi zones shape the urban transportation environment. Sustainable transportation has been under continuous development, in both academic and industry sectors, for at least the last two decades and many of these means are now well established in many cities around the world. However, electrification has been a game-changing concept in this context.

Transport electrification has been declared as a main strategy to address the social and political movement to clean energy. Electric buses and scooters have been introduced to already developed transportation infrastructure. This is, somehow, the first significant contact between the mobility and electricity sectors after electrifying the metro and light trains in cities. The massive electrification of transportation requires significant interactions between mobility and electricity infrastructure.

For example, interaction between electric buses and the power grid, in the form of charging and possible services, should be well addressed in the transition from fossil fuel-based buses to EVs. Reliability in providing mobility services is the top priority of any public transport system. A comprehensive study of mobility demand considering practical constraints such as different driving scenarios, peak demand, and traffic conditions is required to achieve the energy and fleet requirements of a reliable electric bus system. For example, electrification of a bus transport system with 4000 fleet may augment approximately 1.2 gigawatt hours (GWh) electricity demand to the grid considering a typical electric bus with a 300 kilowatt hour (kWh) battery capacity. The problem will become more sophisticated if 175–500 kilowatts (kW) fast chargers are installed to meet tight bus schedules. That means mobility requirements and power grid constraints limit each other.

A sustainable transportation system requires a comprehensive integration of mobility, energy, and social infrastructure, as shown in Figure 4.4. Traditionally, networks of different means of transport have been designed in most cities such that an interconnected mobility network has been achieved. Transition toward e-mobility is probably the main challenge of this sector. This is where the mobility sector meets the energy sector to facilitate charging of these new mobility means. Clearly, renewable and clean energy technologies are required to establish a sustainable charging infrastructure. Last, but not least, the social aspect of these transitions should be well addressed through effective and smooth communication of benefits of these transition to people.

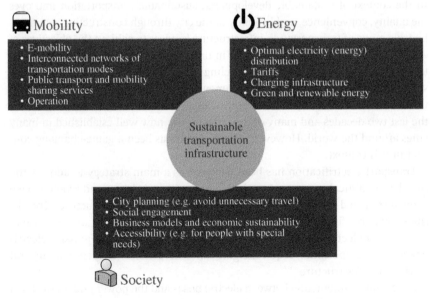

Figure 4.4 Different infrastructure for the transition toward sustainable transportation system.

4.6.1 Governance

Transition to sustainable transportation and keeping it operational in the long term often requires heavy investment in infrastructure upgrades. Traditionally, there has been a role-shrinking agreement between the central and local governments, which has been somehow successful in keeping the infrastructure operational. However, during transition to sustainable transport, investing in major infrastructure is likely to be ineffective unless accompanied by local actions and policies using small-scale investments [59]. In other words, planners and policy-makers should consider the details on the community scale while they are concurrently planning major transportation corridors. There are several stakeholders when it comes to sustainable transportation including mobility service providers and information and computer technology (ICT), urban and transport planning authorities, electricity market operators and aggregators, electric bus, grid, and charging station operators.

Enhancement of the cooperation among these stakeholders in the context of sustainable transportation is one of the key enablers underpinning a successful transition and further operation. The complexity is that many of these entities are working isolated from the others in the current transportation system. Therefore, the first step is to bring them together and start negotiations between them about the operation and planning of future transportation. For example, in a 2021 position paper, the European Network of Transmission System Operators for Electricity (ENTSO-E), called on transmission system operators (TSOs) to govern the optimal transition to EVs [60]. This means that TSOs must look beyond their traditional responsibilities, get engaged with the mobility sector, and adapt and support the wider energy system integration.

4.6.2 Interaction with Electricity Infrastructure

There is a significant interest these days to revisit the sustainable transportation concept in the context of electrification of everything. Many of the transportation means, from vehicles to bicycles and scooters, are going to be electric. That means the future sustainable transport system and the power grid, mainly empowered by renewable energy resources, will be heavily interconnected.

4.6.2.1 Electric Buses and the Power Grid

The capacity of electric bus batteries is normally between 60 and 500 kWh. The size of the battery is chosen based on the travel distance and travel time of electric buses in each route. Despite the design process, these buses generally require on-route fast chargers (up to 500 kW) since their time schedule does not let them leave the service for charging. This is on top of the fast chargers in depots. There-fore, a significant upgrade of the electricity network is required on the scale of increasing the capacity of substations and feeders [61]. Practical constraints should be also considered to address the limitations of the distribution grid where e-mobility normally connects.

The location and size of charging and battery stations pose significant challenges for e-mobility. Optimal placement of charging stations is an active field of research and requires mobility information such as bus routes and schedules. This optimi-zation sometimes requires smart scheduling algorithms for electric bus deploy-ment from the depot and main stops as well as for electric bus recharging. On the other hand, network operators are mainly concerned with the capital costs of installing a fast charger or a depot. Unpredictable and intermittent electric bus charging is another challenge. Any charging activity should be well aligned with the peak time of the energy demand. Optimal daytime charging could be helpful in spreading the load.

4.6.2.2 Operational Strategies

Operational strategies identify the time and duration of operation, sleep, and char-ging of a transportation means based on the required mobility service. For exam-ple, most of the transportation means are normally in the stations during night-time; thus, night-time charging could be the first option. However, it does not line up with solar generation times and is often aligned with electricity network peaks. New charging strategies that maximize renewable energy storage are required. These strategies should utilize the flexibility of each mobility service to manage green charging. For example, opportunity on-route chargers can be supplied by solar if appropriate storage systems are used. Another challenge is the small-energy capacity of EVs compared to fossil fuel-based cars [62]. That means at least for short-term planning, electric buses with large and heavy storage systems should be considered. This adds operational limitations when it comes to challeng-ing conditions, such as hilly routes.

4.6.2.3 Compensation for the Minimum Demand Reduction

As the use of distributed energy resources like solar rooftop photovoltaic and bat-tery storage systems increases, it becomes challenging to maintain a minimum operational demand for the grid [63]. This can impact grid security and reliability. To address this, network operators can implement several strategies such as using storage systems to absorb excess solar generation, curtailing photovoltaic systems during emergencies, and encouraging e-mobility charging during low-mobility demand periods. These strategies can help to increase the minimum operational demand and ensure that EVs charge from clean energy. However, it is crucial for network operators to carefully balance the integration of distributed energy resources with grid reliability and security.

4.6.2.4 Flexible Operation of E-mobility

The demand for electricity and transportation systems peaks at different times, presenting an opportunity to integrate sustainable transportation systems flexibly into the grid. For instance, during the early evening peak in electricity demand, transportation systems slow down, while the opposite happens during peak day-time mobility demand. Integrating sustainable transportation systems can align the fleet's electricity demand with the daily electricity demand profile.

Sustainable transportation planning, such as dynamic route assignment, can also increase flexibility in the transportation system. This can reduce the number of fleet vehicles required and increase the utilization of purchased vehicles, particularly for electric buses. Such planning can help align the electricity demand profile with the transportation system, resulting in a more sustainable and efficient use of resources.

4.6.3 Features of Integrated Sustainable Transportation Networks

For a long time, several transportation networks have been implemented and run, sometimes by different operators, in a city. The main focus of the network designers has been either on vehicles or on individual network design and optimization [64]. For example, the planning concerns of bus-operating companies and city transport service operators have been limited to bus networks in order to improve the network configuration [65], optimize the number of fleets through dynamic bus routing [66, 67], or reduce passenger waiting time [68]. The same scenario happens for other transportation means such as trains and trams. However, in the context of integrated transportation, these networks are identified as fundamental subsystems that need to work collaboratively.

The problem of integrated transportation system brings up the spatial distribution of transportation services as a measure for urban development [69]. The main idea is to guarantee fair accessibility of all urban services, such as education, entertainment, and health, using the transportation system. As a part of the urban management system hierarchy, integrated transport can be managed through hierarchical approaches such as the demand-responsive transit proposed in Ref. [70]. In fact, the problem of integrated transportation planning should be addressed in a layer of the comprehensive, urban planning framework. In this context, some transportation links need improvement, some of them can be removed, while new links should be also defined.

Designing an integrated transportation system in the context of comprehensive urban planning is indeed a large-scale optimization problem. It requires location models for city facilities and can be formulated as either a continuous or discrete optimization problem [71]. A wide range of objective functions can be considered from minimizing the cost of construction of new links to minimizing the travel time between two specific centers [72]. This sophisticated network optimization problem requires advanced methods to solve such as mathematical graph theory [73].

4.6.3.1 Transport Resilience and Sustainability

As cities become more complex in terms of social relations, population, and infrastructure, they become more vulnerable to failures and poor management. Several uncertain factors, such as political instability and cyberattacks, may also happen in this interconnected infrastructure. In addition, new disruptive technologies are emerging in our daily lifestyle, which are sometimes out of the direct control of planners and urban managers. For example, the emergence of EVs impacts the stability of the power distribution grid in a city since unexpected EV charging may happen anywhere at any time.

Resilience of a transportation system can be studied either from the supply-side, i.e., the system planners and operators, or from the customer-side [74]. The focus on the supply side is to recover the quality of service as fast as possible after the failure or attack. It may include different strategies such as establishing redundancies in the system, shifting between different transportation modes, and repair of critical inputs. On the customer-side, the focus is on running the life or business activities as usual using the remained and healthy transport services. Social reactions from the customers may also impact transportation. Studies show a significant reduction in the number of passengers in the United Kingdom and the United States after unexpected serious incidents [75]. This puts the transport system under financial pressure.

While transport resilience focuses on fast system recovery after unexpected events, transport sustainability always looks for socioeconomic equilibrium, i.e., self-sufficiency, as well as ecologically viable solutions with minimum environmental impact. An integrated sustainable transport system should have close interaction and co-evolution with other urban subsystems to contribute to the whole urban management system without compromising other city services. Although resilience and sustainability seem to be totally different concepts, the contrast is not clear in the research environment [76].

The interconnection between transport resilience and sustainability can be shown using Figure 4.5 proposed in Ref. [76]. In fact, the performance of any large-scale hierarchical system can be studied through a bottom-up or top-down approach. The former is more subjective, called active, while the latter is more subjective and passive. Also, the transport development activities can be more formal (rational) than informal. Figure 4.5 shows how the resilience and sustainability of

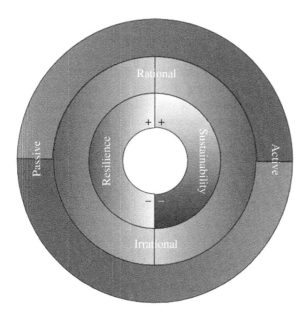

Figure 4.5 Difference between resilience and sustainability of the transport system (*Source:* Adapted from [76]).

transportation can be distinguished considering these concepts. For example, it shows that active and rational approaches result in sustainable transport while passive-rational is related to resilience. Therefore, resilience and sustainability of the transport system are not necessarily aligned to each other.

In addition to transportation means and networks, other aspects related to them, such as stations, should be considered in an integrated sustainable transport system. For example, the land use planning problem has been addressed in Ref. [77] to sustainably use the area around a transport station in Beijing, China. This is indeed a multi-objective optimization problem including different features such as density and cultural diversity of the area, accessibility, and environmental impact. In addition to technical complexity, there are several regulation and legislation challenges. For example, a study in the United Kingdom shows how conflict of responsibilities, such as duplication of procedures, may cause failure in integrating the transportation system sustainably.

4.6.4 Transition to a Sustainable Transportation

Transition from the current unsustainable transportation to a sustainable one is a complicated socio-technical problem [78]. Governance, financing, infrastructure, and neighborhoods are the four pillars identified for this transition [59].

A successful transition requires an interdisciplinary major activity to cover all these four pillars simultaneously. For example, an ASI approach is proposed in Ref. [79] to overcome the encountered barriers in transition to sustainable transportation. The first step in this approach is to "avoid" unnecessary traveling demand by implementing smart city planning. This can be done by reducing the distance between residential locations and places of daily activities or prioritizing local pedestrians or cyclers. "Shifting" the traveling demand from personal vehicles to public transport is a popular approach toward clean transportation. Finally, the infrastructure related to clean transportation, such as EV charging and intelligent transport facilities, should be "improved". Another example in the Mexican bajío shows how the lack of social support can hinder the integration of a 115 M pesos cycling infrastructure project into daily lives [80].

In late 2021, the United States announced a $1.2 trillion Infrastructure Investment and Jobs Act (IIJA) to boost three key areas of transport electrification. The first one, called connectivity, funds highways, roads, and bridges, and invests in public transit and water system upgrades, passenger rail, airports, supply chain ports, and high-quality broadband connectivity [81]. Secondly, the decarbonization part supports the expansion of renewable energy, EV adoption, and emerging technologies such as energy storage, blue hydrogen, geothermal, and advanced nuclear power generation. Finally, the Act supports the workforce by creating 1.5 million jobs per year for the next 10 years [82]. All these areas should progress simultaneously to achieve sustainable green transportation.

4.7 Conclusion

This chapter presents an overview of various aspects of sustainable transportation. It encompasses different perspectives from cogent reasoning for the adoption of sustainable transportation modes to the infrastructure requirements for its well-planned implementation on a global scale. It also provides a deep insight into the existing challenges for the transition from the conventional fossil fuel-based transportation system toward the clean, secure, reliable, and healthy modes of sustainable transportation. It also throws some light on the key benefits in terms of social and economic views of shifting to these modes. It also highlights the core issues in planning for the establishment of proper infrastructure that can meet the increasing demand for mobility. This chapter is quite beneficial for academic researchers and industrial workers to be aware of the intensity of the existing climate crisis in which the conventional fuel-based transportation system plays a significant role. Moreover, it elaborates on the systematic ways to not only improve the current situation of the transportation sector but also plan for future policies to successfully achieve green economy and sustainable development. In view of the discussion on various aspects of sustainable transportation networks in this chapter, some conclusions are extracted as follows:

- In order to achieve sustainability in urban areas, proper planning of a low-emission, responsive, and well-designed transport system is extremely important.
- The major challenges involved in transforming the transport sector into a sustainable system include huge investment in both infrastructure and technology, proper planning, safety, and security risks.
- The implementation of sustainable transport can provide several social, environmental, and economic benefits.
- Urban planning has an important role in the implementation of sustainable transportation infrastructure.
- There is still a lack of government policies and regulations in promoting sustainable transportation.
- The use of EVs and fuel cell-based vehicles in urban areas has significantly reduced emissions in the transportation sector and improved air quality in recent years.
- One of the major causes of accidents and injuries is the lack of dedicated bike lanes and safe pedestrian crossings in many urban areas.
- Adoption of sustainable transport can create many job opportunities such as in the construction, maintenance, and operation of transport systems.
- Cities can use some strategies to promote sustainable transport i.e., implementing sustainable urban planning, investing in public transportation, encouraging active modes, and promoting EVs.
- The main strategy to address the social and political movement to clean energy is transport electrification.
- A sustainable transportation network requires a comprehensive integration of energy, mobility, and social infrastructure.
- TSOs must get engaged with the mobility sector and support the wider energy-system integration.
- The peak time of the energy demand should be considered while planning for any vehicle charging activity.
- Unlike nighttime charging, new charging strategies that maximize renewable energy storage are required.

References

1 EPA United States Environmental Protection Agency (2022). Sources of greenhouse gas emissions. EPA. https://www.epa.gov/ghgemissions/sources-greenhouse-gas-emissions (accessed 6 March 2023).

2 Smart Freight Center (2021). Annual report, leading the way to efficient and zero emission freight and logistics. Amsterdam: SFC. https://shellfoundation.org/app/uploads/2021/03/SFC_-Annual_Report_2020.pdf (accessed 6 March 2023).

3 International Transport Forum (2021). ITF Transport Outlook. OECD. https://www.itf-oecd.org/itf-transport-outlook-2021 (accessed 6 March 2023).

4 McKinnon, A.C. (2010). The potential of economic incentives to reduce CO_2 emissions from goods transport. *DILF Orientering* 47 (1): 24–38.

5 Intergovernmental Panel on Climate Change (2023). Climate change 2021 the physical science basis. https://www.ipcc.ch/report/ar6/wg1/downloads/report/IPCC_AR6_WGI_SPM_final.pdf (accessed 6 March 2023).

6 Nasrin, N., El-Sayed, I., and García, J. (2022). A review of simulation models for CO2 pollution reduction in transportation sector. *2022 IEEE Vehicle Power and Propulsion Conference (VPPC)*, 1–6 (1 November 2022). IEEE.

7 An Australian Government Initiative (2023). Green vehicle guide, driving better choices. Australia. https://www.greenvehicleguide.gov.au/ (accessed 6 March 2023).

8 Madhav University (2018). Vehicular pollution. London: India. https://madhavuniversity.edu.in/vehicular-pollution.html (accessed 6 March 2023).

9 European Environmental Agency (2018). Transport. Copenhagen. https://www.eea.europa.eu/themes/transport/intro (accessed 6 March 2023).

10 Romano, A., Tocchi, D., Tinessa, F. et al. (2022). Analysis of the carbon footprint of freight transport in the mass market retail sector: a case study in Campania (Italy). *2022 IEEE International Conference on Environment and Electrical Engineering and 2022 IEEE Industrial and Commercial Power Systems Europe (EEEIC/I&CPS Europe)*, 1–6 (28 June 2022). IEEE.

11 Manual, S. (2010). *A Guide for Sustainable Urban Development in the 21st Century.* China: Shanghai.

12 Hussain, Z. (2022). Environmental and economic-oriented transport efficiency: the role of climate change mitigation technology. *Environmental Science and Pollution Research* 29 (19): 29165–29182.

13 Change IC (2014). Mitigation of climate change. Contribution of working group III to the fifth assessment report of the intergovernmental panel on climate change, 1454: 147.

14 Dunn, D.R. (2016). Monthly Energy Review.US Energy Information Administration.

15 Taiebat, M., Brown, A.L., Safford, H.R. et al. (2018). A review on energy, environmental, and sustainability implications of connected and automated vehicles. *Environmental Science & Technology* 52 (20): 11449–11465.

16 Chapman, L. (2007). Transport and climate change: a review. *Journal of Transport Geography* 15 (5): 354–367.

17 Edelenbosch, O.Y., McCollum, D.L., Van Vuuren, D.P. et al. (2017). Decomposing passenger transport futures: comparing results of global integrated assessment models. *Transportation Research Part D: Transport and Environment* 55: 281–293.

18 Girod, B., Van Vuuren, D.P., Grahn, M. et al. (2013). Climate impact of transportation A model comparison. *Climatic Change* 118: 595–608.

19 Creutzig, F., Jochem, P., Edelenbosch, O.Y. et al. (2015). Transport: a roadblock to climate change mitigation? *Science* 350 (6263): 911–912.

20 Pietzcker, R.C., Longden, T., Chen, W. et al. (2014). Long-term transport energy demand and climate policy: alternative visions on transport decarbonization in energy-economy models. *Energy* 64: 95–108.

21 McCollum, D.L., Wilson, C., Bevione, M. et al. (2018). Interaction of consumer preferences and climate policies in the global transition to low-carbon vehicles. *Nature Energy* 3 (8): 664–673.

22 Weiss, M., Dekker, P., Moro, A. et al. (2015). On the electrification of road transportation – a review of the environmental, economic, and social performance of electric two-wheelers. *Transportation Research Part D: Transport and Environment* 41: 348–366.

23 Andwari, A.M., Pesiridis, A., Rajoo, S. et al. (2017). A review of Battery Electric Vehicle technology and readiness levels. *Renewable and Sustainable Energy Reviews* 78: 414–430.

24 Weiss, M., Patel, M.K., Junginger, M. et al. (2012). On the electrification of road transport-Learning rates and price forecasts for hybrid-electric and battery-electric vehicles. *Energy Policy* 48: 374–393.

25 Plötz, P., Axsen, J., Funke, S.A., and Gnann, T. (2019). Designing car bans for sustainable transportation. *Nature Sustainability* 2 (7): 534–536.

26 Nykvist, B. and Nilsson, M. (2015). Rapidly falling costs of battery packs for electric vehicles. *Nature Climate Change* 5 (4): 329–332.

27 Zhang, R. and Fujimori, S. (2020). The role of transport electrification in global climate change mitigation scenarios. *Environmental Research Letters* 15 (3): 034019.

28 European Commission. CO_2 emission performance standards for cars and vans. Europe. https://ec.europa.eu/clima/eu-action/transport-emissions/road-transport-reducing-co2-emissions-vehicles/co2-emission-performance-standards-cars-and-vans_en (accessed 6 March 2023).

29 Boxill, S.A. and Yu, L. (2000). *An Evaluation of Traffic Simulation Models for Supporting Its.* Houston, TX: Development Centre for Transportation Training and Research, Texas Southern University.

30 Zhongming, Z., Linong, L., Xiaona, Y., and Wei, L. *The Role of CCUS in Low-Carbon Power Systems.* Paris: IEA.

31 Jacob Teter (2023). Transport. https://www.iea.org/reports/transport (accessed 6 March 2023).

32 U.S. Energy Information Administration (2021). *International Energy Outlook 2021 Narrative*. Washington, DC: U.S. Energy Information Administration https://www. eia.gov/outlooks/ieo/pdf/IEO2021_Narrative.pdf (accessed 6 March 2023).

33 Jaworski, A., Mądziel, M., and Kuszewski, H. (2022). Sustainable public transport strategies – decomposition of the bus fleet and its influence on the decrease in greenhouse gas emissions. *Energies* 15 (6): 2238.

34 Mądziel, M., Campisi, T., Jaworski, A. et al. (2021). Assessing vehicle emissions from a multi-lane to turbo roundabout conversion using a microsimulation tool. *Energies* 14 (15): 4399.

35 United Nations. Population Division. https://www.un.org/development/desa/pd (accessed 6 March 2023).

36 Infrastructure Outlook (2023). Forecasting infrastructure investment needs and gaps. https://outlook.gihub.org (accessed 6 March 2023).

37 Milani, L., Mohr, D., and Sandri, N. (2021). *Built to Last: Making Sustainability a Priority in Transport Infrastructure*. McKinsey.

38 Müller, P., Schleicher-Jester, F., and Schmidt, M.P. (1992). Konzepte Flächenhafter Verkehrsberuhigung in 16 Städten. Gruene Reihe, Fachgebiet Verkehrswesen der Universitaet Kaiserslautern.

39 Transport S (2004). *A Sourcebook for Policy-makers in Developing Cities*. GTZ, Sector.

40 World Health Organization (2009). *Violence, Injury Prevention, World Health Organization. Global Status Report on Road Safety: Time for Action*. World Health Organization.

41 Xia, T., Zhang, Y., Braunack-Mayer, A., and Crabb, S. (2017). Public attitudes toward encouraging sustainable transportation: an Australian case study. *International Journal of Sustainable Transportation* 11 (8): 593–601.

42 Cuthill, N., Cao, M., Liu, Y. et al. (2019). The association between urban public transport infrastructure and social equity and spatial accessibility within the urban environment: an investigation of Tramlink in London. *Sustainability* 11 (5): 1229.

43 Potrč, S., Čuček, L., Martin, M., and Kravanja, Z. (2021). Sustainable renewable energy supply networks optimization – the gradual transition to a renewable energy system within the European Union by 2050. *Renewable and Sustainable Energy Reviews* 146: 111186.

44 Newman, P. and Kenworthy, J. (1999). *Sustainability and Cities: Overcoming Automobile Dependence*. Island Press.

45 Newman, P.W. (2015). Transport infrastructure and sustainability: a new planning and assessment framework. *Smart and Sustainable Built Environment* 4 (2): 140–153.

46 Kristel Staci (2018). Top 5 eco-friendly transportation methods you can feel great about. Blue & Green Tomorrow. https://blueandgreentomorrow.com/transport/

top-5-eco-friendly-transportation-methods-you-can-feel-great-about/ (accessed 6 March 2023).

47 Conserve Energy Future (2023). Modes and benefits of green transportation. https://www.conserve-energy-future.com/modes-and-benefits-of-green-transportation.php (accessed 6 March 2023).

48 EMESA 30 (2023). Sustainable transport and mobility: examples of taking care of the environment. https://www.emesa-m30.com/sustainable-transport-and-mobility/ (accessed 6 March 2023).

49 Gemma Howard-Vyse (2022). The most sustainable modes of transport – is snail travel the answer? Working Abroad. https://www.workingabroad.com/blog/the-most-sustainable-modes-of-transport-is-snail-travel-the-answer/ (accessed 6 March 2023).

50 Gössling, S., Schröder, M., Späth, P., and Freytag, T. (2016). Urban space distribution and sustainable transport. *Transport Reviews* 36 (5): 659–679.

51 Hagen, O.H. and Tennøy, A. (2021). Street-space reallocation in the Oslo city center: adaptations, effects, and consequences. *Transportation Research Part D: Transport and Environment* 97: 102944.

52 Silva, D., Földes, D., and Csiszár, C. (2021). Autonomous vehicle use and urban space transformation: a scenario building and analysing method. *Sustainability* 13 (6): 3008.

53 Henderson, J. and Gulsrud, N.M. (2019). *Street fights in Copenhagen: Bicycle and Car Politics in a Green Mobility City*. Routledge.

54 Lai, C.S., Jia, Y., Dong, Z. et al. (2020). A review of technical standards for smart cities. *Clean Technologies* 2 (3): 290–310.

55 Bamwesigye, D. and Hlavackova, P. (2019). Analysis of sustainable transport for smart cities. *Sustainability* 11 (7): 2140.

56 Nilsson, J.H. (2019). Urban bicycle tourism: path dependencies and innovation in Greater Copenhagen. *Journal of Sustainable Tourism* 27 (11): 1648–1662.

57 Newson, C. and Sloman, L. (2018). The value of the cycling sector to the British economy: a scoping study transport for quality of life. http://s27245.pcdn.co/wp-content/uploads/2018/06/The-Value-of-the-Cycling-Sector-to-the-British-Economy-FINAL2.pdf (accessed 5 April 2023).

58 Yanocha, D. and Mawdsley, S. (2022). Making the economic case for cycling. Institute for Transportation and Development Policy. https://www.itdp.org/wp-content/uploads/2022/06/Making-the-Economic-Case-for-Cycling_6-13-22.pdf (accessed 5 April 2023).

59 Kennedy, C., Miller, E., Shalaby, A. et al. (2005). The four pillars of sustainable urban transportation. *Transport Reviews* 25 (4): 393–414.

60 European Network of Transmission System Operators (ENTSO-E) (2021). Electric vehicle integration into power grids. https://www.entsoe.eu/2021/04/02/electric-vehicle-integration-into-power-grids/ (accessed 6 March 2023).

61 Gao, Z., Lin, Z., and Franzese, O. (2017). Energy consumption and cost savings of truck electrification for heavy-duty vehicle applications. *Transportation Research Record* 2628 (1): 99–109.

62 International Energy Agency (2015). CO2 emissions from fuel combustion. https://www.oecd-ilibrary.org/energy/co2-emissions-from-fuel-combustion-2015_co2_fuel-2015-en (accessed 6 March 2023).

63 Australian Energy Market Operator (AEMO) (2021). Minimum operational demand. https://www.aemo.com.au/-/media/files/learn/fact-sheets/minimum-operational-demand-factsheet.pdf?la=en (accessed 6 March 2023).

64 Taylor, C. and de Weck, O. (2006). Integrated transportation network design optimization. *47th AIAA/ASME/ASCE/AHS/ASC Structures, Structural Dynamics, and Materials Conference 14th AIAA/ASME/AHS Adaptive Structures Conference 7th 2006*, 1912.

65 Shimamoto, H. and Kurauchi, F. (2012). Optimisation of a bus network configuration and frequency considering the common lines problem. *Journal of Transportation Technologies* 2 (3): 220.

66 Koh, K., Ng, C., Pan, D., and Mak, K.S. (2018). Dynamic bus routing: a study on the viability of on-demand high-capacity ridesharing as an alternative to fixed-route buses in Singapore. *2018 21st international conference on intelligent transportation systems (ITSC)*, 34–40 (4 November 2018). IEEE.

67 Jiménez, F. and Román, A. (2016). Urban bus fleet-to-route assignment for pollutant emissions minimization. *Transportation Research Part E: Logistics and Transportation Review* 85: 120–131.

68 Sadrani, M., Tirachini, A., and Antoniou, C. (2022). Vehicle dispatching plan for minimizing passenger waiting time in a corridor with buses of different sizes: model formulation and solution approaches. *European Journal of Operational Research* 299 (1): 263–282.

69 Faludi, A. (2000). The European spatial development perspective-what next? *European Planning Studies* 8 (2): 237–250.

70 Edwards, D., Trivedi, A., Elangovan, A.K., and Dickerson, S. (2011). The network-inspired transportation system: a hierarchical approach to integrated transit. *2011 14th International IEEE Conference on Intelligent Transportation Systems (ITSC)*, 1507–1512 (5 October 2011). IEEE.

71 Bigotte, J.F., Krass, D., Antunes, A.P., and Berman, O. (2010). Integrated modeling of urban hierarchy and transportation network planning. *Transportation Research Part A: Policy and Practice* 44 (7): 506–522.

72 Drezner, Z. and Wesolowsky, G.O. (2003). Network design: selection and design of links and facility location. *Transportation Research Part A: Policy and Practice* 37 (3): 241–256.

73 Háznagy, A., Fi, I., London, A., and Németh, T. (2015). Complex network analysis of public transportation networks: a comprehensive study. *2015 International*

Conference on Models and Technologies for Intelligent Transportation Systems
(MT-ITS), 371–378 (3 June 2015). IEEE.

74 Cox, A., Prager, F., and Rose, A. (2011). Transportation security and the role of
resilience: a foundation for operational metrics. *Transport Policy* 18 (2): 307–317.

75 Rose, A., Oladosu, G., Lee, B., and Beeler-Asay, G. (2009). The economic impacts of
the 2001 terrorist attacks on the World Trade Center: a computable general
equilibrium analysis. *Peace Economics, Peace Science and Public Policy* 16 (2):
Article 6.

76 Zhang, X. and Li, H. (2018). Urban resilience and urban sustainability: what we
know and what do not know? *Cities* 72: 141–148.

77 Ma, X., Chen, X., Li, X. et al. (2018). Sustainable station-level planning: an
integrated transport and land use design model for transit-oriented development.
Journal of Cleaner Production 170: 1052–1063.

78 Geels, F.W. (2019). Socio-technical transitions to sustainability: a review of
criticisms and elaborations of the multi-level perspective. *Current Opinion in
Environment Sustainability* 39: 187–201.

79 Shah, K.J., Pan, S.Y., Lee, I. et al. (2021). Green transportation for sustainability:
review of current barriers, strategies, and innovative technologies. *Journal of
Cleaner Production* 326: 129392.

80 Soliz, A. (2021). Divergent infrastructure: uncovering alternative pathways in
urban velomobilities. *Journal of Transport Geography* 90: 102926.

81 Klingel, J. and von Loesecke, E. (2022). Transportation electrification gets a boost in
the infrastructure investment and jobs act. *Climate and Energy* 38 (12): 16–19.

82 House, W. (2021). President Biden's bipartisan infrastructure law. Washington, DC.
https://www.whitehouse.gov/bipartisan-infrastructure-law (accessed 6
March 2023).

5

Multi-Energy Technologies in Green and Integrated Transportation Networks

M. Edwin[1], M. Saranya Nair[2], and S. Joseph Sekhar[3]

[1] Department of Mechanical Engineering, University College of Engineering, Nagercoil, Anna University Constituent College, Nagercoil, Tamilnadu, India
[2] School of Electronics Engineering, Vellore Institute of Technology, Chennai, Tamilnadu, India
[3] Department of Engineering, University of Technology and Applied Sciences-Shinas, Al-Aqar, Oman

5.1 Introduction

Because of urban growth over the last few years, the proportion of the world inhabitants of cities has risen from 39% in 1970 to 60% in 2022 [1]. The number of licensed automobiles has risen from 200 million in 1960 to 1431 million in 2018. This is due to the expanding population and financial progress, as well as the high requirement of the millennial population for a customized transportation [2]. The existing transportation found a significant positive to make it possible for people to transit to their normal tasks, improve the energy efficiency of transportation services, and reduce the detrimental effects of road transport on the environment, climatic conditions, and health impacts. In addition, sustainable development is described as addressing present needs without compromising those of the future [3]. In the same vein, it is possible to define sustainable transportation as the capacity to meet the accessibility needs of the modern society, while minimizing environmental pollution and preserving the movement of future generations [4]. The different pollution attributes include other harmful emissions in addition to greenhouse gases (GHG), so even the future system with the minimum GHG emissions might not be able to address possible environmental issues [5]. An alternative approach would be to look for new types of fuel for the transportation system [6].

By minimizing carbon pollution and implementing low-carbon systems throughout their life cycle, low-carbon transit infrastructure can be attained. The key

Interconnected Modern Multi-Energy Networks and Intelligent Transportation Systems: Towards a Green Economy and Sustainable Development, First Edition. Edited by Mohammadreza Daneshvar, Behnam Mohammadi-Ivatloo, Amjad Anvari-Moghaddam, and Reza Razzaghi.
© 2024 The Institute of Electrical and Electronics Engineers, Inc.
Published 2024 by John Wiley & Sons, Inc.

technologies typically cover every stage of a transportation infrastructure's life cycle, including the manufacturing of raw materials, the building of the facilities, the maintenance and operation phases, and the end-of-life phases [7]. The automobile industry is at a turning point as a result of the emergence of these new innovations and the growing accessibility of renewable energies. Modern transportation networks are anticipated to be more connected with industrial automation, the energy infrastructure, and information ecosystems even while the techniques and energies that will carry people and commodities are yet unknown. This offers never-before-seen chances to make use of these links and achieve system-level ideal results [8]. Technologies that combine transportation and energy by taking into consideration shifts in the demand for transportation and mobility, the usage of fuel and energy, as well as the effects of actions and regulations, modeling tools may help explain the shift to this future. Understanding the linkages between various systems, the effects of technology and innovative mobility marketing strategies, and the infrastructure required to accommodate changes in the way cars are fueled will be crucial for modeling future mobility and energy systems. A more in-depth comprehension of the spatiotemporal energy usage needs that are unrelated to the existing systems will be necessary, in particular, to investigate the function of alternative fuels [9]. Utilization of renewable energy and effective resource use is necessary for green transportation. The transportation sector might become carbon neutral through conserving energy and reducing emissions in addition to embracing renewable energy, which is the primary option. According to our estimates, the transportation sector's carbon emissions will reach their high in 2030 at 1.33 billion tonnes before falling to 77% from 1.16 billion tonnes in 2019 to 261 million tonnes in 2060 [10].

In conclusion, to the best of their understanding, studies on multi-energy technologies that incorporate green and integrated transportation networks are sparse, but a collaboration of electrical grids, heat power, and energy for transportation might capitalize on the benefits of multi-energy sources. More research is needed to completely understand the ideal design and operation of multi-energy systems that can improve efficiency, flexibility, and carbon footprint. Although resilience has been extensively explored in the literature, outage survival difficulties necessitate more research for a thorough understanding of resilience, particularly in multi-energy technologies in the transportation industry. The situation has an impact on every kind of transportation, including cars driven by individuals, city buses, trains, and aeroplanes. It is the moment to consider economic and technical advancements in order to improve the transportation infrastructure. Researchers desperately require clarifying the system and defining GT after taking into account the seriousness of the issue related to the overall transportation system. This chapter is organized into eight sections. Section 5.1 is introductory. Section 5.2 presents the definition of green transportation. Section 5.3 presents the technological development and managerial integration for green transportation. It includes a smart transportation network, resource-efficient technologies, green technology, hybrid systems, and integrated management for green transportation. Section 5.4 describes the definition and features of an integrated multi-energy system. It comprises major characteristics of an integrated multi-energy system. Section 5.5 presents the electric vehicle integration with renewable energy sources. Section 5.6 describes hybrid fuel cell/battery vehicle systems. Section 5.7 presents the barriers and challenges of integrated multi-energy systems. Finally, Section 5.8 summarizes the conclusions and discusses the scope for future research.

5.2 Definition of Green Transportation

Transportation has a benefit of the societal economic, sociological, and environmental footprint. Transportation should be prioritized in order to meet urban sustainability objectives; this may be accomplished by regulating various components of the transportation network (public transportation), automobile developmental stages, and effective resource patterns [11]. GT is described as "transportation that has less negative effects on human health and the environment than present transportation services" [12]. GT may be defined as a hybrid technology that combines the optimum use of conventional fuels, the most effective use of electric car systems, the utilization of bioenergy as a fuel for automobiles, and improved public transit [13]. Lower risk, reduced traffic issues, enhanced energy and resource sustainability, reduced emissions, and improved assurance of safety are all benefits of a functional GT system. The goal of smart, green, and integrated transportation is to increase the competitiveness of the transportation sectors and develop a resource-efficient, environment and eco-friendly, stable, and fully integrated transportation system for the benefit of the public, the economy, and society [14]. The interconnections between different resource consumption in the urban agricultural system and the GT system, together with the sustainability features of smart hybrid vehicles, have emerged as a new research area of focus in a food-energy-water nexus [15].

5.3 Technological Development and Managerial Integration for Green Transportation

Transportation agencies are under intense pressure to adjust to changing needs and declining income in this era of globalization and the green frontier. Although adaptable, transportation's influence on the environment is heavily reliant on transportation technology to be lessened. In the current situation, agencies look for innovative ways to increase the productivity of their current resources so they can provide the public with delightful travel opportunities. There are various technologies that can help the development of GT [16].

5.3.1 Energy-Efficient Technology

The re-design concept is the basis of energy-efficient technologies. Its fascinating to observe that as time goes on, manufacturers are shifting away from enhancing certain parts of cars and toward fully revamping them to achieve improved energy efficiency [17]. To increase energy efficiency, hybrid combustion technology is being used. Friction also results in the loss of 10–15% of the energy produced by an automobile's IC engine. The performance of an engine may be improved by using lubricants to reduce this drag. To increase energy efficiency, several research facilities are creating brand-new nanofluids or lubricants that might address this frictional problem. Hybrid technologies might be a dependable solution to improve the efficiency of a combustion engine. In order to increase the efficiency of combustion engines, fuel is also essential [18]. Hydrogen fuel cells are another good alternative energy option in combination with bioenergy [19].

5.3.2 Eco-Friendly Technology

Numerous technical advancements, including lighter vehicles, eco-friendly elements and manufacturing techniques, lower friction coefficient, and better transmissions, can increase the performance of the vehicle [5]. Green resources have been used in operation with a noticeable decrease in GHG emissions without compromising safety. In general, a vehicle's weight may be decreased to minimize energy consumption. According to a basic rule of thumb, a 10% decrease in gross weight can minimize energy consumption by 5–7% when combined with good engine technology. BMW's new all-electric compact car, the Intel Xeon, which is constructed from reinforced carbon fiber to save its weight by 30–50% compared to steel, is one such example [20].

5.3.3 Intelligent Transportation System (ITS)

Because of the emergence of innovative technologies including hardware and software systems, navigation technologies, smart sensors, communications, cloud computing, virtual operations, and planning approaches in recent decades, the condition of the transportation network has completely altered. Traffic monitoring, cargo monitoring, systems integration, emergency logistics system, and consumer management platform are important intelligent transportation system (ITS) components. Numerous IT technologies may contribute significantly to risk mitigation, current traffic abatement, and reduced environmental impact along with improved associated with vehicle flow to provide an enjoyable ride for all modalities. Some of them integrate automated toll collecting, data gathering from highways, transportation management systems, automotive information gathering, transit signal priorities, and other technologies [21]. The following are some examples of ITS advances that may be seen in electric vehicles:

Automated vehicles – These vehicles can operate themselves by using advanced sensors and a customized transport network. If the technology is suitable to travel on challenging routes, these are safer [22].

Smart roads can detect how many tyres are being driven on them and provide information to drivers about transportation systems so that they can plan their journey appropriately [23].

Biometric vehicle access–With the use of these technologies, users of vehicles may start and access them using their eyes or biometrics. However, compared to optic scanners, biometric scanners have a higher chance of being accepted [24].

Remote vehicle shutoff is a feature that has been added to cell phones to stop thievery. The use of these technologies can disable stolen autos. By 2020, this remote vehicle shut-down technology is anticipated to be used [25].

Vehicle-to-vehicle communication (peloton technologies) – Fuel and safety are two major issues that the transportation sector faces. To improve safety and save power, large vehicles are connected by peloton, vehicle-to-vehicle connections, and sensor automatic safety systems. For the second vehicle and 4.5% for the main vehicle, fuel savings are anticipated to total 10% [20]

The electronic toll collection system (ETC) is designed to reduce wait times at toll booths. The ETC system is designed to automatically deduct money from authorized automobile users' account while forcing them to end. Taiwan was the first nation in the world to implement electronic tolls on all of its highways [26].

5.3.4 Integrating Systems: Efficiency by Design

High-strength iron, optical fiber composites, metal matrix composites, advanced composite matrix, and magnesium composites may reduce weight by 10–28%, 15–25%, 30–60%, 50–70%, and 30–70%, respectively, during the design process of vehicles. In terms of internal components, strong insulation provides for a concentration drop of 0.4 kg, polyamide polymers at the rear seat, chassis, and contoured light panel decrease weight by 0.3 kg, 0.7 kg, and 0.4 kg, respectively, without sacrificing durability and stability, and traditional combination and insert hybrid automobiles have their own designs. However, a normal hybrid has to have its gas tank filled periodically, necessitating many trips to the gas station. Additionally, it features a power pack that may be used to power the vehicle or launch it. Fuel is the main source of energy. A battery serves as the plug-in hybrid vehicle's primary fuel source, with petrol serving as a backup. The plug-in hybrid technology is more user-friendly when compared to the other two since it can be fueled at home and does not need to be refilled [27]. The British automobile scientist remarked on the logicalness of the race vehicle design, but his ideas apply to many cutting-edge ways of dealing with energy efficiency, such as replacing steel with lightweight materials [28].

5.3.5 Energy Re-using

With this technique, it is anticipated that less fuel will be used and more air will be used. The energy generated during a vehicle's break may be recovered and used again using kinetic energy restoration technologies. For use later, while accelerating, the enhanced energy is typically stored in a reservoir. Multiple applications, including racing automobiles and passenger vehicles, have made use of it.

5.3.6 Solar Impulse Technology

An electric vehicle that uses solar energy as its energy is referred to as a solar car. Photovoltaic systems are used to turn solar energy into electrical energy. With sunrise, the first solar-powered flight took off in 1974 [29]. By completing the first round-the-world voyage powered entirely by solar energy in July 2016, ABB alliance partner Solar Impulse made history. For the use of solar energy for public transport services in 2020, the Tube Way Solar Company has created a solar cargo bike and tube way system [30]. The advantage of Solar Impulse Technology goes beyond sustainability and offers management over the continuously growing cost of electricity.

5.3.7 Integrated Management for Green Transportation

In order to maintain development in the economy, the ecological, and safety, the transportation sector modifies how it is structured and controlled. It is challenging to tackle the problem of urbanization and transportation, and any attempt to achieve sustainability in the transportation network needs to take into account all the many facets of travel demands. The main factor consists of land use efficiency, traffic control, emissions rates, and vehicle growth patterns. Since each of these factors affects the others, it can be challenging to address any one of them on its own. Consequently, it is crucial to address them at the stage of integration [31]. The five key components of the integrated management framework for the GT are benchmark assessment, target formulation, governmental pledges, execution and monitoring, and assessment and adaptation. These five components are schematically shown in Figure 5.1. GT can be effectively applied in several integrated management industries.

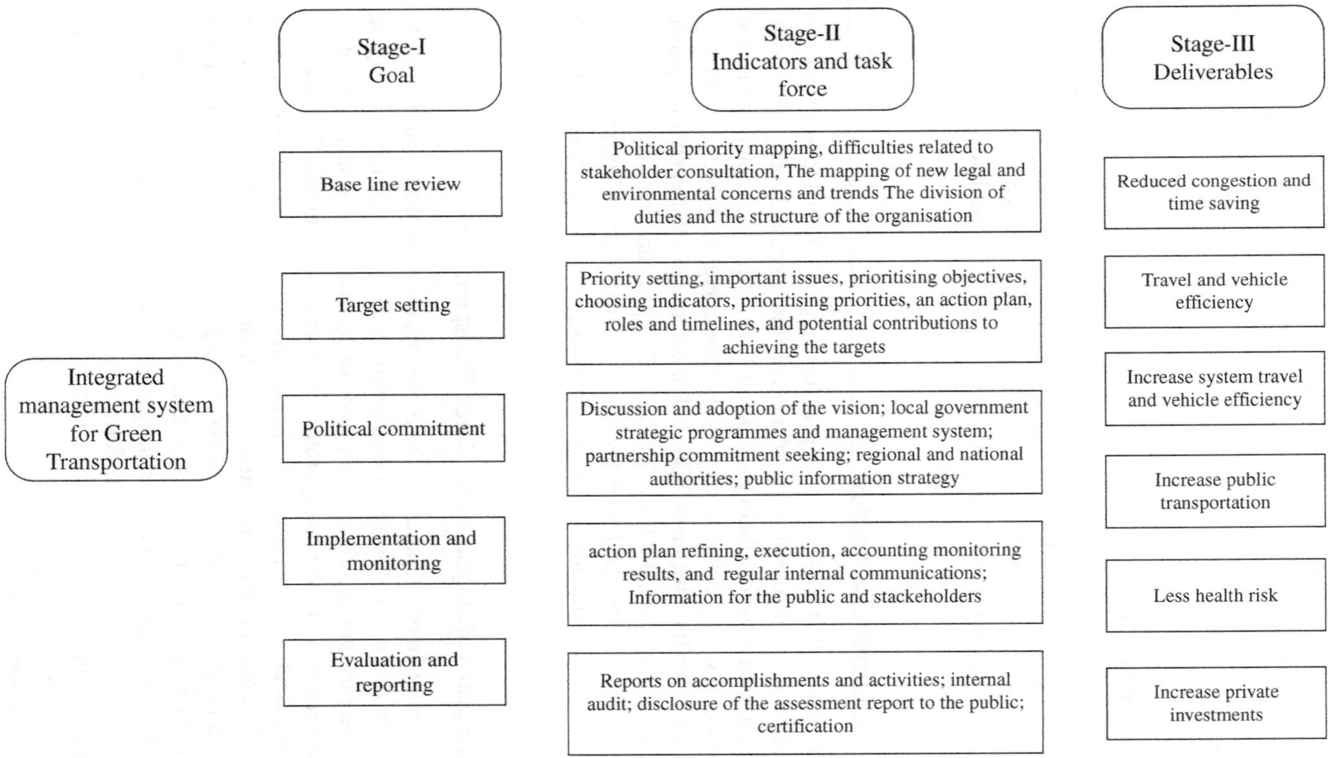

Figure 5.1 Stages involved in integrated management system for green transportation network.

5.3.7.1 Infrastructure Development

In the automotive industry, maintaining the network is a significant difficulty for many emerging regions since it not only takes a significant financial investment but also calls for the right operational and technological abilities. To transform the cityscape and public domain, significant spending on infrastructure for walkers and bicycles is needed. Additional management strategies include carbon restriction, geological formations expanded vehicular access, vehicular traffic regulation, restrictions on automobile parking and accessibility, lowered speed restrictions, and behavioral modification strategies.

5.3.7.2 Alternative Measures in Urban Transportation

The following are alternate energy sources, managerial strategies, and technological advancements to consider for the world's public transportation system:

To encourage the use of natural gas vehicles, such as petroleum products, as an alternative to fossil fuels, such as hydrocarbons, coal, and petroleum; raising public awareness of the benefits of using electric vehicles; to grant hybrid electric automobiles a tax break; to encourage research toward the creation of fuel-cell automobiles; creating a safer hydrogen energy source; creating a tax relief policy to encourage the use of biofuels; developing an inspection and maintenance strategy for an environmental certification scheme; developing and deploying emission control technologies; building bypasses and signals to improve traffic management; and promoting the use of metro rail and monorail travel.

5.4 Definition and Features of Integrated Multi-Energy System

5.4.1 Definition of Integrated Multi-Energy System

Energy systems integration (ESI) is the method of integrating the operations and management of energy technologies across several geographic regions and/or channels to supply dependable, affordable energy services with low environmental impact. Energy system interactions (ESI) involve connections between energy sources (power, heat, and fuel) as well as interconnections with other large-scale applications such as irrigation, transportation, and communications systems constitute an underlying technology for ESI [31]. The structural, organizational, and geographical interactions are where ESI is most useful since there are new possibilities and difficulties for development, implementation, and application that can lead to economic and social advantages. ESI is a multidimensional field that includes human behavior as well as science, engineering, and technology [32]. Compared to "traditional" energy technologies, which respect the multiple energy segments "individually" multi-energy systems (MES), where power generation, refrigeration, fuel sources, and transport interact efficiently at different levels, have a significant chance to enhance technical, economic, and involve performing [33]. The concentration on both broadness and depth at the same time is what makes ESI such a difficult and fascinating field. Through plug-in electric (hybrid) cars and car battery packs, the hitherto completely independent domains of transportation and power may become increasingly connected. However, the customer must embrace this method of transportation [34]. Enhance the energy system's adaptability, for example, by enabling electrically powered heating and cooling loads to naturally possess storage properties to contribute to power source balancing by offering reserve and performance requirements, or by utilizing the adaptable storage technologies found in electric vehicles (EVs) to endorse wind connectivity while supplying cleaner energy for transportation [35].

5.4.2 Major Characteristics of Integrated Multi-Energy System

To emphasize the variety of views and diversity that often constitute MES, four sequences of classification will be carefully pursued. They are as follows [36]:

The geographical viewpoint is used to illustrate whether MES might be meant theoretically or at various degrees of grouping in terms of its constituents. Buildings may be the first level of grouping (where, for example, multiple technology classes providing multiple energy channels combine *with one another*), followed by districts (such as the critically essential situations of district energy technologies), and eventually, territories and even continents.

The objective of the multi-service viewpoint is on delivering several functions or "outcomes" (from multiple forms of energy utilities to the transportation industry) by integrating energy pathways as best as possible, especially at the level of production. A situation that is especially pertinent is the one involving the joint creation of many fuel sources. The multi-fuel approach emphasizes how various sources of energy, from "traditional" fossil fuels to biomass resources and renewable energy sources (RES) for electric and thermal energy, can be incorporated together to meet the best availability of the multi-service consumption in an MES (commonly for both environmental and economic benefits). The channel characteristics discusses the vital function that energy networks play in the evolution of multi-energy systems and how they engage to reduce system costs and maximize sustainability impact. These channels include those for energy, fuel, district heating and cooling, hydrogen, and other types of energy.

A balanced allocation, integrated control, and functional management of multiple energy sources comprise integrated multi-energy. It is an optimization issue with several activities and multiple constraints, thus during the application procedure, special attention should indeed be paid to the requirements indicated below [37]: (i) the availability of resources to meet the integrated multi-energy system's demand for energy forms. Integrated multi-energy systems must take into account all available recourses and be tailored to the area's specific needs. In order to satisfy the need for energy supply, it needs to rationally embrace energy of diverse forms to create combinations that integrate one another. (ii) The potential for different energy levels to interact, integrate, and support one another. The energy streams on the side of the power sources are interchangeable in certain respects in integrated multi-energy. Technically, integrated multi-energy cannot be realized if two sources of energy cannot be replaced. (iii) Integrated multi-energy can provide the load requirement in a cohesive manner. Based on load requirements, integrated multi-energy implementation should be carried out. The layout of energy demand is estimated directly by the load profile. Commercial, domestic, and professional energy usage will all be very different in terms of proportion and features. The user load demand for power, heating, and cooling varies with seasonal and demographic shifts. Various working modes can be made synchronized by the integrated multi-energy. (iv) The use of pertinent technological advancements that can facilitate integrated multi-energy. The intricacy of the global energy system is increased by integrated multi-energy. Due to the intimate coupling between the elements, problems such as malfunctioning and disruption can propagate throughout the subsystems and have an influence on each other. The resilience and reliability of the system are ensured by the energy infrastructure technologies that facilitate the deployment of integrated multi-energy, such as technological integrating, control and monitoring, dispatch planning,and information processing. (v) The business strategy and operational services may satisfy the needs of an integrated multi-energy activity. Users of the integrated multi-energy can flexibly shift between a variety of energy sources, including power, gasoline, coal, and natural gas. It is time for traditionally grown service providers to become multi-energy providers. As the use of an integrated multi-energy system develops, so should the operating framework, market process, oversight methods, commercial model, and access requirements.

5.4.3 Role and Effects of Multi-Energy Conversion Systems in Green and Integrated Transportation Networks

Ever-rising CO_2 emissions from transport are a challenge that no one technology can address alone. Here, a comprehensive strategy for integrating transport into energy planning is put out, employing a variety of strategies to promote sustainable travel. It is stated that while a 100% renewable energy transportation system is feasible, getting there will include major obstacles. Because biomass is a finite resource, itscritical to prevent any impact on the food supply. A multifaceted approach and the integration of the transportation and energy systems are essential. The long-term objective must be taken into account in short-term solutions.

Biofuels are the major focus of the growing global attention on the transportation industry. However, because biomass is a finite resource, it cannot provide a viable mode of transportation by itself. Biofuels will affect food production if used without restriction. Biofuels for transport have a significant role in the long-term planning for 100% renewable energy sources when combined with other equally essential technologies and ideas [38].

About 25% of the world's greenhouse gas emissions are caused by the transportation industry. Fossil fuels predominate, which has a variety of detrimental effects on both people and the environment. Research in alternative automobile engines, such as various electric car models and low-carbon fuels, has grown over the past several years. Fuel cell transportation is a unique subset of electric vehicles. In such automobiles, hydrogen is refueled, preserved, and then utilized to generate power on board rather than charging batteries at charging points. Similar to electricity, hydrogen is a secondary energy carrier that may be created from a variety of main energy sources, such as fossil fuels and renewable energy. Green hydrogen is the common name for hydrogen produced using renewable energy sources. However, high ecological sustainability is taking precedence if hydrogen is to be regarded as a source of energy for usage in the transportation industry.

5.5 Electric Vehicle Integration with Renewable Energy Sources

Due to their minimal carbon pollution and reduced dependency on oil, electric vehicles (EVs) have gained a considerable degree of mainstream adoption during the past 10 years. According to the estimates, there will be more than 36 million EVs worldwide by 2023. However, a significant issue with EVs is that they have a significant adoption rate, which increases terminal and substation traffic and the local grid's excessive energy usage. Integrating energy production, such as RESs, into the EV charging network is one effective way to mitigate the impact [39–42].

High standards for system reliability and relative power regulation have been prompted by the increasing generation capacity of solar and wind power in many places [43, 44]. Its significant that the investigation has mostly concentrated on the possible impact of EVs in easing the integration of RESs in the electricity grid. The research topic, which is focused on how EVs may create reciprocal advantages for the electric energy system with RES and prospective EV consumers, has been created within the context of the smart grid [45]. In particular, the study has concentrated on EV-based approaches for providing additional benefits for wind integration and power storage for PV integration.

5.5.1 Electric Vehicle Integration with Wind Energy

According to the reports, integrating wind energy and electric vehicles into electricity grids makes sense for the delivery of auxiliary functions [46]. EVs should indeed be synchronized for elevated activities, especially supplementary services that frequently lower the operational cost to EV users in the shortterm [47]. Even though EVs have a greater beginning value than ICE vehicles, EV users are expected to pay less overall. The advantages of auxiliary benefits rendered by EVs in the power grid [48]. Their primary focus was on the management of secondary reserves and load frequency control, which is evaluated using computational methods. The article describes that EVs may effectively manage energy imbalance brought on by wind power's fluctuation, doing away with the need for traditional power stations. Huge numbers of distributed EVs and home appliances may be leveraged to provide LFC as secondary reserves in the electric grid. According to the simulation's findings, it can be shown that battery packs experience significant energy excursions that often switch between an empty and full level of charge [49]. It is thought that increasing the objective of voltage regulation in Denmark is unavoidably necessary in order to decrease the unnecessary use of automated reserves [50] and to re-establish their supply. The authors claim that one of the best alternatives for the future replacement of the diminished reserve capacity produced by traditional power plants is employing EVs to provide measuring energy in Denmark. The study examined micro-grid uses as well. Lopes et al. address the application of EV demand management and a droop management approach to stabilize the synchronism in a micro-grid with wind energy in Ref. [51]. It was shown that by utilizing a synchronized EV demand, the degree of penetration of wind energy may be raised even further. In a previous study, power grids and EVs were thought to be ideal controlled loads in a network simulator, but the study neglected to consider potential obstacles such as the need for EV management and the behavior of EV parts during synchronization times.

5.5.2 Electric Vehicle Integration with Solar Energy

In contrast to other studies concentrating on wind power and EVs, the study on the use of solar energy with EVs is substantially more diverse. Significantly, PV electricity production is feasible inside the power system networks at both moderate and low-voltage levels. Moreover, this option further encourages the idea of combining PV energy with EVs [52]. Furthermore, studies show that solar energy may be conveniently conserved in automobile battery packs for later use throughout the day, when radiation from the sun is at its strongest. The concept of "green"-charging [53], for example, is one addition that has helped users grasp the necessity of limiting the pricing of EVs during the irradiation duration. It is proposed the idea that EVs may be fueled on a daily basis in parking spaces located, for example, inside companies. In order to implement the solar-to-vehicle (SV2) technique, EVs may also be fully recharged during business hours. The study also shows that energy produced in each parking space is crucial for the additional generation of enough electricity to meet the needs of EV operators for mobility [54].

Despite the possibility of substantial PV power system adoption, lower voltage is also a possibility. The main restriction under such circumstances is related to fluctuations in voltage measurement along the sources. Additionally, these differences are especially noticeable at times of high output and low demand [55]. Consequently, these occurrences are expected to happen frequently, but mostly in locations where there are a lot of homes with roof-mounted solar panels. On the other hand, several investigations have looked at different options for reducing voltage capability, including grid strengthening, voltage regulation strategies [56], harmonized power flow restriction, and long-term energy storage [57]. Research reveals that the synchronized EV load inside the feeders employing high PV adoption has not been adequately evaluated.

5.6 Hybrid Fuel Cell/Battery Vehicle Systems

New varieties of fuel cell vehicles have been made available for purchase for multiple transportation needs. Cars, buses, submarines, and motorbikes are a few examples of common applications. Due to its relative high maturity, impressive performance, and quick reaction, proton exchange membrane fuel cells (PEMFC) are the fuel source of choice for the majority of applications. The majority of the energy needed to run the vehicle is produced by the PEMFC stacks when hydrogen is utilized as the fuel. In order to achieve a quick reaction to sudden acceleration or deceleration, a power supply is included. Alternative energy sources for fuel cell cars include methanol, ethanol, gasoline, etc., however, they require additional fuel processing components like reformers or partial oxidation reactors (POX) and water-gas shift reactors. The preferred operating temperature for solid oxide fuel cells (SOFC) is normally between 600°C and 1000°C. Due to this, balance of plant is more complicated and SOFC is less adaptable in terms of reboot and energy management. However, an automobile normally needs a power output of 100 kW. SOFC has the edge in terms of efficiency at this size [58]. This SOFC-based power generation has the unique characteristic that the

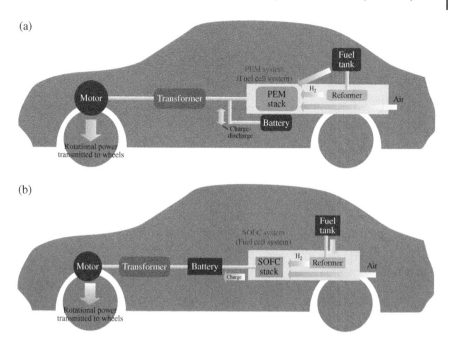

Figure 5.2 Schematic diagram of PEM fuel cell vehicle system (a) and SOFC vehicle system (b). *Source:* Adapted from [58].

SOFC stack feeds the power pack instead of directly powering the motor. Users can therefore receive a driving dynamic that is comparable to that of a fully electric car. A typical PEMFC-based electric vehicle system is shown in Figure 5.2.

5.6.1 PEMFC-Based Fuel Cell Vehicle Systems

Although it is commonly accepted that hydrogen is the end objective for fuel cell vehicles, there continues to be disagreement over how to use it. Even though it will cost more to build refueling stations and store hydrogen, internal compressed hydrogen storage can streamline the fuel cell car system. Figure 5.3a–c displays the system schematics for these three fuel cell cars. These systems attempted to power a midsize automobile car similar to the partnership for a new generation of vehicles (PNGV) with less weight, rolling resistance, and aerodynamic drag [60].

To simulate these three systems, Ogden created a numerical simulation model that included a dynamic PEMFC model, a steam reformer, a POX reactor, a WGS reactor, and a selective catalytic oxidation (SCO) reactor. The power battery can supply additional required power when the fuel cell is unable to run the motor by itself. According to the premise that the battery should be kept close to an

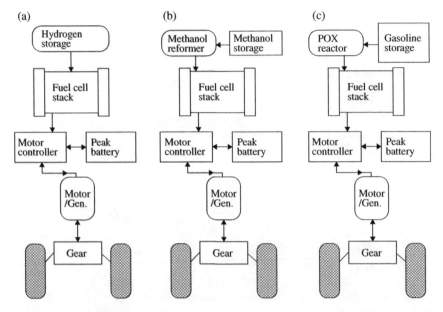

Figure 5.3 Schematics of PEMFC vehicle systems fueled with hydrogen (a), methanol (b), and gasoline (c). *Source:* Adapted from [59].

optimal state of charge, this system will distribute the power demand between the fuel cell and the battery (SOC). According to the PEMFC stack's polarization curve and the efficiency of the fuel processing subsystem, this model determined the amount of fuel that was used. The anode of the PEMFC stack was supplied with pure H_2 for the hydrogen-fueled system. However, whether the system is powered by methanol or gasoline, the incoming H_2 concentration depends on the fuel processing subsystem's output. For methanol-fueled systems, the intake H_2 level is around 75%, whereas for gasoline-fueled systems, it is 40%. Fuels that are liquid at room temperature and normal pressure, such as methanol and gasoline, have a greater volumetric energy density. Hydrocarbon-fueled systems need more fuel reformers, POX reactors, WGS reactors, SCO reactors, and other BOPs such as heat exchangers, burners, compressors, and turbines than H_2-fueled systems need. This mechanism is substantially more intricate as a result. The combined peak power of the power battery pack and PEMFC stack in these three fuel cell vehicle systems is closed. Because they are lightweight, hydrogen-fueled systems require the least peak power. In these unique circumstances, the power battery was employed to provide the additional power requirements. The expenses of the fuel processing subsystems for methanol- and gasoline-fueled systems are comparable. The PEMFC stack's subpar electrochemical performance at low H_2 concentration was largely responsible for the high investment costs for the gasoline-fueled fuel cell vehicle system. The PEMFC stacks and other elements have to be bigger in size and heavier in the gasoline-fueled fuel cell vehicle system in order to achieve the same power output.

In conclusion, hydrogen-fueled fuel cell cars have a more straightforward system architecture, less system weight, greater electrochemical performance, higher fuel economy, and cheaper investment costs. Due to the high volumetric energy density of liquid fuels, fuel cell cars that run on methanol or gasoline have a greater range. Methanol-fueled fuel cell cars among these three fuel cell vehicles became the most economically viable after at least two years of operation, assuming the same maintenance cost and lifetime. After three to four years of operation, gasoline-fueled fuel cell vehicles are more cost-effective than hydrogen-fueled fuel cell vehicles. Of course, if a carbon emission penalty is taken into account, this issue has to be reassessed.

5.6.2 SOFC-Based Fuel Cell Vehicle Systems

Due to its high efficiency and fuel adaptability, a solid oxide fuel cell (SOFC) provides several fixed and mobile application possibilities. The POX reformer and a 12-kW SOFC unit make up the SOFC subsystem. In contrast to hybrid PEMFC/battery vehicle systems, hybrid SOFC/battery vehicle systems directly provide all of the power loads from the power battery. Thus, the driving experience of this hybrid SOFC/battery car is comparable to that of an electric vehicle system that is only powered by batteries. To greatly increase the driving distance, the power battery is directly charged by the SOFC unit.

In general, solid oxide FCs (SOFCs) are not taken into consideration for propulsion in automotive applications due to their high operating temperature and slow start-up times. However, because SOFCs can run on reformed diesel fuel and do not need the establishment of hydrogen fuel facilities, they are being developed for auxiliary power units (APUs). Truckers will be able to power the sleeper cabins using SOFCs thanks to the development of APUs for trucks, which will eliminate the need to idle the diesel engine while the driver is sleeping [61]. Due to their high operating temperature and lengthy start-up times, SOFCs are typically not thought of for rocket engines in vehicle industries [62]. However, since reformed gasoline and diesel can be used in these systems without the development of hydrogen supply facilities, SOFCs for APUs are being established [61]. For instance, SOFC-based APUs are being created to replace the need for idle diesel engines in stationary cars.

An additional choice is a SOFC-battery hybrid system, in which the battery is in charge of the vehicle's quick start-up and resource efficiency during braking, and the SOFC stack is used to either drive or recharge the battery. The SOFC-battery hybrid systems have a lower top speed but comparable driving range and acceleration times to PEMFCs [59].

5.6.3 Present Situation of Fuel Cell Vehicle Technology

More than 35,000 fuel cell cars will be sold by the end of 2023, and Toyota and Hyundai will each sell more than 10,000 fuel cell cars. Future sales of fuel cell vehicles will rise as a result of the conventional vehicle's equal recharging time. As a result, the customer drives the long route without hesitation. In comparison to a conventional vehicle, the fuel cell electric vehicle has a low operating cost [63]. The fuel cell vehicle is being promoted by Indian officials due to its numerous advantages in the context of green and clean energy. In contrast to the rising cost of conventional fuel in the worldwide industry, they claimed that the number of green hydrogen fuel vehicles would rise in the future, allowing every Indian to purchase green fuel. This raises the nation's overall level of independence. The government also finances numerous initiatives to create environmentally friendly hydrogen vehicles. By 2030, India wants to have produced five million tonnes of green hydrogen [64].

5.6.4 Confronts of Fuel Cell Vehicle Technology

Keeping the fuel cell stack's temperature under control and storing hydrogen are the two biggest challenges for fuel cell vehicles. Since hydrogen has a pressure of about 700 bar (10,000 psi), storing it is challenging. Only a few fuel cell electric vehicle models are offered globally, with the Toyota Mirai being the most well-known model. There are a finite number of fueling stations for fuel cell electric vehicles. The Toyota Mirai is a fuel cell electric vehicle that the Indian government has introduced. After the start of the green hydrogen project in 2025, a hydrogen fueling station will be accessible. Compared to electric and conventional vehicles, the cost of a fuel cell vehicle is very high. Car manufacturers keep working to bring down the price of fuel cell electric vehicles. The most crucial factor for fuel cell electric vehicles is public awareness. The Toyota Mirai fuel cell electric vehicle was introduced by the Indian government, and a pilot project there began in March 2023. The project to launch fuel cell cars for people is very beneficial [64].

5.7 Barriers and Challenges

The terms "barrier" and "challenges" refer to things like rules, laws, or policies that prevent anything from occurring as well as difficulties that might be resolved. All stakeholders, including mobility drivers, logistics providers, infrastructure operators, shipowners, environmentalists, public officials, legislators, enterprises, R&D institutions, and academic institutions, must be taken into account when analyzing the barriers and challenges because their respective initiatives and goals differ from one another [65]. To make the goals of all stakeholders more manageable, barriers and challenges are segmented into the compliance, organizational, scientific, economic, and public segments. However, the most recent COVID-19 pandemic have created a number of new difficulties in production processes and logistic support [66].

5.7.1 Societal Barriers and Challenges

Due to rapid urbanization, inhabitants and rising incomes have shifted topographically over the past several years, which have fueled an increase in the road transport system's reliance on cars. This is made worse by the shift to service-oriented jobs, which are more likely to be found in cities, as well as the ineffectiveness of authority at all levels in planning to promote sustainable and inclusive development [67]. The distance between individuals living and working is growing as a result. This has been unable to show how transportation, territory, residential, and employment legislation have been comprehensively incorporated at the nationwide, geographical, and microlevels [68]. Due to fierce competition during a period of rapid urbanization, lower penetration levels and a reluctance to engage in automobiles with increased emission regulations were seen in the case of the transport industry. Increased taxes during organizing could cause perverse incentives and harm domestic transporters' ability to compete. The interruption of the existing policies may raise concerns about a loss of domestic profitability, which could lead to slower adoption of new methods and service practices in the cargo industry segments. Social approval of the transformative process to accomplish the GT objectives is ambiguous because of the uncertainty surrounding the level of engagement between the public and policymakers [69]. Consequently, this is a significant issue for decision-makers because they are unable to put initiatives into effect without public support.

5.7.2 Technological Barriers and Challenges

Scientific knowledge, future technologies, and engineering applications of technological advancements are noted as being key factors in GT's goal of environmentally friendly transportation. Technological innovations can be used to reduce environmental pollution and ensure the sustainability of natural resources. However, the main obstacles and challenges to a successful transformation of technology from conventional to innovative are modification, restricted access to low-carbon technologies, and lack of performance of technology [70]. High-quality energy sources and the use of more environmentally friendly, highly effective transportation methods may improve vehicle performance and lessen environmental problems. The number of vehicles is rising, though, due to current policy decisions. In order to significantly reduce the pollution load caused by vehicle fleets and the use of low-quality fuels, a variety of technologies (such as enhanced automotive vehicles and fuels) should be embraced. It is best to maximize the use of low-carbon fuels rather than liquid fuels. Alternative energy sources and automobiles, such as renewable fuels, biogas, battery-operated cars, hybrid electric vehicles, and hydrogen fuel cell cars, are currently in the market or nearly commercially viable [71]. However, there are a number of complicating factors that contribute to weak sustainability, such as high capital costs, a lack of natural resources, the growth of filling stations, high maintenance costs, and negative public understanding. It is crucial to deploy the new technology with knowledge that is easily available to end users and legislators.

5.7.3 Financial Barriers and Challenges

The economic cause of poor sustainable practices is the cost of going sustainable, as green technologies are too expensive because of decaying infrastructure in both developing and developed nations. It is challenging to implement the necessary changes to move to GT without the backing of a strong infrastructure [72]. Because of the need for required facilities, such as a freight channel and roads, for the effective delivery of GT, developing nations are falling behind [73]. To address these issues and meet GT's current needs, the outdated facilities must be renovated. Government and financial institutions are least interested in undertaking this full or partial rehabilitation because it requires a significant time and financial commitment, especially given the state of the economy in 2008. Transportation regulations should be designed to draw major investment firms.

5.8 Conclusion

Due to the growing attention to adopting sustainability in the global increasing urbanization, the existing transport network should be transitioned into a sustainable and inclusive GT by implementing novel technology and management approaches. This chapter created the conceptual underpinning of a sustainable transportation system from the context by describing GT and identifying limitations and problems in constructing the GT. In recent years, there has been a huge need for green supply chains, green ecotourism, and a sustainable economy in order to accomplish global sustainability objectives. Unfortunately, the current public transportation networks are unable to shift the global transportation system toward green on a technological and environmental level. Yet, this study demonstrated that using multi-energy technologies in green and integrated transportation systems in the current transportation system might have positive effects on the environment as well as the economy. Thus, it can be concluded that in order to transform a current transport network into a GT and advance sustainable growth, multi-energy technology adoption in green and integrated transportation systems is essential. The integration of GT with green ecotourism, green distribution networks, and GT should be done to create a resource-efficient system that will replace the existing cycle and be the futuristic society, even if it is not addressed in this study and is left to future studies. The most effective strategy for creating smart cities and smart tourism, while taking into account the environmental and financial benefits, is green transportation.

References

1 Amegnaglo (ONU) (2018). World urbanization prospects. *Demographic Research* 12: 197–236.

2 Davis, S.C. and Boundy, R.G. (2020). *Transportation Energy Data Book: Edition 38.2.* Energy and Transportation Science Division, Oka Ridge National Laboratory.

3 Sun, D., Zeng, S., Lin, H. et al. (2019). Can transportation infrastructure pave a green way? A city-level examination in China. *Journal of Cleaner Production* 226: 669–678.

4 Li, J., Pan, S.-Y., Kim, H. et al. (2015). Building green supply chains in eco-industrial parks towards a green economy: barriers and strategies. *Journal of Environmental Management* 162: 158–170.

5 Van, F.Y., Klemeš, J.J., Walmsley, T.G., and Perry, S. (2019). Minimising energy consumption and environmental burden of freight transport using a novel graphical decision-making tool. *Renewable and Sustainable Energy Reviews* 114: 109335.

6 Raymand, F., Ahmadi, P., and Mashayekhi, S. (2021). Evaluating a light duty vehicle fleet against climate change mitigation targets under different scenarios up to 2050 on a national level. *Energy Policy* 149: 111942.

7 Li, H., Hao, Y., Xie, C. et al. (2022). Emerging technologies and policies for carbon–neutral transportation. *International Journal of Transportation Science and Technology* 12: 329–334.

8 Muratori, M., Jadun, P., Bush, B. et al. (2020). Future integrated mobility-energy systems: a modeling perspective. *Renewable and Sustainable Energy Reviews* 119: 1–20.

9 Yeh, S., Mishra, G.S., Fulton, L. et al. (2017). Detailed assessment of global transport-energy models' structures and projections. *Transportation Research Part D Transport and Environment* 55: 294–309.

10 Guidebook to Carbon Neutrality in China. *Guidebook to Carbon Neutrality in China.* 2022.

11 Kumar Singh, S., Banerjee, S., and Chakraborty, I. (2020). Importance of traffic and transportation plan in the context of land use planning for Indian cities. *International Journal of Town Planning and Management* 14 (9): 1–6.

12 Björklund, M. (2011). Influence from the business environment on environmental purchasing –drivers and hinders of purchasing green transportation services. *Journal of Purchasing and Supply Management* 17 (1): 11–22.

13 Lee, C.T., Hashim, H., Ho, C.S. et al. (2017). Sustaining the low-carbon emission development in Asia and beyond: sustainable energy, water, transportation and low-carbon emission technology. *Journal of Cleaner Production* 146: 1–13.

14 Shah, K.J., Pan, S.-Y., Lee, I. et al. (2021). Green transportation for sustainability: review of current barriers, strategies, and innovative technologies. *Journal of Cleaner Production* 326: 129392.

15 Elkamel, M., Valencia, A., Zhang, W. et al. (2023). Multi-agent modeling for linking a green transportation system with an urban agriculture network in a food-energy-water nexus. *Sustainable Cities and Society* 89: 104354.

16 Klemeš, J.J., Van, F.Y., and Jiang, P. (2021). COVID -19 pandemic facilitating energy transition opportunities. *International Journal of Energy Research* 45 (3): 3457–3463.

17 IEA Technology Roadmaps (2011). *TechnologyRoadmap: Biofuels for Transport.* OECD Publishing.

18 Kim, M., Li, D., Choi, O. et al. (2017). Effects of supplement additives on anaerobic biogas production. *Korean Journal of Chemical Engineering* 34 (10): 2678–2685.

19 Martins, J. and Brito, F.P. (2020). Alternative fuels for internal combustion engines. *Energies* 13 (16): 4086.

20 Lammert, M.P., Duran, A., Diez, J. et al. (2014). Effect of platooning on fuel consumption of Class 8 vehicles over a range of speeds, following distances, and mass. *SAE International Journal of Commercial Vehicles* 7 (2): 626–639.

21 Maimaris, A. and Papageorgiou, G. (2016). A review of intelligent transportation systems from a communications technology perspective. *2016 IEEE 19th International Conference on Intelligent Transportation Systems (ITSC)*, Rio de Janeiro, Brazil (1–4 November 2016), 54–59. IEEE.

22 Knight, W. (2017). Finally, a driverless car with some common sense. MIT Technology.

23 Smart roads in the making (2017). Innovation and technology lifestyle. The SupercopeReport.

24 Priya, G.R. and Shree, M.M. (2016). Biometric authentication based vehicular safety system. *South Asian Journal of Engineering and Technology* 3 (1): 27–34.

25 Mikulec, R. (2012). Remote shutdown of heavy duty vehicles. Master Thesis.

26 Abdulla, R., Abdillahi, A., and Abbas, M.K. (2018). Electronic toll collection system based on radio frequency identification system. *International Journal of Electrical and Computer Engineering* 8 (3): 1602.

27 EPO (European Patent Office) (2017). Green transportation: five innovations that are driving efficient vehicle technology.

28 Henriksson, F. and Johansen, K. (2016). On material substitution in automotive BIWs – from steel to aluminum body sides. *Procedia CIRP* 50: 683–688.

29 Sai, L., Wei, Z., and Xueren, W. (2017). The development status and key technologies of solar powered unmanned air vehicle. *IOP Conference Series: Materials Science and Engineering* 187: 012011.

30 Bran, B.R., Jin, P.J., Boyce, D. et al. (2012). Perspectives on future transportation research: impact of intelligent transportation system technologies on next-generation transportation modeling. *Journal of Intelligent Transportation Systems* 16 (4): 226–242.

31 O'Malley, M., Kroposki, B., Hannegan, B. et al. (2016). *Energy Systems Integration: Defining and Describing the Value Proposition*. Golden, CO (United States): The International Institute for Energy Systems Integration.

32 O'Malley, M. and Kroposki, B. (2013). Energy comes together: the integration of all systems [Guest Editorial]. *IEEE Power and Energy Magazine* 11 (5): 18–23.

33 Mancarella, P. (2014). MES (multi-energy systems): an overview of concepts and evaluation models. *Energy* 65: 1–17.

34 Lund, H. and Münster, E. (2006). Integrated energy systems and local energy markets. *Energy Policy* 34 (10): 1152–1160.

35 Ruth, M.F. and Kroposki, B. (2014). Energy Systems integration: An evolving energy paradigm. *The Electricity Journal* 27 (6): 36–47.

36 Mancarella, P. (2014). MES (multi-energy systems): an overview of concepts and evaluation models. *Energy* 65: 1–17.

37 Vujanović, M., Wang, Q., Mohsen, M. et al. (2021). Recent progress in sustainable energy-efficient technologies and environmental impacts on energy systems. *Applied Energy* 283: 116280.

38 Sandy Thomas, C.E. (2009). Transportation options in a carbon-constrained world: hybrids, plug-in hybrids, biofuels, fuel cell electric vehicles, and battery electric vehicles. *International Journal of Hydrogen Energy* 34 (23): 9279–9296.

39 Mathiesen, B.V., Lund, H., and Nørgaard, P. (2008). Integrated transport and renewable energy systems. *Utilities Policy* 16 (2): 107–116. https://doi.org/10.1016/j.jup.2007.11.007].

40 Richardson, D.B. (2013). Electric vehicles and the electric grid: a review of modeling approaches, impacts, and renewable energy integration. *Renewable and Sustainable Energy Reviews* 19: 247–254.

41 Mwasilu, F., Justo, J.J., Kim, E.-K. et al. (2014). Electric vehicles and smart grid interaction: a review on vehicle to grid and renewable energy sources integration. *Renewable and Sustainable Energy Reviews* 34: 501–516.

42 Liu, L., Kong, F., Liu, X. et al. (2015). A review on electric vehicles interacting with renewable energy in smart grid. *Renewable and Sustainable Energy Reviews* 51: 648–661.

43 Tan, K.M., Ramachandaramurthy, V.K., and Yong, J.Y. (2016). Integration of electric vehicles in smart grid: a review on vehicle to grid technologies and optimization techniques. *Renewable and Sustainable Energy Reviews* 53: 720–732.

44 Dubarry, M., Devie, A., and McKenzie, K. (2017). Durability and reliability of electric vehicle batteries under electric utility grid operations: bidirectional charging impact analysis. *Journal of Power Sources* 358: 39–49.

45 Pal, B.C. and Jabr, R.A. (2009). Intermittent wind generation in optimal power flow dispatching. *IET Generation Transmission and Distribution* 3 (1): 66–74.

46 Bollen, M., Zhong, J., Zavoda, F. et al. (2010). Power Quality aspects of smart grids. *Renewable Energy and Power Quality Journal* 1 (8): 1061–1066.

47 Kempton, W. and Tomić, J. (2005). Vehicle-to-grid power implementation: from stabilizing the grid to supporting large-scale renewable energy. *Journal of Power Sources* 144 (1): 280–294.

48 Pillai, J.R. and Bak-Jensen, B. (2010). Integration of vehicle-to-grid in the Western Danish power system. *IEEE Transactions on Sustainable Energy* 2: 12–19.

49 Galus, M.D., Koch, S., and Andersson, G. (2011). Provision of load frequency control by PHEVs, controllable loads, and a cogeneration unit. *IEEE Transactions on Industrial Electronics* 58 (10): 4568–4582.

50 Pecas Lopes, J.A., Rocha Almeida, P.M., and Soares, F.J. (2009). Using vehicle-to-grid to maximize the integration of intermittent renewable energy resources in islanded electric grids. *2009 International Conference on Clean Electrical Power*, Capri, Italy (9–11 June 2009), 290–5. IEEE.

51 Peças Lopes, J.A., Polenz, S.A., Moreira, C.L., and Cherkaoui, R. (2010). Identification of control and management strategies for LV unbalanced microgrids with plugged-in electric vehicles. *Electric Power Systems Research* 80 (8): 898–906.

52 Shaukat, N., Khan, B., Ali, S.M. et al. (2018). A survey on electric vehicle transportation within smart grid system. *Renewable and Sustainable Energy Reviews* 81: 1329–1349.

53 Bessa, R.J. and Matos, M.A. (2012). Economic and technical management of an aggregation agent for electric vehicles: a literature survey. *European Transactions on Electrical Power* 22 (3): 334–350.

54 Birnie, D.P. (2009). Solar-to-vehicle (S2V) systems for powering commuters of the future. *Journal of Power Sources* 186 (2): 539–542.

55 Carvalho, P.M.S., Correia, P.F., and Ferreira, L.A.F. (2008). Distributed reactive power generation control for voltage rise mitigation in distribution networks. *IEEE Transactions on Power Apparatus and Systems* 23 (2): 766–772.

56 Tonkoski, R., Lopes, L.A.C., and El-Fouly, T.H.M. (2011). Coordinated active power curtailment of grid connected PV inverters for overvoltage prevention. *IEEE Transactions on Sustainable Energy* 2 (2): 139–147.

57 Demirok, E., González, P.C., Frederiksen, K.H.B. et al. (2011). Local reactive power control methods for overvoltage prevention of distributed solar inverters in low-voltage grids. *IEEE Journal of Photovoltaics* 1 (2): 174–182.

58 Nissan SOFC powered vehicle system runs on bioethanol. (2016) *Fuel Cells Bulletin* 2016 (7): 2–3.

59 Ramadhani, F., Hussain, M.A., Mokhlis, H., and Hajimolana, S. (2017). Optimization strategies for solid oxide fuel cell (SOFC) application: a literature survey. *Renewable and Sustainable Energy Reviews* 76: 460–484.

60 Ogden, J.M., Steinbugler, M.M., and Kreutz, T.G. (1999). A comparison of hydrogen, methanol and gasoline as fuels for fuel cell vehicles: implications for vehicle design and infrastructure development. *Journal of Power Sources* 79 (2): 143–168.

61 Barelli, L., Bidini, G., Ciupăgeanu, D.A. et al. (2020). Stochastic power management approach for a hybrid solid oxide fuel cell/battery auxiliary power unit for heavy duty vehicle applications. *Energy Conversion and Management* 221: 113197.

62 Rechberger, J., Kaupert, A., Hagerskans, J., and Blum, L. (2016). Demonstration of the first European SOFC APU on a heavy duty truck. *Transportation Research Procedia* 14: 3676–3685.

63 Hyundai. (2022). *Hyundai ix35 Fuel Cell*.

64 Gupta, S. and Perveen, R. (2023). Fuel cell in electric vehicle. *Materials Today Proceedings* 79: 434–437.

65 Psaraftis, H.N. (2016). Green maritime logistics: the quest for Win-win solutions. *Transportation Research Procedia* 14: 133–142.

66 Edwin, M., Nair, M.S., and Sekhar, S.J. (2022). A comprehensive review on impacts of COVID-19 in food preservation and cold chain: an approach towards implementing green energy technologies. *Environmental Progress & Sustainable Energy* 41 (5): e13820.

67 Joumard, I., Morgavi, H., and Bourrousse, H. (2017). Achieving strong and balanced regional development in India. *OECD Economics Department Working Papers* 1412: 107–139.

68 Deakin, E. (2020). *Transportation, Land Use, and Environmental Planning*. Elsevier.

69 Browne, D., O'Mahony, M., and Caulfield, B. (2012). How should barriers to alternative fuels and vehicles be classified and potential policies to promote innovative technologies be evaluated? *Journal of Cleaner Production* 35: 140–151.

70 Van Fan, Y., Klemeš, J.J., Perry, S., and Lee, C.T. (2018). An emissions analysis for environmentally sustainable freight transportation modes: distance and capacity. *Chemical Engineering Transactions* 70: 505–510.

71 Li, D., Kim, M., Kim, H. et al. (2018). Evaluation of relationship between biogas production and microbial communities in anaerobic co-digestion. *Korean Journal of Chemical Engineering* 35 (1): 179–186.

72 Pan, S.-Y., Gao, M., Kim, H. et al. (2018). Advances and challenges in sustainable tourism toward a green economy. *Science of The Total Environment* 635: 452–469.

73 Rodrigue, J.P., Comtois, C., and Slack, B. (2016). *The Geography of Transport Systems*, 1–440. Routledge.

6

Flexible Operation of Power-To-X Energy Systems in Transportation Networks

Hessam Golmohamadi

Department of Computer Science, Aalborg University, Aalborg, Denmark

Table of Acronyms

ADRA	Agricultural demand response aggregators
CDRA	Commercial demand response aggregators
COP	Coefficients of performance
DFAFC	Direct formic acid fuel cell
DH	District heating
DHW	Domestic hot water
DMFC	Direct methanol fuel cell
DRP	Demand response program
EV	Electric vehicles
EWH	Electric water heaters
FCEV	Fuel cell electric vehicle
FFV	Fossil fuel vehicle
GHG	Greenhouse gas
H2X	Hydrogen-to-X
HP	Heat pump
IDRA	Industrial demand response aggregators
IHFC	Indirect hydrogen fuel cell
IMFC	Indirect methanol fuel cells
MPC	Model predictive control
Non-TCA	Non-thermostatically controlled appliances

Interconnected Modern Multi-Energy Networks and Intelligent Transportation Systems: Towards a Green Economy and Sustainable Development, First Edition. Edited by Mohammadreza Daneshvar, Behnam Mohammadi-Ivatloo, Amjad Anvari-Moghaddam, and Reza Razzaghi.
© 2024 The Institute of Electrical and Electronics Engineers, Inc.
Published 2024 by John Wiley & Sons, Inc.

P2C	Power-to-chemical
P2D	Power-to-diesel
P2FA	Power-to-formic acid
P2H	Power-to-heat
P2H	Power-to-hydrogen
P2M	Power-to-methane
P2Me	Power-to-methanol
P2T	Power to transport
P2X	Power-to-X
PEV	Plug-in electric vehicle
PV	Photovoltaic
RDRA	Residential demand response aggregators
RMFC	Reformed methanol fuel cell
SNG	Synthetic natural gas
TCA	Thermostatically controlled appliance

6.1 Introduction

6.1.1 Problem Description and Motivation

Climate changes have caused serious concerns about the future of the world [1]. With increasing the average global temperature, human beings and agriculture are exposed to a high risk of extinction. Therefore, it does not matter how capacity is the global fossil fuel resources or when they run out. The matter is that the current exploitation rate of fossil fuels may cause irreversible damage to the environment. To overcome this problem, policymakers have decided to accelerate investment in renewable energy extraction. There is no doubt that future energy systems are independent of fossil fuels and dependent on renewable energies. To decarbonize energy systems, renewable energy should penetrate all energy sectors, including power [2], heat [3], and mobility (transportation) [4]. Therefore, sector coupling is essential to transform renewable energy, e.g., wind energy, into thermal energy in the residential heating system [5] or mobility in electric vehicles (EVs). The power-to-X (P2X) structure is a long-term solution not only to couple different energy sectors but also to provide enough reserve and flexibility [6]. The flexibility potentials of the P2X are reflected in the flexible operation of facilities and demands in different sectors. The facilities include water electrolyzers, hydrogen storage, chemical processes, district heating (DH), and electric and hydrogen vehicles. Therefore, the flexible operation of joint sectors counterbalances the renewable power fluctuation. Consequently, the sustainable P2X structure requires the flexible operation of coupled sectors. This is the main motivation of this study to provide a critical review of the flexibility potentials of the P2X facilities in different sectors.

6.1.2 State of the Art

In the literature, the number of research studies in the domain of P2X plants has increased recently. Many scoping studies have been conducted to investigate the operation of energy sectors in P2X systems. Generally, the P2X transforms renewable energy, e.g., wind, solar, and geothermal, into other forms of energy, e.g., hydrogen, heat, syngas, and liquid chemical fuels, which are easily storable and transformable in energy sectors. Therefore, the P2X is comprised of various energy networks including power networks, heat networks, gas networks, and transportation systems. Some review studies are carried out to discuss the structure of energy management in the mentioned sectors. To provide a general insight into the P2X, Figure 6.1 elaborates on the different energy sectors as intermediary or final P2X products.

First, the power-to-hydrogen (P2H) plants transform renewable power into hydrogen energy. The hydrogen can be stored in the storage, e.g., tanks or salt caverns, or be used as feedstock for downstream sectors. A comprehensive review is presented to discuss methodologies of carbon- and hydrocarbon-assisted water electrolysis in hydrogen production facilities [7]. Technological, financial, and environmental aspects of green hydrogen production plants are surveyed to provide a sustainable operation for large-scale plants [8]. The state of the art of integrated photovoltaic (PV) and hydrogen production facilities is presented [9]. The technical challenges of solar to hydrogen are also reviewed. The challenges and prospects of renewable-based hydrogen production plants are surveyed to decarbonize stationary power applications [10]. A holistic review is provided about the expansion planning of integrated power-hydrogen-gas energy systems [11].

Hydrogen energy can be transformed into syngas, methane, and liquid fuels. In addition, the hydrogen may be injected into gas networks directly. Syngas and synthetic methane are used by other sectors, e.g., industries and heating systems. In Ref. [12], the advanced methodologies are discussed to produce hydrogen or syngas using solar-thermal aided reforming of methane feedstocks. A critical review is conducted for power-to-syngas facilities supplied by various renewable energy resources including hydro, wind, solar, bio, and geothermal energies [13]. Syngas is suggested as a substitute for natural gas in power and mobility sectors to relieve socioeconomic and environmental concerns [14]. The current methodologies of catalytic bi-reforming of methane are reviewed to produce clean syngas [15]. A mini-review is presented to investigate the environmental implications of hydrogen-rich syngas production [16]. In addition to syngas, renewable power

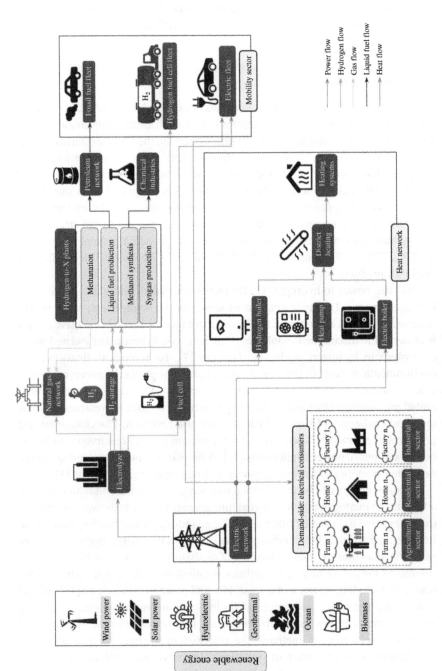

Figure 6.1 General energy structure of power-to-X facilities with different types of energy flow.

is transformed into chemicals and liquid fuels which are easily stored and transformed. Recent technologies are reviewed to produce green liquid fuels, i.e., formic acid, formaldehyde, methanol, gasoline, diesel, and jet fuel [17]. Advanced methodologies of power-to-methanol production are reviewed from a viewpoint of systems engineering [18]. The scoping study discusses commercialization methods to choose methanol as the final production.

In the conventional heating system, fossil fuels are the main energy source of building heating systems for both space heating and hot water consumption. In the P2X system, hydrogen energy can be used in green hydrogen boilers or transformed into power to supply electrical heat pumps (HP). A literature review of power-heat sector coupling is presented to contribute to renewable power integration and energy system decarbonization [19]. An extensive survey is conducted to address the joint heat-to-power flexibility potentials of electrically operated heating systems, including HP, and DH [20]. The flexibility potentials of HP with their barriers and methods of energy efficiency improvement are reviewed [21]. In Ref. [22], 34 large-scale projects are reviewed to address demand response programs (DRPs) of power-to-heat plants. The flexibility opportunities of electricity-thermal systems, including buildings and DH, are surveyed [23].

The future mobility sector benefits from P2X in EVs and hydrogen fuel cell electric vehicles (FCEVs). In EV fleets, the electrical batteries act as the source of electrical storage and flexibility for the power grid [24]. In public parking lots, the aggregated storage capacity of parked EVs provides great flexibility potential for the supply network [25]. The flexibility potentials of EVs in the power distribution grid and the technical, economic, and regulatory barriers to EVs' flexibility integration are reviewed [26]. In addition to EVs, the aggregated flexibility potentials of parked hydrogen vehicles can be integrated into energy hubs [27]. Intelligent parking lots integrated with electrolyzers, hydrogen storage, and fuel cells facilitate the flexible operation of P2X [28]. Flexible interoperation of hydrogen electrolyzers and heavy-duty trucks reduces system operation costs and carbon dioxide emissions [28].

To sum up, Table 6.1 surveys the key objectives of some recent review studies in different energy sectors of the P2X structure. The references are extracted from various publishers to convey the general objectives of multidisciplinary domains including power, thermal, and chemical engineering.

6.1.3 Contributions and Organization

This chapter reviews the flexibility potentials of different energy sectors in the P2X structure. The energy sectors include power-to-hydrogen, power-to-chemicals, power-to-syngas, power-to-heat, and power-to-mobility. In addition, flexibility opportunities for electrical demands are surveyed in residential, commercial,

Table 6.1 Classification of some recent scoping studies in the transportation and P2X sectors.

Sector	Key objective(s)	Publisher	Reference
Power-to-hydrogen	1) Integration of wind power to water electrolyzers 2) Electric network support 3) Cost reduction and CO_2 mitigation	Renewable and Sustainable Energy Reviews	[29]
	1) Integrating hydrogen energy into hybrid energy systems 2) Survey of computer software 3) Review of environmental and sociopolitical issues	Applied Energy	[30]
	1) Discuss the key role of hydrogen in the future energy transition 2) Impact of hydrogen energy on climate	Energy Conversion and Management	[31]
Power-to-syngas	1) Review of pros and cons of co-electrolysis to generate syngas 2) General applications of syngas in traffic/transportation and chemical industry	Angewandte Chemie: Journal of the German Chemical Society	[32]
	1) Electrolysis and methanation technologies 2) Efficiencies and economics of power-to-gas chains	Renewable Energy	[33]
	1) Renewable energy-based power-to-gas concepts 2) Comparison between environmental impacts, operation cost, and efficiency	International Journal of Hydrogen Energy	[34]
Power-to-chemical	Providing an overview of the existing and emerging power-to-chemicals and fuels, including ammonia, ethylene, methanol, ethanol, and formic acid	Technologies for Integrated Energy Systems and Networks	[35]
	Extensive review of electrochemical energy storage to increase penetration of green energies	Chemical Reviews	[36]
	Review of thermochemical and power-to-gas technologies to produce advanced fuels	Renewable and Sustainable Energy Reviews	[37]

Power-to-heat	Review of supervisory controls to unlock heat-power flexibility of heat pumps	*Journal of Process Control*	[38]
	1) Survey of power-heat flexibility of residential buildings 2) Address quantification methods and metrics of energy flexibility	*Advances in Applied Energy*	[39]
	1) Review of requirements on demand-side flexibility in electricity markets. 2) Classifying building energy flexibilities based on power grid requirements	*Energy*	[40]
Power-to-transportation	Extensive review of political, economic, social, technical, legislative, and environmental aspects of EV integration	*International Journal of Electrical Power & Energy Systems*	[41]
	Review second-life batteries as electrical storage for power peak shaving in commercial and industrial sectors	*Journal of Energy Storage*	[42]
	Review of energy management strategies of electric buses in public transport to decarbonize urban areas	*Renewable and Sustainable Energy Reviews*	[43]

industrial, and agricultural sectors. The key objectives and applications of energy flexibility are discussed. To the best of the authors' knowledge, no comprehensive review study has been conducted to elaborate on the flexibility potentials of different energy sectors of the P2X. Therefore, the main contributions of the current study can be stated as follows:

1) Providing an extensive review of flexibility potentials of different energy sectors in the P2X structure, including power-to-hydrogen, power-to-chemicals, power-to-syngas, power-to-heat, and power-to-mobility, as well as electricity demand, i.e., residential, commercial, industrial, and agricultural sectors.
2) Classifying the key objectives and applications of the flexibility opportunities in different energy sectors of the P2X system with key results and the studied region.

The rest of the chapter is organized as follows. In Section 6.2, the flexibility potentials of Power-to-Hydrogen facilities are discussed. Section 6.3 addressed the flexibility opportunities of Power-to-Methan plants. The flexible operation of power-to-chemical, including diesel, formic acid, and methanol, is explained in Section 6.4. The power-to-heat sector is described in Section 6.5. Section 6.6 illustrates the flexibility potential of the mobility sector. In Section 6.7, flexibility is discussed in four electricity demand sectors. Finally, Section 6.8 concludes the review study.

6.2 Power to Hydrogen

The power-to-hydrogen (P2H) sector transforms renewable power into hydrogen energy through water electrolysis. In water electrolysis, renewable power is used to decompose water (H_2O) into hydrogen (H_2) and oxygen (O_2). The electrolysis process takes place in an electrolysis cell which is comprised of anode/cathode electrodes, a catalyst, and a membrane. Generally, the electrolysis is carried out based on three main technologies as follows:

1) Alkaline water electrolysis [44]
2) Polymer-electrolyte membrane [45]
3) Solid oxide electrolysis [46]

Water electrolysis is the core part of P2H plants. The main product is H_2, which is a storable energy commodity. The by-product is O_2, which can be further utilized by downstream industries. The chemical reaction of the water electrolysis can be formulated as follows:

$$2H_2O \rightarrow 2H_2 + O_2 \tag{6.1}$$

Figure 6.2 describes the energy paradigm of the P2H plants. In this figure, the inlet and outlet energy and materials are stated. As the graph reveals, the outlet hydrogen energy can be used in three ways:

1) Store in hydrogen storage and/or salt caverns
2) Inject into the gas network directly
3) Export to (hydrogen-to-X) H2X facilities

First of all, the produced hydrogen can be stored in hydrogen storage. Generally, hydrogen storage is in the form of storage tanks and salt caverns. There is a direct relationship between the flexible operation of the electrolyzer and storage capacity. To ensure constant operation of the electrolyzers, an adequate storage capacity is required. The "adequate storage capacity" is determined considering the lull duration of renewable energies. The main challenge of hydrogen tanks is the large dimensions. Increasing the storage volume of the hydrogen, the dimension and investment cost of the storage tanks increase noticeably. To overcome this problem, salt caverns are the practical solution to store hydrogen underground. The salt caverns have a geometrical volume from 100,000 to 1,000,000 m^3 [47]. Salt caverns suffer from local constraints. It means that the salt caverns are normally created on salt bed deposits. The stored hydrogen can be used by (1) fuel cells to reconvert to power and (2) H2X plants to convert to other types of energy commodities.

The produced hydrogen can be injected into the natural gas network directly. The amount of hydrogen injected into the gas network is limited. The reason is that the mixture of hydrogen reduces the gas quality. Meanwhile, it increases the loss of the natural gas network. In the literature, it is suggested that the acceptable amount of hydrogen share in the gas network should not exceed 10%. In some studies, a higher share of hydrogen blending is investigated. The results show that the injection point plays a key role in the upper threshold of hydrogen blending. In this way, as the hydrogen injection point approaches the pipeline outlet of the gas network, the loss of the natural gas network decreases up to 36.5% [48].

In the literature, many studies have been conducted to unlock the flexibility potentials of P2H plants in response to renewable power fluctuations. Table 6.2 surveys the key objectives and flexibility purposes of some recent studies for P2H plants. In Ref. [57], the P2X is suggested to counterbalance the wind power fluctuation and decarbonize the transportation sector. The study proposes a new energy system that is comprised of combined cooling, heat, and power (CCHP), wind power, and hydrogen electrolysis systems. The daily production capacity of the hydrogen electrolysis reaches 500 and 260 kg in winter and summer, respectively. Besides, the total energy system efficiency increases to 72%. The levelized costs of hydrogen models in China are discussed in the research study [58]. The

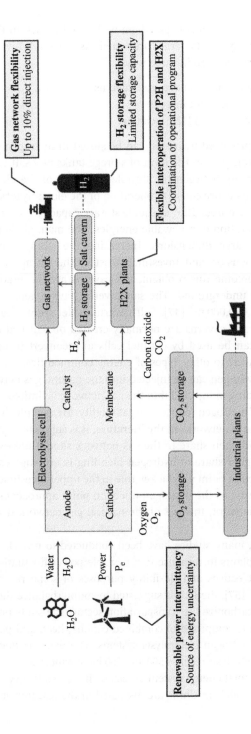

Figure 6.2 Energy paradigm of P2H plants with the flexibility potential.

Table 6.2 Key objectives and flexibility purposes of P2H plants in recent studies.

Reference	Main objective(s) and flexibility potentials	Demand sectors	Key components	Region	Key results
[49]	Flexibility for electricity and hydrogen markets	Industrial sector	Integrated electricity and hydrogen markets	Netherlands	A carbon price of 150–700 €/ton is needed for beneficial investment in hydrogen production facilities.
[50]	1) Stabilize wind and solar market values 2) Flexible operation of hydrogen electrolysis	Flexible electricity demand	1) Electricity market 2) Wind and solar units 3) H_2 storage unit	Germany	Flexible hydrogen production causes market values in Europe above 19 ± 9 €/MWh and 27 ± 8 €/MWh for solar and wind energy in 2050
[51]	1) Lower electricity costs 2) Reduce carbon dioxide emissions 3) Flexible operation of electrolysis	Transportation (EVs)	1) Hydrogen fuel cell 2) EVs 3) Electricity market	PLEXOS power system, Western US	By increasing hydrogen production flexibility, the cost of hydrogen and electricity generation decreases
[52]	1) Optimizes investment and hourly dispatch of energy units 2) Lowers total system costs 3) Reduces system emissions 4) Decarbonize transport and industry 5) Flexible power and hydrogen production	Fixed demand	1) Wind and solar units 2) Battery storage 3) CO_2 transport and storage 4) Thermal power plant	Germany	1) Increases the optimal wind and solar share by 50% 2) Lowers total system costs by 8% 3) Reduces system emissions from 45 to 4 $kgCO_2$/MWh 4) Produces clean hydrogen equivalent to about 90% of total electricity demand
[53]	1) Economic competitiveness of flexible H_2 production 2) Mitigates renewable energy curtailment 3) Integrate fluctuating renewable energies into the power system 4) Flexible hydrogen production	Flexible electricity demand	1) Power system 2) Renewable energy generation	1) France 2) European power system	The approach with very high penetration rates of decarbonized energy in Europe can produce low-carbon hydrogen with a competitive price of €3–5/kgH_2.

(Continued)

Table 6.2 (Continued)

Reference	Main objective(s) and flexibility potentials	Demand sectors	Key components	Region	Key results
[54]	1) Reduction of operational costs and emissions. 2) Flexibility of electrical and thermal demands. 3) Flexible production and storage of hydrogen.	1) Water desalination units 2) Commercial buildings	1) Seawater desalination units 2) Reverse osmosis (RO) technology 3) Organic Rankine cycle (ORC) for generating electrical power 4) Parking lot and EVs 5) Hydrogen tank	Qeshm Island, Iran	1) Total cost of the energy hub is decreased by 11.08% 2) Electrical and thermal power generation increased up to 56.03% and 70.53%
[55]	1) Increases the interrelation of biofuel supply with renewable power 2) Improve system flexibility 3) Minimize the system cost 4) Increase flexibility in the system to respond to the imbalances between energy demand and supply	1) Residential sector 2) Transportation 3) Industrial demands	1) CHP plants 2) Hydrotreated pyrolysis oil production plant 3) Rooftop PV systems 4) Power-to-hydrogen storage 5) Heat pumps 6) EV	Västerås, Sweden	Increase the share of renewables in energy supply by 6%
[56]	1) Maximize flexible operation for electricity, hydrogen, ammonia, and heating 2) Flexible operation of an integrated renewable multi-generation system	1) Industrial sector 2) District heating	1) Hydrogen and ammonia as short-term and long-term energy storage mediums 2) Solar System 3) Wind power unit	10 cities in China	Increase energy and exergy efficiencies by 50% and 73.1%

study concludes that the cost of the hydrogen model for coal-to-hydrogen is 13.1–19.4 RMB/kg, which is 20.5–61% lower than the hydrogen production electrolyzed by renewable energies. In Sweden, the feasibility of hydrogen production is investigated in nuclear power plants to hedge against the inherent inflexibility of the power plant. Besides, hydrogen production for industrial applications is discussed [59]. Molten salt heat storage is addressed to improve the operational flexibility of the power grid whose electricity and heat demands are supplied by hydrogen production. The results confirm a 4.4% reduction in the operational cost of the power grid [60]. In Ref. [61], the power and hydrogen networks are integrated to supply fuel cell EVs with the purpose of maximizing daily profit. The simulation results show that the integrated network increases the daily profit by up to 25.32% and decreases the renewable energy curtailment by up to 36.4%. In Germany, the P2H system is discussed to decarbonize energy-intensive industries, e.g., copper production, through on-site hydrogen production. Based on the result, not only are the heavy industries decarbonized but also the by-product of oxygen brings economic benefit [62]. In Ref. [63], a multi-energy hub system is proposed to deliver power, heating, cooling, and hydrogen. The study aims to maximize the total energy efficiency of the energy system using fuel cells. The results reveal that the suggested structure increases energy efficiency by up to 87.43%. In a smart multi-carrier energy system, hydrogen storage is considered a flexible resource to counterbalance renewable power fluctuations. The power is converted to hydrogen energy in low-price hours (surplus of renewable energy), and then electrical power is generated by hydrogen-based fuel cells during high-price hours (deficit of renewable energy) [64]. In Ref. [65], a multi-carrier storage system is comprised of thermal storage, ice banks, and hydrogen storage to supply electricity, heating, and cooling demands. Minimizing the operational cost of the energy system, the simulation results confirm up to a 5% reduction in the total system operation cost. The P2H and lithium batteries are evaluated to hedge against wind power fluctuations [66]. The study concludes that the P2H is more beneficial for profitability, reliability, and full-load operation. In contrast, the lithium battery is advantageous in reducing the lost wind with a lower environmental impact. To conclude this section, Table 6.3 classifies the flexibility potentials of hydrogen storage and consumption in the P2H facilities.

6.3 Power to Methane

Power to methane (P2M) is an energy sector in which the original renewable power is transformed into synthetic gas energy through the methanation process. Methanation is an exothermic chemical reaction in which carbon dioxide and hydrogen are converted to methane through hydrogenation. Methanation is a

Table 6.3 Applications of hydrogen storage and production in the P2H plants.

Main application(s)		Main objective(s)	Potential flexible demand	Key result(s)	Reference
Storage	**H₂ storage**	1) Providing flexible hydrogen production sources for vehicles 2) Sector coupling between hydrogen, gas, and power 3) Reduce operation costs and emissions	1) FCEVs 2) Hydrogen-based industries	Reduction cost and emission by 8.04% and 9.8%	[67]
		1) Providing heat and power demand response 2) Minimize the operation cost.	1) FCVs 2) Parking lot of PEVs 3) Fuel cell micro-CHP	Reduce the whole system operation cost by 76.35%	[68]
		1) Integrate flexibility to combined heat and power and photovoltaic systems 2) Optimization of emissions and power import into the regional systems.	1) District heating 2) Regional electricity demand	Reduce power imports and emissions by 53%	[69]
	Salt cavern	Reducing emissions within an integrated energy systems context	General electrical and thermal demands	The type of H₂ storage and salt caverns has a strong dependency on the RES availability and needs for smooth charging and discharging.	[70]
		1) Decrease power supply curtailment 2) Optimize operation costs	Electricity demands	2030 and beyond seasonal electricity storage reduces the system cost by 10–25%	[71]
		Provide flexibility, reliability, and lower operational costs for energy systems	1) FCVs 2) PEVs	1) FCVs are more flexible and cheaper for backup power than PEVs. 2) PEVs are appropriate for local balancing power, e.g., office buildings.	[72]

Gas network 5–10%	Injection of extra hydrogen to transmission pipelines of gas network	NA	Maximum fraction of hydrogen in natural gas should not exceed 15–20%; otherwise, gas quality decreases.	[73]
	1) Counterbalance photovoltaic power generation 2) Power and gas sectors coupling	Gas network demand	1) Place of the gas injection point determines the share of hydrogen blending. 2) High solar irradiation injects 20–30% hydrogen into a gas network which is an unacceptable blend.	[74]
	Facilitates the gas and electricity sector coupling	Gas network demand	Methanation is an alternative solution to increase the hydrogen injection in the gas network and facilitates the gas and power sectors' coupling.	[75]
Fuel cells	Zero-carbon fuels in marine applications	Ship electrical demands	1) Cost analysis: fuel cells ships have 27–43% higher costs than a diesel-powered ship 2) Green ammonia-fuel cells reduce up to 84% in CO_2-eq emissions in comparison with diesel-powered ships.	[76]
	Flexible energy operation of urban districts and public transportation	1) Urban residential district 2) Mobility sector	1) Annual efficiency of the system is within 55–70% 2) Daily efficiency of hydrogen and electricity changes the mobility demand between 25–51%.	[77]
	Integrate renewable energy into the residential sector	Residential sector	Fuel cells are highly applicable in residential markets as alternative RES	[78]

critical process to create synthetic natural gas (SNG) through the following chemical reaction [79]:

$$CO_2 + 4H_2 \rightleftharpoons CH_4 + 2H_2O \tag{6.2}$$

The methanation methods are classified into two different approaches: (i) Thermochemical methanation [80] and (ii) biological methanation [81]. In the former, the methanation process is done using a transition metal catalyst with an efficiency of 70–85% [82]. In the latter, the methanation is carried out using microorganisms or biogas with an efficiency of 75–98% [83].

Figure 6.3 describes the different sections of P2M facilities. Based on the diagram, the main share of produced methane is injected into the natural gas network. Generally, natural gas is comprised of more than 75% methane [84]. Therefore, in contrast to hydrogen, there is no limitation to inject methane into the gas pipelines. For regions with well-developed natural gas networks, pipelines are the best solution. In intercontinental applications, the gas network may not be practical due to oceans and deep seas. In this condition, compression and liquefaction facilities are applied to convert methane into pressurized and liquefied methane, respectively. Consequently, the volume decreases noticeably, especially for the liquid methane. The liquid/pressurized methane can be transported by trucks, trains, and ships. It is worth mentioning that the liquefication/compression facilities consume a certain percentage of methane energy [85]. The compression and liquefaction plants consume 7.2–13% [86] and 10–25% [87] of methane energy, respectively. It means that the total efficiency of the P2M facilities decreases for compressed and liquid methane.

The produced hydrogen of P2H plants is the main feedstock of the methanation facilities. In the P2H sector, two technical factors limit the flexible operation of the electrolyzers: (i) the limited share of the hydrogen blending to natural gas networks and (ii) the limited storage capacity of hydrogen tanks. Consequently, the methanation facilities can increase the flexibility potential of the P2H plants. In the literature, many studies have been conducted to facilitate the flexible operation of methanation plants. In Ref. [88], a mathematical model is suggested for power-to-renewable methane in regions with high penetration of wind energy and biogas plants. The problem aims to provide power grid balancing in the Baltic States, including Estonia, Latvia, and Lithuania. The results show that the current wind power potential can increase mechanized biogas by up to 48.4%. The dynamic model of power-to-synthetic natural gas is presented in Ref. [89]. The study investigates the impact of hydrogen generation, storage, and buffering devices on the operation of the P2M plants. The role of P2M plants is studied in the security and operation of the networked energy hubs, including power and gas grids [90]. The results confirm that the suggested approach can overcome gas shortages in critical hours. Moreover, it reduces wind generation curtailment.

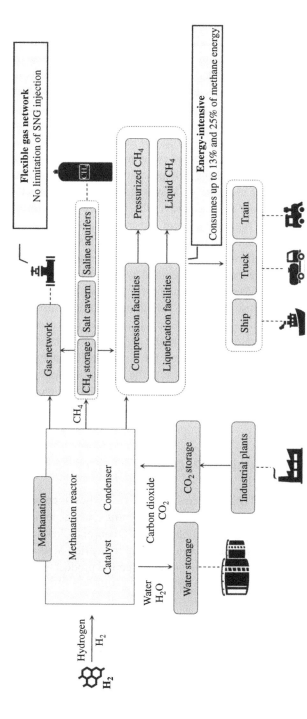

Figure 6.3 Energy paradigm of P2M facilities.

The economic analysis is carried out on a fully renewable P2M system [91]. The study concludes that the Levelized cost is between 0.24 and 0.30 €/kWh of SNG. Also, it conveys that the by-products, e.g., oxygen, can increase the competitiveness of the P2M plants. In Ref. [92], the P2M plant is suggested as an auxiliary operation of thermal power plants. Based on the study, the competitiveness of the P2M depends mainly on the interconnectedness of other technologies. The analysis reveals that the payback period for the P2M facilities ranges between 3.8 and 5.1 years. A poly-generation system is proposed for hydrogen production and electric power generation [93]. A combined-cycle power plant is modeled to utilize methane. The results show that the poly-generation system can reach an exergy efficiency of 83.1%. In Ref. [94], a bio-P2M system is discussed to convert the biogas carbon dioxide to methane. Based on the results, if the bio-P2M is integrated with renewable energies, the suggested system can meet the 2026 EU goal of greenhouse gas (GHG) reduction. The research study [95] proposes liquid CO_2 energy storage as a flexible regulator for P2M facilities to hedge against the uncertainties associated with renewable power, gas demand, and electricity prices. In this way, CO_2 energy storage ensures continuous power and gas production. A bi-level profit-driven approach is suggested for P2M systems to maximize the benefit of electricity, natural gas, and hydrogen sectors [96]. In the upper level, the energy value and price bids are submitted in the day-ahead markets for the three energy commodities. In the lower level, the multi-carrier energy markets are characterized. To have a general insight into the flexibility potentials of methane storage and gas network injection, Table 6.4 explains some recent studies for P2M facilities.

6.4 Power to Chemical (P2C)

In the power-to-chemical (P2C) process, the electrical power is converted to liquid energy commodity or chemical energy. In the literature, power-to-liquid [104] and power-to-fuel [105] have a close relation to the P2C concept. Chemical energy normally is used in the transportation and chemical industries. Methanol, synthetic diesel, and formic acid are the most important chemical products of the P2C. In the following, the three liquid fuels are described.

6.4.1 Power to Diesel (P2D)

In the power-to-diesel (P2D) plants, hydrogen and carbon dioxide are reacted to a mixture of hydrocarbons as the following reaction [85]:

$$nCO_2 + (3n + 1)H_2 \rightleftharpoons C_nH_{2n+2} + 2nH_2O \tag{6.3}$$

Table 6.4 Flexibility opportunities and objectives of P2M plants for methane storage and gas network injection.

Methane output		Key flexibility objective(s)	Region	Key result(s)	Reference
Storage	Salt cavern	Provide balancing between the electricity and the gas markets	Germany	Feed-in-tariffs of 100 and 130 €/MW for hydrogen and methane promote gas storage.	[97]
		Increase the stability of the salt cavern to balance the intermittent renewable power	China	1) In salt caverns, a higher content of interlayers reduces the capacity of gas storage. 2) Also, the flexibility of gas operation increases and the service period of the gas storage is lengthened.	[98]
		Increase the economic potential of energy storage with high shares of intermittent renewable electricity	Austria and Germany	1) Daily cycles have the highest impact on storage revenues. 2) Day-storage cycles with a ratio of capacity to power rate of 7–8 are the optimal economic layout	[99]
	Saline aquifers	Integration of renewables into electricity grids through large-scale storage balancing supply and demand	Portugal	Critical criteria for selecting underground reservoirs include land surface constraints, subsurface, general geology of the area, candidate rock types, structural and tectonic factors, seismicity risks, hydrogeological and geothermal issues, containment issues, and geotechnical factors.	[100]
		1) Investigation of the role of carbon capture and storage technologies on energy balance 2) Integration of power-to-gas into energy networks	Italy	1) The CO_2 capture causes a less than 1% efficiency reduction 2) Synthetic natural gas plant reaches near-zero life cycle emissions.	[101]
		Economical and environmental analysis of storage on traditional and renewable power plants	Germany	The greenhouse gas emissions from 1 kWh of electricity generated by first-generation carbon capture and storage power plants reduce by 68–87%	[102]

(Continued)

Table 6.4 (Continued)

Methane output		Key flexibility objective(s)	Region	Key result(s)	Reference
Gas network	**Pipe storage**	Provide temporal arbitrage and balancing energy between power and gas markets	Germany	1) P2G system used for bridging the balancing markets for power and gas is not profitable. 2) For temporal arbitrage and balancing energy, pipe storage is preferred. 3) For cost-effective pipe storage, feed-in-tariffs of around 100 and 130 €/MW for hydrogen and methane are needed.	[97]
	Pipe injection	Review challenges and opportunities of syngas injection into gas networks	The UK Worldwide	1) The governmental support and power balancing markets play a key role in the commercialization of P2G. 2) The efficiency of P2G and hydrogen injection gas grids will increase by 2030	[103]

There are two widely accepted technologies to produce liquid fuel including methanol synthesis [106] and the Fischer-Tropsch reaction [107]. Diesel is an easily transportable energy commodity that can be transported via pipe networks or fuel transportation fleets, e.g., railway, ship, and trucks.

In the literature, some studies have been conducted to produce diesel fuel from renewable hydrogen. In Ref. [108], an experimental study is conducted to produce diesel fuel from biomass. Also, the study integrated wind energy into hydrogen production facilities, P2G, and biomass-to-liquid processes. The study concludes that wind energy can be stored in chemical commodities with high energy density. Considering diesel as a transportable energy carrier, a research study shows that renewable energy in North Africa can be transformed into Hydrogen, diesel, or gasoline and transported to Europe by pipe networks and/or tankers (ships) [109]. Therefore, liquid fuel is an economic intermediary energy commodity to transfer renewable energy from North Africa to the center of Europe at the price of 54–119 €/MWh. In another study, synthetic diesel is suggested as a means of renewable energy storage [110]. The study concludes that the suggested system can supply between 0.9% and 32% of the annual diesel demand for road transportation with 100% emission savings. However, the production cost of synthetic diesel is not competitive enough in comparison to conventional diesel. Synthetic diesel plays a key role in energy system decarbonization. The research study suggested dimethyl ether as a synthetic diesel for road transportation with low CO_2, NO_x, and particulate matter emissions [111]. It is shown that synthetic fuel can decarbonize the agricultural sector. To have a net-zero local energy system, a dairy farm is modeled with microturbines, solar panels, batteries, and synthetic fuel production facilities [112]. The results reveal that the electricity generation of the farm is profitable; adversely, the synthetic fuel production causes high costs that make it nonprofitable. In Sweden, synthetic fuel is used to integrate renewable energies into the transportation sector [113]. Comparing hydrogen, methane, methanol, and diesel in the transportation sector, the results show that methane and diesel have great potential for cost competitiveness in the near term. In contrast, hydrogen is price-competitive for longer terms. A hybrid energy system is suggested to supply transportation fuel in Finland with renewable energy [114]. The hybrid system is comprised of PV, wind generation units, electrolysis, and hydrogen-to-liquids plants. Based on the results, the combined renewable electricity-diesel value chains are competitive for crude oil prices within a minimum price range of about 79–135 US$/barrel regarding costs associated with capital, oxygen, and CO_2 emissions. Although synthetic fuels are used to supply industries and the transportation sector, a study is carried out in the Maldives to decarbonize the energy systems by substituting diesel generations with 100% renewable energies, including floating offshore solar PVs, wave power, and offshore wind [115].

6.4.2 Power-to-Formic Acid (P2FA)

In power-to-formic acid (P2FA), hydrogen and carbon dioxide are reacted chemically to produce formic acid as the following formulation [116]:

$$CO_2 + H_2 \rightleftharpoons HCOOH \tag{6.4}$$

Formic acid is a nontoxic environmentally friendly hydrogen energy with low flammability under ambient conditions that show great applicability in the mobility sector [117]. Generally, formic acid has applications as an energy carrier and a chemical combination. In the former, formic acid is used to carry hydrogen energy. In the latter, formic acid is used in heavy industries, chemical industries, food industries, and pharmacological manufacturing for non-energy applications. The current study concentrates on the energy applications of hydrogen energy carriers; therefore, the latter is excluded. Considering formic acid as hydrogen energy, formic acid impacts the energy systems with the following objectives:

1) supplying the transportation sector with green hydrogen energy;
2) storing renewable electricity in the form of a green hydrogen carrier;
3) decarbonization of energy systems and industries.

First, formic acid supplies the transportation sector, including vehicles, buses, and trucks, with green energy. Formic acid can generate electricity in two ways: (i) direct formic acid fuel cell (DFAFC) [118] and (ii) indirect hydrogen fuel cell (IHFC) through dehydrogenation reactions [119]. The DFAFC is suggested for the mobility sector to replace traditional fossil fuels. In Ref. [120], a 25 kW formic acid-to-power system technology is proposed to supply city buses or serve as a standalone carbon-neutral electricity generator unit. Storage technologies are developed to store hydrogen energy in the form of formic acid [121]. Formic acid has a high volumetric hydrogen storage capacity of 53 g H_2/L: therefore, it takes advantage of low-cost and large-scale hydrogen storage and transportation in contrast to energy-intensive hydrogen liquefaction and compression [122]. Formic acid not only provides green energy to the transportation sector but also causes a net reduction of carbon dioxide due to the captured CO_2 during the synthesis process, i.e., Eq. (6.4). A technical report carried out by the EU Commission shows that the capacity of the carbon dioxide utilization plant depends on the available renewable power, rather than on the product's demand [123]. To sum up, Figure 6.4 explains how the green formic acid decarbonizes the mobility sector.

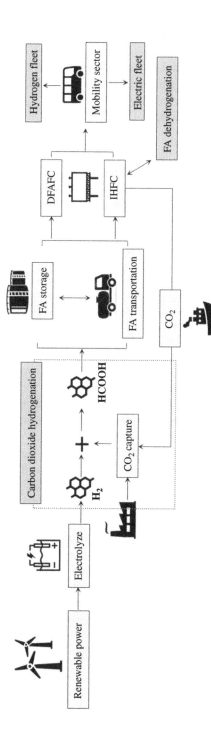

Figure 6.4 Integration of formic acid into the transportation sector, "FA stands for formic acid."

6.4.3 Power to Methanol (P2Me)

In the power-to-methanol (P2Me) plants, hydrogen and carbon dioxide are reacted to produce methanol and water using the following formulation [124]:

$$CO_2 + 3H_2 \rightleftharpoons CH_3OH + H_2O \tag{6.5}$$

The hydrogen energy of methanol normally supplies the mobility sector in three ways: (i) methanol blending with fossil fuel vehicles (FFVs), (ii) direct methanol fuel cells (DMFC), and (iii) indirect methanol fuel cells (IMFC). In methanol blending, alcohol-blended fuels are used in fossil fuel-based engines to reduce emission production [125]. In the DMFC, the chemical energy of liquid methanol is transformed into electric power through an electrochemical energy conversion process [126]. In the IMFC, also called reformed methanol fuel cell (RMFC), the liquid methanol is reformed before feeding the fuel cells.

The flexible operation of P2Me plants plays a key role in the integration of intermittent renewable energy into chemical production. The research study shows that the flexible operation of chemical processes can reduce the methanol production cost by 20–35% [127]. In addition to the flexible operation of the P2Me facilities, the flexible operation of the electrolysis system and power grid should be integrated to hedge against intermittent renewable power. The dynamic simulation of methanol synthesis reveals that the energy efficiency of the P2Me varies between 75.4% and 90.2% under different load change operations [128].

A research study is conducted to assess the production and transportation costs of switching from fossil fuels to renewable methanol [129]. The results reveal that the production cost of renewable methanol can reach the market price of fossil fuels by 2030. In Ref. [130], wind energy and captured CO_2 are discussed in six countries to decarbonize the transportation-power nexus and reduce gasoline consumption. A chemically recuperated gas turbine is suggested to be fed by renewable methanol from power-to-liquid facilities. The integrated gas turbine, hydrogen electrolyzer, and methanol production plants can store a surplus of renewable energy with very low CO_2 emission and an overall power-to-power efficiency of 0.23 [131]. In the other study, a novel system is proposed to produce solar methanol by hybridizing natural gas with an energy efficiency of 62% [132]. When the system is integrated with a power generation unit, the net solar-to-electricity efficiency reaches 23%. To elaborate more on the energy management of power-to-liquid plants, Table 6.5 classifies the key characteristics of flexible operation, objectives, and outcomes.

Table 6.5 Energy management objectives of power-to-chemical plants with key results.

Reference	Main objective(s)	RES	Key point(s)	Region[a]	Key numerical results
[131]	Store RES surplus and produce electricity with very low CO_2 emissions	General renewable	Recovering exhaust heat of chemically recuperated gas turbine	Italy	1) Overall efficiency of power to power is 0.23 2) Production of 350 kg/h of renewable methanol from a renewable power capacity of 600–750 kW
[133]	Determine the efficiency of converting renewable power to methanol	Wind	Direct CO_2 air capture	Netherlands	1) Efficiency of converting wind power to methanol is 50% 2) Capital cost breakdown: 45% for electrolyzers, 50% for the CO_2 air capture installation, and 5% for the methanol synthesis system
[134]	Determine the environmental opportunity cost of P2Me	1) Solar 2) Wind	Mitigate climate change by carbon dioxide capture	The US	Use RES to produce methanol from CO_2 only if the grid intensity is below 67 gCO_2/kWh
[130]	1) Decarbonize the electricity and transportation 2) Displace conventional fossil fuels	Wind	Carbon capture offers greater environmental benefits	1) The US 2) UK 3) France 4) Germany 5) Poland 6) China	Decarbonizing the power grid using RES has greater environmental benefits, compared to producing methanol from the hydrogenation of captured CO_2.
[135]	Minimize the system's operational cost or maximize its green hydrogen production	1) Wind 2) Solar	Energy management system addressing P2X demand response and flexibility of electrolysis plant	Denmark	1) Potential reduction of 51.5–61.6% for the total system operational cost 2) Increase in the share of green hydrogen by 10.4–37.6%
[136]	Increase penetration of solar energy into the chemical industry (solar methanol production)	Solar	Highly efficient use of solar power in the industrial sector	Germany	The efficiency of solar methanol plants is higher than that of the conventional solar power plants

[a] The region of case studies may refer to the partial or whole part of the study region. This is because the project may be comprised of different case studies in different countries/regions.

6.5 Power to Heat (P2H)

Power-to-heat (P2H) refers to energy-converting technologies that generate thermal energy for heating purposes including space heating and domestic hot water (DHW) consumption. In heat networks, the heat energy is distributed to consumers through DH systems. The heat sources of the DH can be power or fuel. Regarding power, electric boilers and HP convert electric power directly to thermal energy. In the latter, the heat source can be natural gas, syngas, and chemical fuels. Therefore, there is a strong interdependency between the heat and power networks to counterbalance the renewable power fluctuations. Figure 6.5 explains the flexibility potentials of P2H for both electric and hydrogen energy consumption.

Regarding the power source, the total energy efficiency of HP is much higher than that of electric heaters. Although the investment cost of HP is currently higher than that of electric heaters, saving energy can compensate for the high initial investment within a few years depending on energy prices. The HP operations are normally optimized by heat controllers. The heat controller can unlock the flexibility potentials of space heating and DHW consumption in response to the energy price or flexibility needs of the upstream networks. In Ref. [137], a model predictive control (MPC) is suggested to unlock the flexibility of the residential heating system in response to day-ahead electricity prices. The flexibility of residential heating systems stems from the upper and lower thresholds of indoor

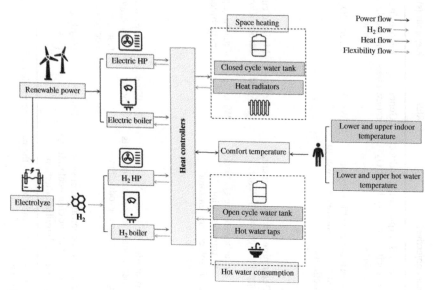

Figure 6.5 Integration of flexibility of buildings heating system to the P2H sector.

air temperature and hot water. The research study suggested a stochastic structure for the MPC to incorporate the electricity price uncertainties of three successive electricity market floors, including the day-ahead, intraday, and balancing markets [138]. In addition to individual HP, many residential buildings are supplied by DH through mixing loop control. Therefore, a mixing loop controller is discussed to optimize the operation of hot water valves [139]. The on-off states of the mixing loop valves are optimized to unlock the flexibility of space heating in response to variable energy prices. In Ref. [140], a two-layer stratified hot water tank model is presented for electric boilers to integrate flexibility potentials of residential heating systems into a low voltage distribution grid in Northern Jutland, Denmark. The role of flexible thermal units in voltage control and grid congestion mitigation is also addressed. A research study was conducted in Norway to investigate the role of electricity price scenarios on the flexible operation of electric boilers integrated with thermal storage as the backup system of HPs [141]. The result reveals that electric boilers contribute to 17% of the heat supply in the case study.

In addition to the electric HP, the gas pumps are operated by hydrogen energy. The technical operation, coefficients of performance (COP), and exergy efficiency of chemical HP are discussed [142]. A thermo-economic cost analysis was carried out for a small prototype of the isopropanol-acetone-hydrogen chemical HP in China [143]. The simulation results revealed that the COP and exergy efficiency of the system were 24.3% and 42.3%, respectively. Although this type of HP is normally used to recover industrial waste heat, it can supply the DH systems in urban areas.

6.6 Power to Transport (P2T)

By increasing the penetration of EVs in private and public transportation, road maps and traffic routes are attracting attention in energy system studies [144]. Nowadays, the transportation sector is not supplied by fossil fuels alone. Electricity and hydrogen energies are displacing traditional fuels to meet climate change requirements [145]. In the mobility sector, there are three types of vehicle fleets as follows:

1) FFVs, supplied by gasoline and natural gas;
2) plug-in electric vehicles (PEVs), supplied by electric power;
3) FCEVs, supplied by hydrogen- and formic acid-based energy.

Figure 6.6 sketches the energy paradigm of the P2T sector. In the transportation sector, the FFVs are the main consumers of P2D products [146]. Synthetic diesel is used for conventional vehicle fleets. In recent years, the FFVs have been retired

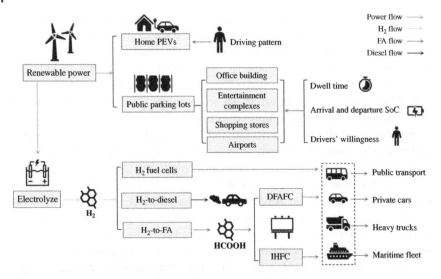

Figure 6.6 General flexibility potentials of the mobility sector.

and replaced by PEVs and FCEVs. The PEVs can provide noticeable demand flexibility for power systems [147] and the P2T sector. The PEVs are mobile electrical batteries that can be charged/discharged when a renewable power excess/shortage occurs on the supply side. The PEVs are discussed in private and public parking lots [148]. In large-scale parking lots, the parking lot operator can coordinate the charging/discharging operation of the PEVs [149]. In the private parking lots, an intermediary entity, e.g., a demand response aggregator, is required to integrate the flexibility potentials of individual PEVs in response to the flexibility requirements of the supply networks. Public parking lots exhibit considerable electric storage capacity to counterbalance renewable power fluctuations [150]. Smart charging stations play a pivotal role in the flexible operation of PEVs [151]. In the home parking lots, the occupancy pattern, driving patterns, and PEV owners' behavior affect the charging/discharging operation of PEVs [152]. In public parking lots, the traffic pattern depends on the application of the parking lot. For official parking, the electrical storage is normally accessible during working hours on weekdays. Therefore, the time duration and capacity of available electric storage depend on the number of office staff and the length of working hours. For the parking of an entertainment complex, the electric storage capacity is normally accessible outside of working hours with a lower time length [153]. Dwell time, arrival/departure State of Charge (SoC), and drivers' willingness to participate in the DRPs are the main factors affecting the flexibility potential of public parking lots. The FCEVs are new hydrogen energy consumers in the mobility sector [154]. The

FCEVs can be supplied by hydrogen and formic acid. In the hydrogen FCEVs, the hydrogen energy is transformed directly into electric power [155]. In the formic acid FCEVs, the electric power is generated by direct and indirect fuel cells in which the formic acid is converted to electricity directly or through dehydrogenation reactions.

6.7 Power Demand Flexibility

In power systems, electricity consumers play a pivotal role in meeting flexibility requirements. The electricity consumers are discussed in four sectors, including residential, industrial, agricultural, and commercial sectors. Responsive demands can integrate demand flexibility into power systems when a power shortage occurs or system reliability is jeopardized due to failure or unforeseen maintenance.

Figure 6.7 explains the various flexibility potentials of electricity consumers in power systems. In residential buildings, the HEMS unleashes the flexibility potential of household appliances [156]. Household appliances are divided into three main categories: (i) thermostatically controlled appliances (TCA) [157], (ii) non-thermostatically controlled appliances (Non-TCAs) [158], and (iii) uncontrollable appliances. The first group includes HP, HVAC, refrigerators, and electric water heaters (EWH). The second group addresses wet appliances, e.g., washing machines, dishwashers, and tumble dryers. The uncontrollable appliances normally refer to TV, Radio, and basic lighting systems which are not subject to DRPs. In Norway, the flexibility potentials of EWH are calculated for different levels of flexible power and activation time. The result indicates that the highest flexibility potential is approximately 54% for a duration of 61 minutes at 8 a.m. [159]. The flexibility potentials of 10 residential buildings are discussed in Madrid, Spain, to integrate HP into solar-assisted DH with thermal storage [160]. The flexibility potentials of HVAC systems are optimized using the Genetic Algorithm in Portugal to reduce operation costs and interaction with the power grid [161]. Home refrigerators with flexible chamber temperatures are discussed as flexible resources in the HEMS [158]. The refrigerator's flexibility stems from the limited gap between the upper and lower thresholds of the chamber temperature. Wet appliances are a practical solution to shift energy consumption out of peak demand. In France, a practical study on 100 French households shows that wet appliances can provide peak shaving for the supply side at a net-zero cost [162]. It is worth mentioning that rooftop PV sites can increase the flexibility opportunities of residential buildings significantly. This way, household energy consumption can be supplied by solar power during high solar irradiation hours and be supplied partially by the stored energy in electrical batteries out of daylight hours [163].

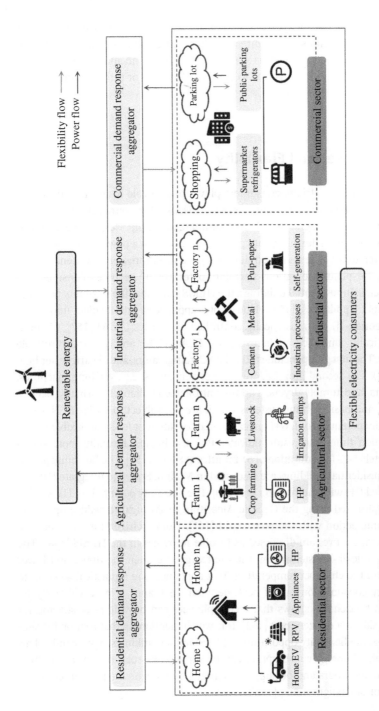

Figure 6.7 General flexibility potentials of four sectors of electricity consumers in power systems.

In the industrial sector, energy-intensive industrial processes provide great flexibility for power systems. In cement manufacturing plants, some processes, e.g., crushers and cement mills, are interruptible: therefore, they can be turned off/on when a power shortage/surplus occurs in the electricity market [164]. Metal smelting plants are comprised of different smelting pots. The operation processes of the smelting pots are normally uninterruptible, but the power consumption can be turned up/down by variable voltage controllers and tap changer transformers in response to the flexibility requirements of the supply side [165]. Oil refinery industries are consumers of power, heat, and steam supplied by power and gas grids. They have self-generation facilities, e.g., gas turbines and heat boilers, to supply domestic energy needs. The generation facilities can provide power and gas flexibility for the energy networks when an energy shortage happens in the power/gas grid [166].

In the agricultural sector, demand flexibility can be addressed for crop farming and livestock farms. In crop farming, water irrigation pumps show great flexibility potential. In Ref. [167], a robust optimization model is proposed to schedule the power consumption of water pumps out of peak hours. As a result, peak shaving and valley filling are provided for the power system. The variable frequency drive of water pumps can turn up/down the power consumption in response to excess/ deficit of renewable power in the electricity market [168]. Tower water storage is a practical solution to draw water from an underground well during low electricity price hours: then the crops can be irrigated by gravity or less power consumption during high electricity price hours [169]. In New Zealand, six large-scale dairy farms are studied to shift demand out of peak periods and balance grid demand. Based on the results, flattening demand reduces electricity costs for the milking process by 3.3% [170]. In Ref. [171], Poultry Demand-Side Management is suggested in Bangladesh to facilitate the use of solar energy in poultry farms and decrease the dependency on the power grid.

In the commercial sector, supermarket refrigerators exhibit great flexibility potential. In Denmark, a research study was conducted to improve the flexibility of supermarket refrigeration systems based on Danfoss A/S. The study concludes that the estimation of food temperature plays a key role in the improvement of DRPs [172]. The optimal DRPs of commercial buildings in a heavily urbanized area are studied [173]. To deploy the DRPs, different commercial buildings, including hotels, shopping malls, and offices, are discussed in Huangpu district, Shanghai. In Iran, a research study investigated the flexibility potential of commercial EV parking lots to supply the demand of shopping stores, office buildings, food courts, and entertainment complexes [153]. To extract the demand flexibility in the four sectors, expert knowledge is required. Therefore, based on Figure 6.7, the flexibility potential of the demand sectors is integrated into power systems by intermediary agents called Industrial Demand Response Aggregators (IDRA)

[164], Agricultural Demand Response Aggregators (ADRA) [167], Residential Demand Response Aggregators (RDRA) [165], and Commercial Demand Response Aggregators (CDRA) [153].

6.8 Conclusion

In this chapter, the flexibility potentials of eight different energy sectors of power-to-X facilities are discussed. The energy sectors include power-to-hydrogen, power-to-methane, power-to-diesel, power-to-methanol, power-to-formic acid, power-to-heat, power-to-transport, and electric power demands.

In the power-to-hydrogen plants, hydrogen storage, i.e., storage tanks and salt caverns, plays a key role in flexible operation. Although the share of hydrogen injection into the gas network is limited to 10%, the flexible interoperation of the gas network and hydrogen electrolyzers can counterbalance the renewable power fluctuations. In the power-to-methan plants, storage tanks, salt caverns, and saline aquifers play a pivotal role in the flexible operation. Also, there is no limitation to inject methane into gas grids: therefore, the gas networks increase flexibility opportunities. For regions without gas network access, energy-intensive liquefaction and compression processes are applied which decrease the energy content by up to 25%. Power-to-chemicals, including power-to-methanol and power-to-Formic Acid, decarbonize the mobility sector. Green methanol and formic acid can supply EVs through direct and indirect fuel cells. Also, methanol can be blended with fossil fuels to decrease GHG emissions. Therefore, the flexible operation of the power-to-chemical depends strongly on the flexible operation of the mobility sector. Regarding the power-to-heat, the thermal demands can be supplied by electricity, e.g., HP and electric boilers, or green hydrogen boilers. The main flexibility of heat demand stems from the flexible bound of buildings' air temperature and DHW consumption. In the power-to-mobility sector, EVs are supplied by direct electricity and hydrogen energy, i.e., hydrogen, formic acid, and methanol. This way, there are strong ties between traffic control, energy management of parking lots, and flexible operation of the power-to-chemical and power-to-transport sectors.

Power demands are addressed in four sectors, including residential, industrial, agricultural, and commercial sectors. In the residential sector, the home energy management system unlocks the flexibility of household appliances, e.g., HP, refrigerators, and wet appliances. In the industrial sector, the power consumption of energy-intensive industrial processes, e.g., cement crushers and metal smelting pots, can be turned off/on or up/down in response to the flexibility requirements of the power systems. In the agricultural sector, the water irrigation pumps, water tank towers, and on-farm solar sites provide great flexibility potential for the power system. In the commercial sector, supermarket refrigerators and public parking lots are addressed as great sources of flexibility potential. To sum up, the key points of the chapter can be summarized as follows:

1) The flexibility potentials of different energy sectors in the P2X structure enable the integration of renewable energies into energy systems in the forms of green hydrogen, chemicals, syngas, heat, and power.
2) The storage capability of various P2X sectors makes it possible to store renewable power in different forms of energy commodities.
3) The flexibility potentials and storage capabilities of the P2X energy sectors mitigate the volatility and intermittency of renewable energies.
4) The flexibility potentials of the transportation sector, particularly PEVs and FCEVs, play a pivotal role in decarbonizing energy systems.

References

1 Puertas, R. and Marti, L. (2021). International ranking of climate change action: an analysis using the indicators from the Climate Change Performance Index. *Renewable and Sustainable Energy Reviews* 148: 111316. https://doi.org/10.1016/j.rser.2021.111316.

2 Maeder, M., Weiss, O., and Boulouchos, K. (2021). Assessing the need for flexibility technologies in decarbonized power systems: a new model applied to Central Europe. *Applied Energy* 282: 116050. https://doi.org/10.1016/j.apenergy.2020.116050.

3 Rüdisüli, M., Romano, E., Eggimann, S., and Patel, M.K. (2022). Decarbonization strategies for Switzerland considering embedded greenhouse gas emissions in electricity imports. *Energy Policy* 162: 112794. https://doi.org/10.1016/j.enpol.2022.112794.

4 Burandt, T., Xiong, B., Löffler, K., and Oei, P.-Y. (2019). Decarbonizing China's energy system – modeling the transformation of the electricity, transportation, heat, and industrial sectors. *Applied Energy* 255: 113820. https://doi.org/10.1016/j.apenergy.2019.113820.

5 Edtmayer, H., Nageler, P., Heimrath, R. et al. (2021). Investigation on sector coupling potentials of a 5th generation district heating and cooling network. *Energy* 230: 120836. https://doi.org/10.1016/j.energy.2021.120836.

6 Buffo, G., Marocco, P., Ferrero, D. et al. (2019). Power-to-X and power-to-power routes. In: *Solar Hydrogen Production* (ed. F. Calise, M.D. D'Accadia, M. Santarelli, et al.), 529–557. Academic Press https://doi.org/10.1016/B978-0-12-814853-2.00015-1.

7 Ju, H., Badwal, S., and Giddey, S. (2018). A comprehensive review of carbon and hydrocarbon assisted water electrolysis for hydrogen production. *Applied Energy* 231: 502–533. https://doi.org/10.1016/j.apenergy.2018.09.125.

8 El-Emam, R.S. and Özcan, H. (2019). Comprehensive review on the techno-economics of sustainable large-scale clean hydrogen production. *Journal of Cleaner Production* 220: 593–609. https://doi.org/10.1016/j.jclepro.2019.01.309.

9 Chatterjee, P., Ambati, M.S.K., Chakraborty, A.K. et al. (2022). Photovoltaic/photo-electrocatalysis integration for green hydrogen: a review. *Energy Conversion and Management* 261: 115648. https://doi.org/10.1016/j.enconman.2022.115648.

10 Maestre, V.M., Ortiz, A., and Ortiz, I. (2021). Challenges and prospects of renewable hydrogen-based strategies for full decarbonization of stationary power applications. *Renewable and Sustainable Energy Reviews* 152: 111628. https://doi.org/10.1016/j.rser.2021.111628.

11 Klatzer, T., Bachhiesl, U., and Wogrin, S. (2022). State-of-the-art expansion planning of integrated power, natural gas, and hydrogen systems. *International Journal of Hydrogen Energy* 47 (47): 20585–20603. https://doi.org/10.1016/j.ijhydene.2022.04.293.

12 Agrafiotis, C., von Storch, H., Roeb, M., and Sattler, C. (2014). Solar thermal reforming of methane feedstocks for hydrogen and syngas production – a review. *Renewable and Sustainable Energy Reviews* 29: 656–682. https://doi.org/10.1016/j.rser.2013.08.050.

13 Choe, C., Cheon, S., Gu, J., and Lim, H. (2022). Critical aspect of renewable syngas production for power-to-fuel via solid oxide electrolysis: integrative assessment for potential renewable energy source. *Renewable and Sustainable Energy Reviews* 161: 112398. https://doi.org/10.1016/j.rser.2022.112398.

14 Kanwal, S., Mehran, M.T., Hassan, M. et al. (2022). An integrated future approach for the energy security of Pakistan: replacement of fossil fuels with syngas for better environment and socio-economic development. *Renewable and Sustainable Energy Reviews* 156: 111978. https://doi.org/10.1016/j.rser.2021.111978.

15 Mohanty, U.S., Ali, M., Azhar, M.R. et al. (2021). Current advances in syngas ($CO + H_2$) production through bi-reforming of methane using various catalysts: a review. *International Journal of Hydrogen Energy* 46 (65): 32809–32845. https://doi.org/10.1016/j.ijhydene.2021.07.097.

16 Ayodele, B., Mustapa, S.I., Tuan Abdullah, T.A.R., and Salleh, S.F. (2019). A mini-review on hydrogen-rich syngas production by thermo-catalytic and bioconversion of biomass and its environmental implications. *Frontiers in Energy Research* 7: https://doi.org/10.3389/fenrg.2019.00118.

17 de Vasconcelos, B. and Lavoie, J.-M. (2019). Recent advances in power-to-X technology for the production of fuels and chemicals. *Frontiers in Chemistry* 7: https://doi.org/10.3389/fchem.2019.00392.

18 Andika, R., Nandiyanto, A.B.D., Putra, Z.A. et al. (2018). Co-electrolysis for power-to-methanol applications. *Renewable and Sustainable Energy Reviews* 95: 227–241. https://doi.org/10.1016/j.rser.2018.07.030.

19 Bloess, A., Schill, W.-P., and Zerrahn, A. (2018). Power-to-heat for renewable energy integration: a review of technologies, modeling approaches, and flexibility potentials. *Applied Energy* 212: 1611–1626. https://doi.org/10.1016/j. apenergy.2017.12.073.

20 Golmohamadi, H., Larsen, K.G., Jensen, P.G., and Hasrat, I.R. (2022). Integration of flexibility potentials of district heating systems into electricity markets: a review. *Renewable and Sustainable Energy Reviews* 159: 112200. https://doi.org/ 10.1016/j.rser.2022.112200.

21 Wang, Y., Wang, J., and He, W. (2022). Development of efficient, flexible and affordable heat pumps for supporting heat and power decarbonisation in the UK and beyond: review and perspectives. *Renewable and Sustainable Energy Reviews* 154: 111747. https://doi.org/10.1016/j.rser.2021.111747.

22 Gjorgievski, V.Z., Markovska, N., Abazi, A., and Duić, N. (2021). The potential of power-to-heat demand response to improve the flexibility of the energy system: an empirical review. *Renewable and Sustainable Energy Reviews* 138: 110489. https:// doi.org/10.1016/j.rser.2020.110489.

23 Luc, K.M., Heller, A., and Rode, C. (2019). Energy demand flexibility in buildings and district heating systems – a literature review. *Advances in Building Energy Research* 13 (2): 241–263. https://doi.org/10.1080/17512549.2018.1488615.

24 Jangdoost, A., Keypour, R., and Golmohamadi, H. (2020). Optimization of distribution network reconfiguration by a novel RCA integrated with genetic algorithm. *Energy Systems* https://doi.org/10.1007/s12667-020-00398-5.

25 Alinejad, M., Rezaei, O., Kazemi, A., and Bagheri, S. (2021). An optimal management for charging and discharging of electric vehicles in an intelligent parking lot considering vehicle owner's random behaviors. *Journal of Energy Storage* 35: 102245. https://doi.org/10.1016/j.est.2021.102245.

26 Gonzalez Venegas, F., Petit, M., and Perez, Y. (2021). Active integration of electric vehicles into distribution grids: barriers and frameworks for flexibility services. *Renewable and Sustainable Energy Reviews* 145: 111060. https://doi.org/10.1016/j. rser.2021.111060.

27 Nasir, M., Rezaee Jordehi, A., Matin, S.A.A. et al. (2022). Optimal operation of energy hubs including parking lots for hydrogen vehicles and responsive demands. *Journal of Energy Storage* 50: 104630. https://doi.org/10.1016/j. est.2022.104630.

28 Razipour, R., Moghaddas-Tafreshi, S.-M., and Farhadi, P. (2019). Optimal management of electric vehicles in an intelligent parking lot in the presence of hydrogen storage system. *Journal of Energy Storage* 22: 144–152. https://doi.org/ 10.1016/j.est.2019.02.001.

29 Apostolou, D. and Enevoldsen, P. (2019). The past, present and potential of hydrogen as a multifunctional storage application for wind power. *Renewable and Sustainable Energy Reviews* 112: 917–929. https://doi.org/10.1016/j. rser.2019.06.049.

30 Eriksson, E.L.V. and MacAgray, E. (2017). Optimization and integration of hybrid renewable energy hydrogen fuel cell energy systems – a critical review. *Applied Energy* 202: 348–364. https://doi.org/10.1016/j.apenergy.2017.03.132.

31 Capurso, T., Stefanizzi, M., Torresi, M., and Camporeale, S.M. (2022). Perspective of the role of hydrogen in the 21st century energy transition. *Energy Conversion and Management* 251: 114898. https://doi.org/10.1016/j. enconman.2021.114898.

32 Foit, S.R., Vinke, I.C., de Haart, L.G.J., and Eichel, R.-A. (2017). Power-to-syngas: an enabling technology for the transition of the energy system? *Angewandte Chemie International Edition* 56 (20): 5402–5411. https://doi.org/10.1002/ anie.201607552.

33 Götz, M., Lefebvre, J., Mörs, F. et al. (2016). Renewable power-to-gas: a technological and economic review. *Renewable Energy* 85: 1371–1390. https://doi. org/10.1016/j.renene.2015.07.066.

34 Ozturk, M. and Dincer, I. (2021). A comprehensive review on power-to-gas with hydrogen options for cleaner applications. *International Journal of Hydrogen Energy* 46 (62): 31511–31522. https://doi.org/10.1016/j. ijhydene.2021.07.066.

35 Schaidle, J.A., Grim, R.G., Tao, L. et al. (2022). Power conversion technologies: the advent of power-to-gas, power-to-liquid, and power-to-heat. In: *Technologies for Integrated Energy Systems and Networks*, 41–70. Wiley https://doi.org/10.1002/ 9783527833634.ch3.

36 Yang, Z., Zhang, J., Kintner-Meyer, M.C.W. et al. (2011). Electrochemical energy storage for green grid. *Chemical Reviews* 111 (5): 3577–3613. https://doi.org/ 10.1021/cr100290v.

37 Wu, B., Lin, R., O'Shea, R. et al. (2021). Production of advanced fuels through integration of biological, thermo-chemical and power to gas technologies in a circular cascading bio-based system. *Renewable and Sustainable Energy Reviews* 135: 110371. https://doi.org/10.1016/j.rser.2020.110371.

38 Péan, T.Q., Salom, J., and Costa-Castelló, R. (2019). Review of control strategies for improving the energy flexibility provided by heat pump systems in buildings. *Journal of Process Control* 74: 35–49. https://doi.org/10.1016/j. jprocont.2018.03.006.

39 Li, H., Wang, Z., Hong, T., and Piette, M.A. (2021). Energy flexibility of residential buildings: a systematic review of characterization and quantification methods and applications. *Advances in Applied Energy* 3: 100054. https://doi.org/10.1016/j. adapen.2021.100054.

40 Tang, H., Wang, S., and Li, H. (2021). Flexibility categorization, sources, capabilities and technologies for energy-flexible and grid-responsive buildings: state-of-the-art and future perspective. *Energy* 219: 119598. https://doi.org/10.1016/j.energy.2020.119598.

41 Capuder, T., Miloš Sprčić, D., Zoričić, D., and Pandžić, H. (2020). Review of challenges and assessment of electric vehicles integration policy goals: integrated risk analysis approach. *International Journal of Electrical Power & Energy Systems* 119: 105894. https://doi.org/10.1016/j.ijepes.2020.105894.

42 Lee, J.W., Haram, M.H.S.M., Ramasamy, G. et al. (2021). Technical feasibility and economics of repurposed electric vehicles batteries for power peak shaving. *Journal of Energy Storage* 40: 102752. https://doi.org/10.1016/j.est.2021.102752.

43 Manzolli, J.A., Trovão, J.P., and Antunes, C.H. (2022). A review of electric bus vehicles research topics – methods and trends. *Renewable and Sustainable Energy Reviews* 159: 112211. https://doi.org/10.1016/j.rser.2022.112211.

44 Kuckshinrichs, W., Ketelaer, T., and Koj, J.C. (2017). Economic analysis of improved alkaline water electrolysis. *Frontiers in Energy Research* 5: https://doi.org/10.3389/fenrg.2017.00001.

45 Gouda, M.H., Elnouby, M., Aziz, A.N. et al. (2020). Green and low-cost membrane electrode assembly for proton exchange membrane fuel cells: effect of double-layer electrodes and gas diffusion layer. *Frontiers in Materials* 6: https://doi.org/10.3389/fmats.2019.00337.

46 Nechache, A. and Hody, S. (2021). Alternative and innovative solid oxide electrolysis cell materials: a short review. *Renewable and Sustainable Energy Reviews* 149: 111322. https://doi.org/10.1016/j.rser.2021.111322.

47 Donadei, S. and Schneider, G.-S. (2016). Compressed air energy storage in underground formations. In: *Storing Energy* (ed. T.M.B.T.-S.E. Letcher), 113–133. Oxford: Elsevier https://doi.org/10.1016/B978-0-12-803440-8.00006-3.

48 Zhou, D., Yan, S., Huang, D. et al. (2022). Modeling and simulation of the hydrogen blended gas-electricity integrated energy system and influence analysis of hydrogen blending modes. *Energy* 239: 121629. https://doi.org/10.1016/j.energy.2021.121629.

49 Li, X. and Mulder, M. (2021). Value of power-to-gas as a flexibility option in integrated electricity and hydrogen markets. *Applied Energy* 304: 117863. https://doi.org/10.1016/j.apenergy.2021.117863.

50 Ruhnau, O. (2022). How flexible electricity demand stabilizes wind and solar market values: the case of hydrogen electrolyzers. *Applied Energy* 307: 118194. https://doi.org/10.1016/j.apenergy.2021.118194.

51 Zhang, C., Greenblatt, J.B., Wei, M. et al. (2020). Flexible grid-based electrolysis hydrogen production for fuel cell vehicles reduces costs and greenhouse gas emissions. *Applied Energy* 278: 115651. https://doi.org/10.1016/j.apenergy.2020.115651.

52 Cloete, S. and Hirth, L. (2020). Flexible power and hydrogen production: finding synergy between CCS and variable renewables. *Energy* 192: 116671. https://doi. org/10.1016/j.energy.2019.116671.

53 Caumon, P., Lopez-Botet Zulueta, M., Louyrette, J. et al. (2015). Flexible hydrogen production implementation in the French power system: expected impacts at the French and European levels. *Energy* 81: 556–562. https://doi.org/10.1016/j. energy.2014.12.073.

54 Kafaei, M., Sedighizadeh, D., Sedighizadeh, M., and Fini, A.S. (2022). An IGDT/ Scenario based stochastic model for an energy hub considering hydrogen energy and electric vehicles: a case study of Qeshm Island, Iran. *International Journal of Electrical Power & Energy Systems* 135: 107477. https://doi.org/10.1016/j. ijepes.2021.107477.

55 Daraei, M., Campana, P.-E., Avelin, A. et al. (2021). Impacts of integrating pyrolysis with existing CHP plants and onsite renewable-based hydrogen supply on the system flexibility. *Energy Conversion and Management* 243: 114407. https:// doi.org/10.1016/j.enconman.2021.114407.

56 Wen, D. and Aziz, M. (2022). Flexible operation strategy of an integrated renewable multi-generation system for electricity, hydrogen, ammonia, and heating. *Energy Conversion and Management* 253: 115166. https://doi.org/ 10.1016/j.enconman.2021.115166.

57 Li, N., Zhao, X., Shi, X. et al. (2021). Integrated energy systems with CCHP and hydrogen supply: a new outlet for curtailed wind power. *Applied Energy* 303: 117619. https://doi.org/10.1016/j.apenergy.2021.117619.

58 Fan, J.-L., Yu, P., Li, K. et al. (2022). A levelized cost of hydrogen (LCOH) comparison of coal-to-hydrogen with CCS and water electrolysis powered by renewable energy in China. *Energy* 242: 123003. https://doi.org/10.1016/j. energy.2021.123003.

59 Tang, O., Rehme, J., Cerin, P., and Huisingh, D. (2021). Hydrogen production in the Swedish power sector: considering operational volatilities and long-term uncertainties. *Energy Policy* 148: 111990. https://doi.org/10.1016/j. enpol.2020.111990.

60 Wei, F., Sui, Q., Li, X. et al. (2021). Optimal dispatching of power grid integrating wind-hydrogen systems. *International Journal of Electrical Power & Energy Systems* 125: 106489. https://doi.org/10.1016/j.ijepes.2020.106489.

61 Chen, H., Song, J., and Zhao, J. (2021). Synergies between power and hydrogen carriers using fuel-cell hybrid electrical vehicle and power-to-gas storage as new coupling points. *Energy Conversion and Management* 246: 114670. https://doi.org/ 10.1016/j.enconman.2021.114670.

62 Röben, F.T.C., Schöne, N., Bau, U. et al. (2021). Decarbonizing copper production by power-to-hydrogen: a techno-economic analysis. *Journal of Cleaner Production* 306: 127191. https://doi.org/10.1016/j.jclepro.2021.127191.

63 Park, C., Jung, Y., Lim, K. et al. (2021). Analysis of a phosphoric acid fuel cell-based multi-energy hub system for heat, power, and hydrogen generation. *Applied Thermal Engineering* 189: 116715. https://doi.org/10.1016/j. applthermaleng.2021.116715.

64 Agabalaye-Rahvar, M., Mansour-Saatloo, A., Mirzaei, M.A. et al. (2021). Economic-environmental stochastic scheduling for hydrogen storage-based smart energy hub coordinated with integrated demand response program. *International Journal of Energy Research* 45 (14): 20232–20257. https://doi.org/10.1002/er.7108.

65 Mansour-Saatloo, A., Mirzaei, M.A., Mohammadi-Ivatloo, B., and Zare, K. (2020). A risk-averse hybrid approach for optimal participation of power-to-hydrogen technology-based multi-energy microgrid in multi-energy markets. *Sustainable Cities and Society* 63: 102421. https://doi.org/10.1016/j.scs.2020.102421.

66 Zhang, Y., Wang, L., Wang, N. et al. (2019). Balancing wind-power fluctuation via onsite storage under uncertainty: power-to-hydrogen-to-power versus lithium battery. *Renewable and Sustainable Energy Reviews* 116: 109465. https://doi.org/10.1016/j.rser.2019.109465.

67 Zhou, L., Zhang, F., Wang, L., and Zhang, Q. (2022). Flexible hydrogen production source for fuel cell vehicle to reduce emission pollution and costs under the multi-objective optimization framework. *Journal of Cleaner Production* 337: 130284. https://doi.org/10.1016/j.jclepro.2021.130284.

68 Mansour-Saatloo, A., Ebadi, R., Mirzaei, M. et al. (2021). Multi-objective IGDT-based scheduling of low-carbon multi-energy microgrids integrated with hydrogen refueling stations and electric vehicle parking lots. *Sustainable Cities and Society* 74: 103197. https://doi.org/10.1016/j.scs.2021.103197.

69 Daraei, M., Campana, P.E., and Thorin, E. (2020). Power-to-hydrogen storage integrated with rooftop photovoltaic systems and combined heat and power plants. *Applied Energy* 276: 115499. https://doi.org/10.1016/j. apenergy.2020.115499.

70 Gabrielli, P., Poluzzi, A., Kramer, G.J. et al. (2020). Seasonal energy storage for zero-emissions multi-energy systems via underground hydrogen storage. *Renewable and Sustainable Energy Reviews* 121: 109629. https://doi.org/10.1016/j. rser.2019.109629.

71 Fasihi, M. and Breyer, C. (2020). Baseload electricity and hydrogen supply based on hybrid PV-wind power plants. *Journal of Cleaner Production* 243: 118466. https://doi.org/10.1016/j.jclepro.2019.118466.

72 Farahani, S.S., Bleeker, C., van Wijk, A., and Lukszo, Z. (2020). Hydrogen-based integrated energy and mobility system for a real-life office environment. *Applied Energy* 264: 114695. https://doi.org/10.1016/j.apenergy.2020.114695.

73 Kuczyński, S., Łaciak, M., Olijnyk, A. et al. (2019). Thermodynamic and technical issues of hydrogen and methane-hydrogen mixtures pipeline transmission. *Energies* 12 (3): https://doi.org/10.3390/en12030569.

74 Cavana, M., Mazza, A., Chicco, G., and Leone, P. (2021). Electrical and gas networks coupling through hydrogen blending under increasing distributed photovoltaic generation. *Applied Energy* 290: 116764. https://doi.org/10.1016/j.apenergy.2021.116764.

75 Romeo, L.M., Cavana, M., Bailera, M. et al. (2022). Non-stoichiometric methanation as strategy to overcome the limitations of green hydrogen injection into the natural gas grid. *Applied Energy* 309: 118462. https://doi.org/10.1016/j.apenergy.2021.118462.

76 Perčić, M., Vladimir, N., Jovanović, I., and Koričan, M. (2022). Application of fuel cells with zero-carbon fuels in short-sea shipping. *Applied Energy* 309: 118463. https://doi.org/10.1016/j.apenergy.2021.118463.

77 Buffo, G., Ferrero, D., Santarelli, M., and Lanzini, A. (2020). Energy and environmental analysis of a flexible Power-to-X plant based on Reversible Solid Oxide Cells (rSOCs) for an urban district. *Journal of Energy Storage* 29: 101314. https://doi.org/10.1016/j.est.2020.101314.

78 Zakaria, Z., Kamarudin, S.K., Abd Wahid, K.A., and Abu Hassan, S.H. (2021). The progress of fuel cell for Malaysian residential consumption: energy status and prospects to introduction as a renewable power generation system. *Renewable and Sustainable Energy Reviews* 144: 110984. https://doi.org/10.1016/j.rser.2021.110984.

79 Emilio, P. and de Miranda, V. (ed.) (2019). Chapter 5.3.3 - Application of hydrogen by use of chemical reactions of hydrogen and carbon dioxide. In: *Science and Engineering of Hydrogen-Based*, 279–289. Academic Press https://doi.org/10.1016/B978-0-12-814251-6.00013-7.

80 Calbry-Muzyka, A.S. and Schildhauer, T.J. (2020). Direct methanation of biogas – technical challenges and recent progress. *Frontiers in Energy Research* 8: https://doi.org/10.3389/fenrg.2020.570887.

81 Rusmanis, D., O'Shea, R., Wall, D.M., and Murphy, J.D. (2019). Biological hydrogen methanation systems – an overview of design and efficiency. *Bioengineered* 10 (1): 604–634. https://doi.org/10.1080/21655979.2019.1684607.

82 Koj, J.C., Wulf, C., and Zapp, P. (2019). Environmental impacts of power-to-X systems – a review of technological and methodological choices in Life Cycle Assessments. *Renewable and Sustainable Energy Reviews* 112: 865–879. https://doi.org/10.1016/j.rser.2019.06.029.

83 Vo, T.T.Q., Xia, A., Rogan, F. et al. (2017). Sustainability assessment of large-scale storage technologies for surplus electricity using group multi-criteria decision analysis. *Clean Technologies and Environmental Policy* 19 (3): 689–703. https://doi.org/10.1007/s10098-016-1250-8.

84 Simon, A.L. (ed.) (1975). Chapter 6: natural gas. In: *Energy Resources*, 67–73. Pergamon https://doi.org/10.1016/B978-0-08-018750-1.50011-6.

85 Hermesmann, M., Grübel, K., Scherotzki, L., and Müller, T.E. (2021). Promising pathways: the geographic and energetic potential of power-to-x technologies based on regeneratively obtained hydrogen. *Renewable and Sustainable Energy Reviews* 138: 110644. https://doi.org/10.1016/j.rser.2020.110644.

86 Bossel, U. and Eliasson, B. Energy and the hydrogen economy. https://afdc.energy.gov/files/pdfs/hyd_economy_bossel_eliasson.pdf

87 Park, J., You, F., Cho, H. et al. (2020). Novel massive thermal energy storage system for liquefied natural gas cold energy recovery. *Energy* 195: 117022. https://doi.org/10.1016/j.energy.2020.117022.

88 Zoss, T., Dace, E., and Blumberga, D. (2016). Modeling a power-to-renewable methane system for an assessment of power grid balancing options in the Baltic States' region. *Applied Energy* 170: 278–285. https://doi.org/10.1016/j.apenergy.2016.02.137.

89 Uchman, W., Skorek-Osikowska, A., Jurczyk, M., and Węcel, D. (2020). The analysis of dynamic operation of power-to-SNG system with hydrogen generator powered with renewable energy, hydrogen storage and methanation unit. *Energy* 213: 118802. https://doi.org/10.1016/j.energy.2020.118802.

90 Gholizadeh, N., Vahid-Pakdel, M.J., and Mohammadi-ivatloo, B. (2019). Enhancement of demand supply's security using power to gas technology in networked energy hubs. *International Journal of Electrical Power & Energy Systems* 109: 83–94. https://doi.org/10.1016/j.ijepes.2019.01.047.

91 Morgenthaler, S., Ball, C., Koj, J.C. et al. (2020). Site-dependent levelized cost assessment for fully renewable Power-to-Methane systems. *Energy Conversion and Management* 223: 113150. https://doi.org/10.1016/j.enconman.2020.113150.

92 Straka, P. (2021). A comprehensive study of Power-to-Gas technology: technical implementations overview, economic assessments, methanation plant as auxiliary operation of lignite-fired power station. *Journal of Cleaner Production* 311: 127642. https://doi.org/10.1016/j.jclepro.2021.127642.

93 Fan, J. and Zhu, L. (2015). Performance analysis of a feasible technology for power and high-purity hydrogen production driven by methane fuel. *Applied Thermal Engineering* 75: 103–114. https://doi.org/10.1016/j.applthermaleng.2014.10.013.

94 Bekkering, J., Zwart, K., Martinus, G. et al. (2020). "Farm-scale bio-power-to-methane: comparative analyses of economic and environmental feasibility," *International Journal of Energy Research*, 44 (3): 2264–2277. https://doi.org/10.1002/er.5093.

95 Qi, M., Park, J., Landon, R.S. et al. (2022). Continuous and flexible Renewable-Power-to-Methane via liquid CO2 energy storage: revisiting the techno-economic potential. *Renewable and Sustainable Energy Reviews* 153: 111732. https://doi.org/10.1016/j.rser.2021.111732.

96 Pan, G., Gu, W., Lu, Y. et al. (2021). Accurate modeling of a profit-driven power to hydrogen and methane plant toward strategic bidding within multi-type markets. *IEEE Transactions on Smart Grid* 12 (1): 338–349. https://doi.org/10.1109/TSG.2020.3019043.

97 Budny, C., Madlener, R., and Hilgers, C. (2015). Economic feasibility of pipe storage and underground reservoir storage options for power-to-gas load balancing. *Energy Conversion and Management* 102: 258–266. https://doi.org/10.1016/j.enconman.2015.04.070.

98 Zhang, X., Liu, W., Jiang, D. et al. (2021). Investigation on the influences of interlayer contents on stability and usability of energy storage caverns in bedded rock salt. *Energy* 231: 120968. https://doi.org/10.1016/j.energy.2021.120968.

99 Kloess, M. and Zach, K. (2014). Bulk electricity storage technologies for load-leveling operation – an economic assessment for the Austrian and German power market. *International Journal of Electrical Power & Energy Systems* 59: 111–122. https://doi.org/10.1016/j.ijepes.2014.02.002.

100 Matos, C.R., Carneiro, J.F., and Silva, P.P. (2019). Overview of large-scale underground energy storage technologies for integration of renewable energies and criteria for reservoir identification. *Journal of Energy Storage* 21: 241–258. https://doi.org/10.1016/j.est.2018.11.023.

101 Bassano, C., Deiana, P., Vilardi, G., and Verdone, N. (2020). Modeling and economic evaluation of carbon capture and storage technologies integrated into synthetic natural gas and power-to-gas plants. *Applied Energy* 263: 114590. https://doi.org/10.1016/j.apenergy.2020.114590.

102 Viebahn, P., Daniel, V., and Samuel, H. (2012). Integrated assessment of carbon capture and storage (CCS) in the German power sector and comparison with the deployment of renewable energies. *Applied Energy* 97: 238–248. https://doi.org/10.1016/j.apenergy.2011.12.053.

103 Quarton, C.J. and Samsatli, S. (2018). Power-to-gas for injection into the gas grid: what can we learn from real-life projects, economic assessments and systems modelling? *Renewable and Sustainable Energy Reviews* 98: 302–316. https://doi.org/10.1016/j.rser.2018.09.007.

104 Herz, G., Rix, C., Jacobasch, E. et al. (2021). Economic assessment of Power-to-Liquid processes – influence of electrolysis technology and operating conditions. *Applied Energy* 292: 116655. https://doi.org/10.1016/j.apenergy.2021.116655.

105 Candelaresi, D. and Spazzafumo, G. (2021). 10 - Power-to-fuel potential market. In: *Power to Fuel* (ed. G. Spazzafumo), 239–265. Academic Press https://doi.org/10.1016/B978-0-12-822813-5.00009-6.

106 Chakraborty, J.P., Singh, S., and Maity, S.K. (2022). Advances in the conversion of methanol to gasoline. In: *Hydrocarbon Biorefinery* (ed. S.K. Maity, K. Gayen, and T.K.B.T.-H.B. Bhowmick), 177–200. Elsevier https://doi.org/10.1016/B978-0-12-823306-1.00008-X.

107 van Vliet, O.P.R., Faaij, A.P.C., and Turkenburg, W.C. (2009). Fischer–Tropsch diesel production in a well-to-wheel perspective: a carbon, energy flow and cost analysis. *Energy Conversion and Management* 50 (4): 855–876. https://doi.org/10.1016/j.enconman.2009.01.008.

108 Müller, S., Groß, P., Rauch, R. et al. (2018). Production of diesel from biomass and wind power – energy storage by the use of the Fischer-Tropsch process. *Biomass Conversion and Biorefinery* 8 (2): 275–282. https://doi.org/10.1007/s13399-017-0287-1.

109 Timmerberg, S. and Kaltschmitt, M. (2019). Hydrogen from renewables: supply from North Africa to Central Europe as blend in existing pipelines – potentials and costs. *Applied Energy* 237: 795–809. https://doi.org/10.1016/j.apenergy.2019.01.030.

110 Samavati, M., Martin, A., Santarelli, M., and Nemanova, V. (2018). Synthetic diesel production as a form of renewable energy storage. *Energies* 11 (5): https://doi.org/10.3390/en11051223.

111 Styring, P., Dowson, G.R.M., and Tozer, I.O. (2021). Synthetic fuels based on dimethyl ether as a future non-fossil fuel for road transport from sustainable feedstocks. *Frontiers in Energy Research* 9: https://doi.org/10.3389/fenrg.2021.663331.

112 Fuchs, C., Meyer, D., and Poehls, A. (2022). Production and economic assessment of synthetic fuels in agriculture – a case study from northern Germany. *Energies* 15 (3): https://doi.org/10.3390/en15031156.

113 Larsson, M., Grönkvist, S., and Alvfors, P. (2015). Synthetic fuels from electricity for the swedish transport sector: comparison of well to wheel energy efficiencies and costs. *Energy Procedia* 75: 1875–1880. https://doi.org/10.1016/j.egypro.2015.07.169.

114 Fasihi, M., Bogdanov, D., and Breyer, C. (2016). Techno-economic assessment of power-to-liquids (PtL) fuels production and global trading based on hybrid PV-wind power plants. *Energy Procedia* 99: 243–268. https://doi.org/10.1016/j.egypro.2016.10.115.

115 Keiner, D., Salcedo-Puerto, O., Immonen, E. et al. (2022). Powering an island energy system by offshore floating technologies towards 100% renewables: a case for the Maldives. *Applied Energy* 308: 118360. https://doi.org/10.1016/j.apenergy.2021.118360.

116 Navlani-García, M., Mori, K., Salinas-Torres, D. et al. (2019). New approaches toward the hydrogen production from formic acid dehydrogenation over Pd-based heterogeneous catalysts. *Frontiers in Materials* 6: https://doi.org/10.3389/fmats.2019.00044.

117 Eppinger, J. and Huang, K.-W. (2017). Formic acid as a hydrogen energy carrier. *ACS Energy Letters* 2 (1): 188–195. https://doi.org/10.1021/acsenergylett.6b00574.

118 Aslam, N.M., Masdar, M.S., Kamarudin, S.K., and Daud, W.R.W. (2012). Overview on direct formic acid fuel cells (DFAFCs) as an energy sources. *APCBEE Procedia* 3: 33–39. https://doi.org/10.1016/j.apcbee.2012.06.042.

119 Chatterjee, S., Dutta, I., Lum, Y. et al. (2021). Enabling storage and utilization of low-carbon electricity: power to formic acid. *Energy & Environmental Science* 14 (3): 1194–1246. https://doi.org/10.1039/D0EE03011B.

120 van Putten, R., Wissink, T., Swinkels, T., and Pidko, E.A. (2019). Fuelling the hydrogen economy: scale-up of an integrated formic acid-to-power system. *International Journal of Hydrogen Energy* 44 (53): 28533–28541. https://doi.org/10.1016/j.ijhydene.2019.01.153.

121 Onishi, N., Iguchi, M., Yang, X. et al. (2019). Development of effective catalysts for hydrogen storage technology using formic acid. *Advanced Energy Materials* 9 (23): 1801275. https://doi.org/10.1002/aenm.201801275.

122 Dutta, I., Chatterjee, S., Cheng, H. et al. (2022). Formic acid to power towards low-carbon economy. *Advanced Energy Materials* 12 (15): 2103799. https://doi.org/10.1002/aenm.202103799.

123 Perez Fortes, M. and Tzimas, E. (2016). *Techno-economic and environmental evaluation of CO₂ utilisation for fuel production.* Synthesis of methanol and formic acid. Publications Office of the European Union, Luxembourg. https://doi.org/10.2790/89238 (print),10.2790/981669 (online).

124 Borisut, P. and Nuchitprasittichai, A. (2019). Methanol production via CO2 hydrogenation: sensitivity analysis and simulation – based optimization. *Frontiers in Energy Research* 7: https://doi.org/10.3389/fenrg.2019.00081.

125 Iliev, S. (2021). A comparison of ethanol, methanol, and butanol blending with gasoline and its effect on engine performance and emissions using engine simulation. *Processes* 9 (8): https://doi.org/10.3390/pr9081322.

126 Park, S.-J. and Seo, M.-K. (2011). Solid-liquid interface. In: *Interface Science and Composites* (ed. S.-J. Park and T. Seo), 147–252. Elsevier https://doi.org/10.1016/B978-0-12-375049-5.00003-7.

127 Chen, C. and Yang, A. (2021). Power-to-methanol: the role of process flexibility in the integration of variable renewable energy into chemical production. *Energy Conversion and Management* 228: 113673. https://doi.org/10.1016/j.enconman.2020.113673.

128 Cui, X., Kær, S.K., and Nielsen, M.P. (2022). Energy analysis and surrogate modeling for the green methanol production under dynamic operating conditions. *Fuel* 307: 121924. https://doi.org/10.1016/j.fuel.2021.121924.

129 Schorn, F., Breuer, J.L., Samsun, R.C. et al. (2021). Methanol as a renewable energy carrier: an assessment of production and transportation costs for selected global locations. *Advances in Applied Energy* 3: 100050. https://doi.org/10.1016/j.adapen.2021.100050.

130 Al-Qahtani, A., González-Garay, A., Bernard, A. et al. (2020). Electricity grid decarbonisation or green methanol fuel? A life-cycle modelling and analysis of today's transportation-power nexus. *Applied Energy* 265: 114718. https://doi.org/10.1016/j.apenergy.2020.114718.

131 Tola, V. and Lonis, F. (2021). Low CO2 emissions chemically recuperated gas turbines fed by renewable methanol. *Applied Energy* 298: 117146. https://doi.org/10.1016/j.apenergy.2021.117146.

132 Liu, X., Hong, H., Zhang, H. et al. (2020). Solar methanol by hybridizing natural gas chemical looping reforming with solar heat. *Applied Energy* 277: 115521. https://doi.org/10.1016/j.apenergy.2020.115521.

133 Bos, M.J., Kersten, S.R.A., and Brilman, D.W.F. (2020). Wind power to methanol: renewable methanol production using electricity, electrolysis of water and CO_2 air capture. *Applied Energy* 264: 114672. https://doi.org/10.1016/j.apenergy.2020.114672.

134 Ravikumar, D., Keoleian, G., and Miller, S. (2020). The environmental opportunity cost of using renewable energy for carbon capture and utilization for methanol production. *Applied Energy* 279: 115770. https://doi.org/10.1016/j.apenergy.2020.115770.

135 Klyapovskiy, S., Zheng, Y., You, S., and Bindner, H.W. (2021). Optimal operation of the hydrogen-based energy management system with P2X demand response and ammonia plant. *Applied Energy* 304: 117559. https://doi.org/10.1016/j.apenergy.2021.117559.

136 von Storch, H., Roeb, M., Stadler, H. et al. (2016). On the assessment of renewable industrial processes: case study for solar co-production of methanol and power. *Applied Energy* 183: 121–132. https://doi.org/10.1016/j.apenergy.2016.08.141.

137 Golmohamadi, H., Guldstrand Larsen, K., Gjøl Jensen, P., and Riaz Hasrat, I. (2021). Optimization of power-to-heat flexibility for residential buildings in response to day-ahead electricity price. *Energy and Buildings* 232: 110665. https://doi.org/10.1016/j.enbuild.2020.110665.

138 Golmohamadi, H., Larsen, K.G., Jensen, P.G., and Hasrat, I.R. (2021). Hierarchical flexibility potentials of residential buildings with responsive heat pumps: a case study of Denmark. *Journal of Building Engineering* 41: 102425. https://doi.org/10.1016/j.jobe.2021.102425.

139 Golmohamadi, H. and Larsen, K.G. (2021). Economic heat control of mixing loop for residential buildings supplied by low-temperature district heating. *Journal of Building Engineering* 103286. https://doi.org/10.1016/j.jobe.2021.103286.

140 Sinha, R., Bak-Jensen, B., Pillai, J., and Moller-Jensen, B. (2018). *Unleashing flexibility from electric boilers and heat pumps in Danish residential distribution network. CIGRE 2018.* Paris, France.

141 Trømborg, E., Havskjold, M., Bolkesjø, T.F. et al. (2017). Flexible use of electricity in heat-only district heating plants. *International Journal of Sustainable Energy Planning and Management* 12: 29–46.

142 Guo, J., Huai, X., Li, X., and Xu, M. (2012). Performance analysis of Isopropanol–Acetone–Hydrogen chemical heat pump. *Applied Energy* 93: 261–267. https://doi. org/10.1016/j.apenergy.2011.12.073.

143 Xu, M., Cai, J., Guo, J. et al. (2017). Technical and economic feasibility of the Isopropanol-Acetone-Hydrogen chemical heat pump based on a lab-scale prototype. *Energy* 139: 1030–1039. https://doi.org/10.1016/j.energy.2017.08.043.

144 Wang, T., Tang, T.-Q., Huang, H.-J., and Qu, X. (2021). The adverse impact of electric vehicles on traffic congestion in the morning commute. *Transportation Research Part C: Emerging Technologies* 125: 103073. https://doi.org/10.1016/j. trc.2021.103073.

145 Sacchi, R., Bauer, C., Cox, B., and Mutel, C. (2022). When, where and how can the electrification of passenger cars reduce greenhouse gas emissions? *Renewable and Sustainable Energy Reviews* 162: 112475. https://doi.org/10.1016/j. rser.2022.112475.

146 Hänggi, S., Elbert, P., Bütler, T. et al. (2019). A review of synthetic fuels for passenger vehicles. *Energy Reports* 5: 555–569. https://doi.org/10.1016/j. egyr.2019.04.007.

147 Zhao, S., Li, K., Yang, Z. et al. (2022). A new power system active rescheduling method considering the dispatchable plug-in electric vehicles and intermittent renewable energies. *Applied Energy* 314: 118715. https://doi.org/10.1016/j. apenergy.2022.118715.

148 Karimi-Arpanahi, S., Jooshaki, M., Pourmousavi, S.A., and Lehtonen, M. (2022). Leveraging the flexibility of electric vehicle parking lots in distribution networks with high renewable penetration. *International Journal of Electrical Power & Energy Systems* 142: 108366. https://doi.org/10.1016/j.ijepes.2022.108366.

149 Shafie-khah, M., Heydarian-Forushani, E., Osório G.J. et al. (2016). Optimal behavior of electric vehicle parking lots as demand response aggregation agents. *2016 IEEE Power and Energy Society General Meeting (PESGM)*, p. 1. https://doi. org/10.1109/PESGM.2016.7741167.

150 Daryabari, M.K., Keypour, R., and Golmohamadi, H. (2020). Stochastic energy management of responsive plug-in electric vehicles characterizing parking lot aggregators. *Applied Energy* 279: 115751. https://doi.org/10.1016/j. apenergy.2020.115751.

151 Cañigueral, M. and Meléndez, J. (2021). Flexibility management of electric vehicles based on user profiles: the Arnhem case study. *International Journal of Electrical Power & Energy Systems* 133: 107195. https://doi.org/10.1016/j. ijepes.2021.107195.

152 Sørensen, Å.L., Lindberg, K.B., Sartori, I., and Andresen, I. (2021). Analysis of residential EV energy flexibility potential based on real-world charging reports and smart meter data. *Energy and Buildings* 241: 110923. https://doi.org/10.1016/j. enbuild.2021.110923.

153 Daryabari, M.K., Keypour, R., and Golmohamadi, H. (2021). Robust self-scheduling of parking lot microgrids leveraging responsive electric vehicles. *Applied Energy* 290: 116802. https://doi.org/10.1016/j.apenergy.2021.116802.

154 Pollet, B.G., Staffell, I., Shang, J.L., and Molkov, V. (2014). 22 - Fuel-cell (hydrogen) electric hybrid vehicles. In: *Alternative Fuels and Advanced Vehicle Technologies for Improved Environmental Performance* (ed. R. Folkson), 685–735. Woodhead Publishing https://doi.org/10.1533/9780857097422.3.685.

155 Rodrigues de Moraes, D., Oliveira Soares, L., Hernández-Callejo, L., and Mancebo Boloy, R.A. (2022). DIR-FCEV powered by different fuels – part I: well-to-wheel analysis for the Brazilian and Spanish contexts. *International Journal of Hydrogen Energy* 47 (38): 17069–17081. https://doi.org/10.1016/j.ijhydene.2022.03.175.

156 Alrumayh, O. and Bhattacharya, K. (2019). Flexibility of residential loads for demand response provisions in smart grid. *IEEE Transactions on Smart Grid* 10 (6): 6284–6297. https://doi.org/10.1109/TSG.2019.2901191.

157 Zhao, B., Zeng, L., Li, B. et al. (2020). Collaborative control of thermostatically controlled appliances for balancing renewable generation in smart grid. *IEEJ Transactions on Electrical and Electronic Engineering* 15 (3): 460–468. https://doi. org/10.1002/tee.23075.

158 Golmohamadi, H., Keypour, R., Bak-Jensen, B., and Radhakrishna Pillai, J. (2019). Optimization of household energy consumption towards day-ahead retail electricity price in home energy management systems. *Sustainable Cities and Society* 47: 101468. https://doi.org/10.1016/j.scs.2019.101468.

159 Lakshmanan, V., Sæle, H., and Degefa, M.Z. (2021). Electric water heater flexibility potential and activation impact in system operator perspective – Norwegian scenario case study. *Energy* 236: 121490. https://doi.org/10.1016/j. energy.2021.121490.

160 Abokersh, M.H., Saikia, K., Cabeza, L.F. et al. (2020). Flexible heat pump integration to improve sustainable transition toward 4th generation district heating. *Energy Conversion and Management* 225: 113379. https://doi.org/ 10.1016/j.enconman.2020.113379.

161 Majdalani, N., Aelenei, D., Lopes, R.A., and Silva, C.A.S. (2020). The potential of energy flexibility of space heating and cooling in Portugal. *Utilities Policy* 66: 101086. https://doi.org/10.1016/j.jup.2020.101086.

162 Vellei, M., Le Dréau, J., and Abdelouadoud, S.Y. (2020). Predicting the demand flexibility of wet appliances at national level: the case of France. *Energy and Buildings* 214: 109900. https://doi.org/10.1016/j.enbuild.2020.109900.

163 Golmohamadi, H. and Keypour, R. (2018). A bi-level robust optimization model to determine retail electricity price in presence of a significant number of invisible solar sites. *Sustainable Energy, Grids and Networks* 13: 93–111. https://doi.org/10.1016/j.segan.2017.12.008.

164 Golmohamadi, H., Keypour, R., Bak-Jensen, B. et al. (2020). Robust self-scheduling of operational processes for industrial demand response aggregators. *IEEE Transactions on Industrial Electronics* 67 (2): 1387–1395. https://doi.org/10.1109/TIE.2019.2899562.

165 Golmohamadi, H., Keypour, R., Bak-Jensen, B., and Pillai, J.R. (2019). A multi-agent based optimization of residential and industrial demand response aggregators. *International Journal of Electrical Power & Energy Systems* 107: 472–485. https://doi.org/10.1016/j.ijepes.2018.12.020.

166 Golmohamadi, H. and Asadi, A. (2020). Integration of joint power-heat flexibility of oil refinery industries to uncertain energy markets. *Energies (Basel)* 13 (18): https://doi.org/10.3390/en13184874.

167 Golmohamadi, H. Operational scheduling of responsive prosumer farms for day-ahead peak shaving by agricultural demand response aggregators. *International Journal of Energy Research* https://doi.org/10.1002/er.6017.

168 Golmohamadi, H. and Asadi, A. (2020). A multi-stage stochastic energy management of responsive irrigation pumps in dynamic electricity markets. *Applied Energy* 265: 114804. https://doi.org/10.1016/j.apenergy.2020.114804.

169 Golmohamadi, H. (2020). Agricultural demand response aggregators in electricity markets: structure, challenges and practical solutions- a tutorial for energy experts. *Technology and Economics of Smart Grids and Sustainable Energy* 5 (1): 17. https://doi.org/10.1007/s40866-020-00091-7.

170 Dew, J.J.W., Jack, M.W., Stephenson, J., and Walton, S. (2021). Reducing electricity demand peaks on large-scale dairy farms. *Sustainable Production and Consumption* 25: 248–258. https://doi.org/10.1016/j.spc.2020.08.014.

171 Zeyad, M., Masum Ahmed, S.M., Hossain, E. et al. (2021). Optimization of a solar PV power plant with poultry demand side management (PoDSM) for poultry farm. *2021 International Conference on Computational Performance Evaluation (ComPE)*, 73–78. https://doi.org/10.1109/ComPE53109.2021.9751839.

172 Pedersen, R., Schwensen, J., Biegel, B. et al. (2017). Improving demand response potential of a supermarket refrigeration system: a food temperature estimation approach. *IEEE Transactions on Control Systems Technology* 25 (3): 855–863. https://doi.org/10.1109/TCST.2016.2583958.

173 Chen, T., Cui, Q., Gao, C. et al. (2021). Optimal demand response strategy of commercial building-based virtual power plant using reinforcement learning. *IET Generation, Transmission and Distribution* 15 (16): 2309–2318. https://doi.org/10.1049/gtd2.12179.

7

Integration of Electric Vehicles into Multi-energy Systems

Samaneh Sadat Sajjadi, Ali Moradi Amani, Nawazish Ali, and Mahdi Jalili

School of Engineering, STEM College, RMIT University, Melbourne, Australia

Abbreviations

AND	Advanced distribution networks
CCHP	Combined cooling, heat, and power
CHP	Combined heating and power
DMG	Distribute multigeneration
DNSP	Distribution Network Service Provider
DR	Demand response
EH	Energy hub
EV	Electric vehicle
EVs	Electric vehicles
G2V	Grid-to-vehicle
HEMS	Home energy management systems
HP	Heat pump
IES	Integrated energy station
LFC	Load frequency control
MES	Multi-energy system
MESs	Multi-energy systems
OF	Objective function
PV	Photovoltaic
P2G	Power-to-grid
RES	Renewable energy sources
SMEMS	Smart multi-energy management system
SMES	Smart multi-energy system

Interconnected Modern Multi-Energy Networks and Intelligent Transportation Systems: Towards a Green Economy and Sustainable Development, First Edition. Edited by Mohammadreza Daneshvar, Behnam Mohammadi-Ivatloo, Amjad Anvari-Moghaddam, and Reza Razzaghi.
© 2024 The Institute of Electrical and Electronics Engineers, Inc.
Published 2024 by John Wiley & Sons, Inc.

SoC	State of charge
VGI	Vehicle-grid integration
VPP	Virtual power plant
V2B	Vehicle-to-building
V2G	Vehicle-to-grid
V2H	Vehicle-to-home
V2I	Vehicle-to-infrastructure
V2L	Vehicle-to-load
V2N	Vehicle-to-network
V2X	Vehicle-to-everything
WPP	Wind power producer
WT	Wind turbine

7.1 Introduction

More than 50% of the world's population resides in urban areas [1]. Developed countries in Europe and North America already have more than 70% and 80%, respectively, of their population classified as urban dwellers. This continuous trend of urbanization has led to an increase in sustainability challenges in cities. For example, cities currently consume approximately 75% of all resources and account for around 70% of all greenhouse gas emissions [2]. Securing a clean, affordable, and reliable energy supply is one of the primary sustainable development challenges of cities for which, integrating renewable energy sources (RES)-based electricity into the energy mix, using innovative concepts such as Smart Grid, is a potential solution. Although, electricity only satisfies a portion of a city's energy needs[3], there is an increasing interest in expanding its role in sustainable urban development to shape future smart cities.

Multi-Energy System (MES) is gaining increasing attention as a promising approach to tackle the growing energy consumption needs while mitigating greenhouse gas emissions and promoting sustainable energy systems. MES integrates multiple energy sources, such as electricity, natural gas, and renewable energy, with various energy storage technologies, such as batteries, thermal storage, and hydrogen storage, to provide a reliable and efficient energy supply [4]. Different energy sources in MES are coupled by an energy hub (EH) or integrated energy station (IES) [5–7]. MES has different definitions in the literature [8]. For example, it has been considered as an innovative integration of energy sectors, including gas, water, electricity, cooling, heating, and transport, at different stages to enable the creation of more services and profitable streams [9].

An MES requires harmony in the operation and planning of the integration of energy systems across various routes and geographic locations that offer reliable and sustainable energy delivery with minimal impact on the environment [10]. In addition, MESs offer optimization capabilities in the form of energy conversion to meet the specific requirements of consumers [11, 12]. It can provide significant advantages such as operational flexibility, sustainable energy utilization, and improved reliability. MES comprises various complex components, such as combined heating and power (CHP) systems, combined cooling, heat, and power (CCHP) systems, heat pumps (HP), boilers, refrigerators, thermal and electricity storage systems, and power-to-gas (P2G) converters [13].

EH is the state-of-the-art solution for MES modeling. It enables the conversion, conditioning, and storage of multiple carriers, thus making it easier to model and operate MES [14–16].The incorporation of RES like photovoltaic (PV) and wind power provides advantageous prospects in the system operation of EH and has also attracted significant attention globally [17]. Furthermore, electric vehicles (EVs) are being recognized as a potential answer to enhance flexibility in MES due to their rapid response, economic viability, and eco-friendly nature. This has led to increased interest in operating an MES structure that incorporates renewable-based generation systems and EVs, leveraging the advantages of the smart grid perspective [12].

The integration of EVs can significantly enhance the overall effectiveness and sustainability of the MES. EV-grid integration requires careful consideration of the charging infrastructure and energy management strategies to optimize performance. This integration is important because changes in electricity load resulting from the adoption of EVs can affect the operation of energy conversion systems like CHP, gas-fired generators, boilers, and others. In addition, by leveraging the bidirectional power flow capabilities of EVs, they can be utilized as a means of energy storage to balance energy demand and supply [18, 19].

Many studies have primarily focused on examining the interconnection between EVs and power grids, disregarding the significant interdependencies between alternative-fuel-powered modes of transportation or mechanical energy storage and the broader energy infrastructure. For example, a multi-objective optimization approach is proposed in Ref. [20] for modeling the components of advanced distribution networks (ADNs) and identification of the optimal location and size of charging stations. Wang et al. [21] use a stochastic approach to develop an approach for expanding the planning of ADNs that considered shared charging stations and the charging behavior of users. The authors in Ref. [22] propose a planning model for distributed generators, charging stations, and charging demand, which identified the distribution curves indicating the likelihood of EVs parking time and durations. However, these studies made the common assumption that charging demand could be predicted in advance using pre-collected data or probability distributions, neglecting the consequences of flow patterns in traffic networks.

In Ref. [23], operational challenges associated with interconnected transportation and distribution systems are investigated, highlighting the potential security risks of an unguided traffic flow. Following that, a model for expansion planning based on scenarios for an ADN coupled with a transportation system is presented in Ref. [24], emphasizing the significance of accounting for the interconnectedness of the two systems. Additionally, a two-stage robust optimization method is employed to effectively synchronize the integrated networks, taking into account the uncertainties associated with renewable energy and demand.

The critical topic of the smart charging of EVs involves a set of smart feature sets that enable a versatile and energy-efficient ecosystem for charging and discharging. This is achieved by facilitating communication between the EVs and the power grid. The rapid acceleration of electrical equipment manufacturing with higher power ratings is imperative to facilitate ultra-fast charging. Powerful converters and integrated control approaches for intelligent charging coordination at high power levels are examples of how the charging time of EVs can be reduced effectively [25]. Smart charging can not only increase power transmission efficiency but also helps to mitigate the electricity demand from the grid. Additionally, charging EVs using solar and wind power can promote products that are designed, manufactured, and utilized in a manner that promotes sustainability.

Smart bi-directional charging, also called Vehicle-to-Grid (V2G) can bring a new role for EVs as ancillary service providers to the grid, e.g. for frequency control [26].By offering these services, smart charging has the potential to slash energy costs of owners associated with EVs by up to 60%. [27]. Furthermore, EV batteries serve a critical function in the charging process of EVs and as energy storage for grid integration. In the V2G mode, during periods of high demand, EVs are commonly linked to the grid to strengthen grid performance and stability.

In light of the explanations provided above, despite significant research on EV and MES separately, there is potential to explore optimal strategies for integrating EVs within MES. This could include investigating the most efficient and effective ways to integrate EVs into existing MES through energy management and system optimization. Furthermore, comprehensive studies are needed for modeling approaches that consider the dynamic interactions among EVs, RES, storage systems, and the grid. Developing robust and holistic modeling approaches that capture the complexities of MES with EV integration could be an area for further research. Moreover, while there may be research into the technical aspects of integrating EVs within MESs, there could be a research gap in terms of exploring the economic and policy aspects of this integration. This could include investigating the economic viability of different integration strategies, evaluating the impact of policies and regulations on the integration of EV within MES, and assessing the potential economic benefits and challenges associated with this integration. In addition to this, still research gaps exist in terms of real-world implementation studies that demonstrate the practicality and feasibility of integrating EV within MES. Examining case studies or pilot projects that have implemented or tested different integration strategies in real-world settings could provide valuable insights into the challenges, opportunities, and lessons learned from actual implementation efforts.

This chapter provides a review of the EV-MES integration. The novelty of this book chapter lies within its comprehensive review of original research works associated with the modeling and management of MES integrating EV routing and charging. It highlights the interdependencies between the energy and transport infrastructure in the MES, particularly in the context of transport electrification and the challenges and opportunities it presents. By addressing the interdependencies and optimization opportunities within the MES, this chapter provides a holistic approach to understanding the integration of EVs into MESs and offers insights into the potential benefits, challenges, and solutions for achieving sustainable and efficient energy systems in the context of transport electrification. The comprehensive review of original research works in this area adds value to the existing literature and provides valuable information for researchers, and policymakers interested in the field of MES and EV integration.

Section 7.2 reviews the structure of MESs, the general aspects and assessment of MES modeling, the concept of EHs and their role in the modeling and implementing of MES, and the challenges in MES modeling. Section 7.3 investigates the integration of EVs with MESs, including integration with RES and the power grid. It also reviews the different strategies for charging/discharging of EVs. Section 7.3 provides the conclusion and future recommendations.

7.2 Multi-energy Systems Structure

Integrating multiple energy carriers like electricity, natural gas, hydrogen, and thermal energy in an MES is a promising approach to enhance energy efficiency and decrease the expenses related to energy supply. However, with the availability of various energy converters and storage devices with distinct characteristics, the selection of appropriate types and capacities of these devices and their effective management pose significant challenges in designing optimal MES structures. Due to the remarkable progress in information and communication technologies, coupled with the proliferation of smart devices such as smart meters/sensors, actuators, switchgears, communication and metering infrastructures, and embedded smart energy management systems, the development of MES has evolved into smart multi-energy System (SMES) [28, 29]. Within this smart environment, devices from each network can exchange relevant information and identify potential synergies in a coordinated manner, resulting in customized and ideal resolution for individual sectors as well as the entirety of the SMES simultaneously. An SMES structure is a complex energy system that integrates multiple energy sources and storage technologies to meet energy demands in an efficient, sustainable, and reliable manner. It is developed to improve efficiency in the use of various energy sources and storage technologies, including renewable and non-renewable sources, while minimizing greenhouse gas emissions and reducing energy costs. The diagram in Figure 7.1 illustrates the structure of a typical SMES that is in use at the community level. This SMES connects to the electricity, natural gas, and district heat distribution networks via input ports, and the output ports provide a range of multi-energy services, such as power, heating, and cooling, all simultaneously. The power, heating, and cooling hubs function as energy nodes, responsible for gathering and distributing power, heat, and cold energy. At any given time, the energy at each hub remains in balance, which means that the energy input matches the energy output. This arrangement allows for a clear visualization of energy flows and the links between energy converters and storage devices. The SMES structure consists of four main components:

- **Energy sources:** The SMES structure includes multiple energy sources such as solar, wind, biomass, geothermal, hydro-, and fossil fuels. The system uses these energy sources to produce electricity and heat to meet the energy demands of various applications.
- **Energy storage:** The SMES structure includes various energy storage systems like batteries, hydrogen, pumped hydro-, and compressed air energy storage. These technologies store excess energy generated during off-peak hours and discharge it during peak periods, thereby ensuring a continuous and stable energy supply.
- **Energy conversion:** The SMES structure includes various energy conversion technologies such as fuel cells, combustion engines, and turbines. In other word, the energy converters include photovoltaic, wind turbine (WT), transformers, gas turbines, gas boilers, heat exchangers, electric chillers, and absorption chillers. These technologies convert stored energy into electricity and heat for various applications.
- **Energy management:** The SMES structure includes advanced energy management systems that monitor and control energy flows between different components of the system. These systems optimize the use of energy sources and storage technologies to meet energy demands in the most optimal and economically feasible way.

In the context of the main components in a MES, diverse energy requirements of end-users can be met using various sources, including renewable generators such as PV and WT, gas turbines, storage systems, or the power network. Similarly, heating demands can be addressed through options such as gas turbines, gas

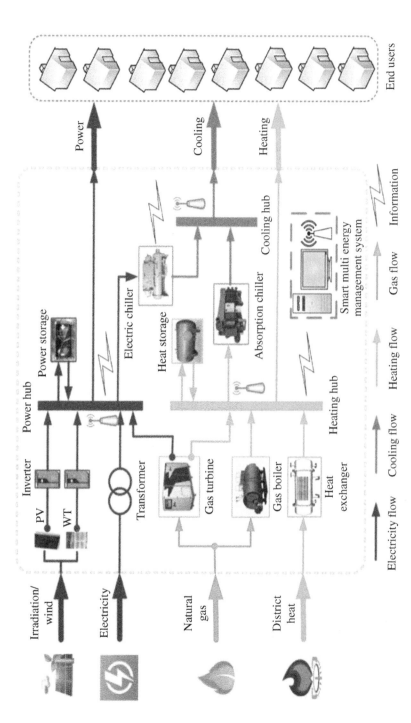

Figure 7.1 The typical configuration of SMES. *Source:* [28] / with permission of Elsevier.

boilers, heat storage systems, or district heat networks. Cooling demands can be fulfilled using electric or absorption chillers. The MES also possesses a bidirectional communication infrastructure, smart energy meters, and a smart multienergy management system (SMEMS). The SMEMS collects data on electricity and natural gas prices, device status, weather patterns, PV and WT power outputs, and power, heating, and cooling loads, among other factors. Utilizing this data, the SMEMS develops an optimal energy management and dispatch approach, sending control signals to each component to efficiently coordinate the operation of the entire energy system.

The SMES structure is also designed to achieve several benefits, including:

- **Energy security:** The incorporation of multiple energy sources and storage technologies reduces dependence on a single energy source, such as fossil fuels. This makes the system less vulnerable to supply disruptions or price volatility associated with a single energy source. The SMES structure can also escalate the reliability of energy supply by providing backup energy storage during peak demand periods or grid outages.
- **Cost-effectiveness:** The integration of multiple energy sources and storage technologies can optimize the use of energy resources and reduce the need for expensive grid infrastructure. For example, by using energy storage technologies to accumulate energy when demand is low, the SMES can help to decrease the need for peak generation capacity and lower overall energy costs. Additionally, by reducing energy waste through better energy management, the SMES can lower energy costs for both the system and the end-users.
- **Sustainability:** By promoting the use of RES and energy storage technologies, the SMES can help to mitigate these environmental impacts and set the stage for a cleaner, more sustainable energy future. Additionally, the use of these technologies can help to reduce the dependence on finite resources and can ensure a more reliable and sustainable energy supply for future generations.

7.2.1 General Aspects of MES Modeling

The analysis and evaluation of how various energy systems interconnect and influence the overall performance of the energy system is known as the modeling and assessment of MES. This field encompasses the integration of diverse energy sources and the assessment of the trade-offs involved in their use within the energy grid. MES modeling and assessment involve a comprehensive study of various energy sources, including solar, wind, hydro-, fossil fuels, and nuclear energy. This analysis considers the energy demand and its variability, including the energy requirements of different sectors such as industry, residential, and commercial. Additionally, the study focuses on energy storage options like batteries, pumped hydro-, and thermal energy storage to manage the variability in energy generation and demand. The modeling of the energy network, including the interconnections between different energy sources and the energy grid, is also a critical aspect. The economic viability of different energy sources and the cost of integrating them into the energy grid is evaluated, along with an assessment of the environmental impact, including emissions and land use. By evaluating various energy sources and their interconnections with the energy grid, MES modeling and assessment can identify the most effective and economical approaches for integrating different energy sources. Moreover, it can assess the impacts of diverse energy policies on the energy system and provide crucial information for decision-makers in prioritizing investments in different energy sources and infrastructure. In addition to this, modeling of MES involves the use of mathematical and computational methods to simulate the behavior of the energy system and its components. The principal objective of MES modeling is to achieve a thorough comprehension of the interrelations among the various energy sources and the energy grid, while evaluating the trade-offs between them. Below are some of the key elements of MES modeling:

- **Energy demand modeling:** This modeling encompasses the examination of the patterns and fluctuations of energy demand, considering the energy necessities of various sectors, including industrial, residential, and commercial. By employing energy demand models, one can evaluate the energy requirements of a specific location and determine the effects of energy-conservation measures on energy demand. To estimate the energy requirements of a specific location, energy demand models use data from multiple sources such as weather forecasts, demographic information, and economic trends. By analyzing this information, energy demand models can accurately predict the demand of different sectors at different times. Moreover, this modeling plays a pivotal role in planning and managing the energy supply of a region. By accurately estimating the demand, energy providers can optimize the design and operation of energy generation systems as well as energy storage systems to ensure that the grid can reliably meet the demand of consumers. Furthermore, this part of MES modeling allows for the optimization of energy systems design and operation and helps evaluate the impact of energy-conservation measures on energy demand. These models can be used to estimate the impact of building retrofits or energy-efficient appliances on the demand of residential or commercial sectors.
- **Energy generation modeling:** This part of the MES modeling includes modeling various energy sources, such as solar, wind, hydro-, and nuclear energy, to estimate their energy yield and evaluate the impact of weather conditions on energy generation. In energy generation modeling, a variety of factors are taken into account, including solar radiation, wind speed, water flow, and reactor

power output. By analyzing these factors, energy generation models can estimate the energy yield of different sources under various conditions. These models can be used to optimize the design and operation of generation systems and to evaluate the feasibility of incorporating new or alternative energy sources into the grid. Weather conditions have a substantial impact on the generation of renewable sources. Energy generation models also take into account different weather scenarios, including seasonal, daily, and hourly variations, to estimate the energy output of renewable sources accurately. This helps to ensure that the energy grid can reliably fulfill the demands of the consumers and can accommodate fluctuations in supply.

- **Energy storage modeling:** Energy storage modeling is also one of the fundamental aspects of MES modeling and assessment, providing critical insights into the performance and efficiency of energy storage systems, which are vital components of a reliable and sustainable energy system. It involves modeling energy storage systems like batteries and pumped hydro- storage to estimate their capacity and efficiency. Energy storage models are necessary for assessing the ability of energy storage systems to manage fluctuations in generation and demand, which is of particular significance for variable RES like wind and solar. Energy storage modeling enables the identification of the most cost-effective and efficient energy storage technologies that can be integrated into the grid to support renewable sources. This helps in minimizing the need for expensive and environmentally damaging backup power sources, such as natural gas plants, during times of low generation or high demand. This part of MES modeling also enables the assessment of the impact of several factors on the performance of storage systems, such as the ambient temperature, battery or storage system degradation, and the depth of discharge. This information is crucial for optimizing the storage system design and operation, maximizing their benefits, and minimizing costs.

- **Energy network modeling:** Network modeling is a crucial part of the modeling and evaluation of MES. It involves creating a model that encompasses all relevant aspects of the energy infrastructure, including the interconnections between different energy sources and the grid, to assess the performance of the energy infrastructure and its ability to meet the demands of consumers. Energy network models simulate various scenarios of energy generation, storage, and distribution to identify potential issues in the grid and develop solutions to mitigate these issues. For example, an energy network model can assess the impact of integrating a new energy source, such as a wind farm, into the existing energy infrastructure, helping to identify the optimal location for the wind farm and the required upgrades to the grid to ensure a reliable energy supply. Furthermore, network models can estimate the cost of integrating different energy sources into the grid by analyzing the cost of generation, storage, and distribution. The model can help identify the most cost-effective solutions for meeting demand, which is useful for decision-makers in determining investment priorities and developing energy policies.

- **Economics modeling:** MES economic modeling involves analyzing the economic feasibility of various energy sources and the associated costs of integrating them into the energy grid. Economic models estimate the cost of energy generation and storage and analyze the impact of energy policies on the energy system. Economic modeling accounts for several factors, such as capital, operational, and maintenance costs, fuel costs, and financing costs, to estimate the overall cost of energy generation and storage for different energy sources. The costs can differ, depending on the energy source type, location, and the level of technology employed for energy generation and storage. The models can also assess the impacts of various energy policies on the energy system. For example, economic models evaluate the impact of carbon pricing or subsidies on energy generation and storage costs, which is valuable for policymakers to identify successful interventions to promote renewable energy and reduce greenhouse gas emissions. Furthermore, economic models identify the most cost-effective solutions for meeting energy demand. By analyzing the costs of energy generation, storage, and distribution, these models help determine the most economically feasible combination of energy sources and the most efficient methods for integrating them into the grid.

The operational and planning optimization of MES is often a complex process due to the diverse range of potential MES types that can be envisioned. Advanced modeling techniques and tools are necessary to address this complexity. It is crucial to capture both the complexity of interactions between multiple sources of energy occurring within the MES and the interactions with the external environment. Power system operators must recognize that distributed multi-generation (DMG) and multi-energy loads, previously considered passive by power system operators, must be integrated into the power system operation to achieve a sustainable energy system while minimizing costs. Because many DMG systems operate at small or even micro-scale levels (For example, micro-CHP generators that are installed in residential homes), general aggregation concepts have been proposed to integrate distributed energy resources into power system operation and planning [30]. Microgrids [31, 32] and virtual power plants (VPP) [33, 34] are two such concepts that can be implemented in MES. On the other hand, the EH concept has been particularly established to technically model generic MES [15, 35]. The significance of these aggregation notions is underscored from an economic standpoint, indicating a necessity to shift away from the traditional scheme of constructing large-scale power plants subject to regulated rates of return, and instead develop new business models for these novel, distributed systems.

7.2.2 Energy Hub Concept

Researchers have shown an increasing interest in utilizing the EH approach for implementing MES [36–40]. In Ref. [36], an EH was defined as a computer-simulated environment where energy carriers are transformed into load demand. Various technologies are available inside the box for energy conversion and storage, as depicted in Figure 7.2. In MES modeling, an EH is a central facility that integrates multiple energy sources and energy carriers to manage energy supply and demand more efficiently. EH uses advanced optimization algorithms and control strategies to balance the flow of energy across different sectors, such as residential, commercial, industrial, and transportation. They can incorporate various energy sources, including RES like solar, wind, and geothermal, as well as non-renewable sources like fossil fuels and nuclear power. These sources are integrated into the EH through various energy carriers, such as electricity, natural gas, hydrogen, and thermal energy [41]. One of the primary functions of EH is energy storage, which, by using technology like batteries, pumped hydro, or compressed air, can accumulate surplus energy during periods of low demand and distribute it when demand is high. This helps balance the energy supply and demand, reducing the need for fossil fuel-based peaking power plants and minimizing energy waste. EH also distributes energy to different sectors. For example, a microgrid can be used to supply electricity to a local community, while a district heating system can supply thermal energy to buildings. By optimizing energy distribution, EH can help reduce transmission losses and increase energy efficiency. Designing and operating an EH requires interdisciplinary energy engineering, economics, and policy expertise. Energy engineers design the technical components of the EH, like energy storage systems, conversion technologies, and distribution networks. Economists evaluate the economic viability of it and analyze the costs and benefits of different energy sources and technologies. Policy experts assess the regulatory framework and incentives needed to support the development of an EH.

As shown in Figure 7.2, the EH is comprised of several components, including input and output, conversion, and storage units. In essence, it is an extension of power system network nodes. The EH framework includes inputs such as natural gas, electricity, and district heat networks, as well as energy conversion components including a natural gas boiler, CHP, transformer, heat exchanger, electric heat pump, and absorption chiller. The EH also includes battery/thermal storage devices [42]. Finally, the EH outputs include heat, cool, and electricity. In this regard, an EH can be broken down into the following four basic areas. The first is energy generation, which encompasses producing energy from various sources, including RES like solar, wind, and geothermal, fossil fuels like coal and natural gas, and nuclear energy. The second component is energy storage, which involves using energy storage technologies like batteries, pumped hydro-, or compressed air

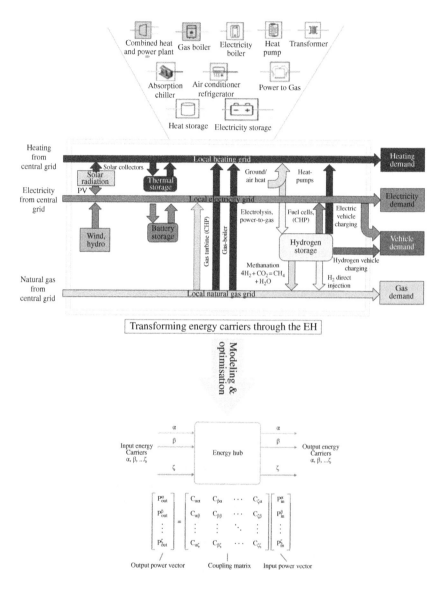

Figure 7.2 Configuration of EH with all-available components. *Source:* Adapted from [8, 37].

to store excess energy during periods of low demand and release it during periods of high demand. Energy storage systems employed in EH may include heat storage, hot water tanks, hydrogen tanks, batteries, ice storage, and flywheels. The third component is energy conversion, which utilizes energy conversion technologies like boilers, chillers, CHP units, heat pumps, fuel cells, electrolysers, or power electronics to convert energy from one form to another (e.g., from electricity to hydrogen or from thermal energy to electricity). These conversion technologies enable the management of different energy resources entering the system and balance supply and demand. Finally, energy distribution involves the distribution of energy across various sectors such as residential, commercial, industrial, and transportation. Energy distribution technologies employed in EH may include microgrids, district heating and cooling systems, and hydrogen pipelines. The hub energy demands for end-users may include electricity, heat, cooling, gas, water, and hydrogen. The energy vectors used as inputs to the EH system may come from any of these sources.

MES planning and management involves the challenge of determining the facilities, systems, or networks required for energy supply, storage, and conversion technologies to effectively fulfill energy needs. Recent studies suggest that a coordinated approach for planning and managing MESs is more desirable compared to the current practice of uncoordinated development. Deterministic and probabilistic models are used to employ the EH scheme for evaluating the management of MESs. Moreover, planning objective functions (OFs) are used to make decisions when planning an EH. The OFs refer to the objectives that guide decision-making during the planning process. These objectives can be broadly grouped into three categories: technical, economic, and environmental. The main OFs that are commonly used include minimizing investment costs, operational costs, and greenhouse gas emissions. These OFs are typically related to the cost and environmental impacts of different energy conversion, storage, and supply technologies [8]. Nevertheless, the majority of studies that focus on planning micro-energy systems tend to emphasize economic and technical objectives, with little consideration for environmental objectives. This is a significant gap, as ecological consequences of energy production and consumption systems are of increasing concern globally. Therefore, there is a need to incorporate environmental objectives into planning models for EH. Furthermore, few studies have compared the advantages of management and planning an EH that synergizes various energy carriers (such as electricity, heat, and gas) with independent energy carrier planning. To address this gap, some objective functions have been proposed for EH planning, including the following:

- **Minimizing investment and operating cost:** This OF designs an energy system that can produce the desired amount of energy with the lowest possible investment in equipment and infrastructure, and the lowest possible operating costs. To this end, a careful analysis of the different components that make up the energy system, including energy generation, transmission, and storage, is required. It can involve choosing equipment and technologies that require less maintenance and have a longer life span, as well as optimizing the operation of the energy system to reduce energy waste and increase efficiency.

- **Minimizing lifecycle cost (LCC):** This involves assessing the total cost of a project over its entire lifecycle, including design, construction, operation, and maintenance costs. The aim is to optimize the LCC by selecting the most cost-effective technologies and designing the system to minimize overall costs.
- **Maximizing the share of RES penetration:** This objective is driven by the need to mitigate the effects of climate change by lowering greenhouse gas emissions, as well as to increase energy security and diversify the energy mix. This OF involves assessing factors such as the availability and reliability of renewable sources, their cost and competitiveness relative to other energy sources, and the potential impacts on the grid and energy storage requirements. Additionally, it may be needed to consider the policy and regulatory frameworks that support the development of renewable sources, such as feed-in tariffs or renewable energy targets.
- **Minimizing energy costs and emissions:** This goal is strongly linked to the previous objective of minimizing lifecycle costs, as reducing energy costs and emissions often requires a focus on energy efficiency, which can result in long-term cost savings. One approach to minimizing energy costs and emissions is to promote the adoption of RES, which can be used to displace fossil fuel-based energy sources that are associated with high emissions. Another approach to achieving this objective is to increase the efficiency of the energy system by reducing energy waste, such as through improved insulation and building design, and by optimizing the use of energy in buildings, industry, and transportation. This can involve the use of energy efficient appliances and equipment, as well as the progression of smart grids and energy management systems that can contribute to a better balance between energy supply and demand and also reduce energy waste.
- **Maximizing system reliability and profits:** System reliability gauges the system's ability to deliver energy consistently and without interruption, while profits refer to the financial benefits of operating the system. To achieve this objective, a range of factors that can impact system reliability and profits, such as the availability and quality of energy sources, the efficiency of energy converters and storage devices, the cost of maintenance and repair, and the demand for energy should be considered. Incorporating advanced control systems and predictive maintenance strategies can help identify potential issues before they become major problems, allowing for timely repairs and maintenance that can prevent downtime and ensure consistent energy delivery. Additionally, incorporating backup power sources or redundant systems can be considered to further improve system reliability. Maximizing profits can also involve strategies such as pricing energy at different rates based on demand, implementing

demand-response programs that incentivize customers to reduce energy usage during peak demand periods, or integrating energy storage devices to take advantage of fluctuations in energy prices.

- **Maximizing social welfare:** This objective considers various factors, such as energy access, affordability, and equity. It should be considered that the benefits of the energy system are distributed fairly across different segments of society. Moreover, maximizing social welfare also involves considering the environmental impact of the energy system. Planners need to design the energy system in a way that minimizes its negative impact on the environment while still providing the necessary energy services to society.

By considering these OFs, a more comprehensive understanding of the costs, benefits, and environmental impacts of different EH designs and configurations can be developed. This can help to inform more sustainable and efficient energy planning decisions and ultimately contribute to a more sustainable energy future.

In addition to those mentioned earlier, the EH planning involves dealing with constraints that limit the possible values of decision variables, such as operational or structural constraints. Operational constraints refer to the power system's limitations, with consideration of power balance, ramping rate thresholds, wind power accessibility, and transmission flow limits. Gas system constraints are limitations for a gas system, incorporating nodal balance equations, well-flow capacity, nodal pressure, compressor constraints, and pipeline limits. Moreover, decision variables determine the output demand of an EH. To plan an EH and solve operational challenges, it is essential to identify decision variables, such as binary variables, to indicate if an energy converter or storage device is selected, or continuous variables to represent energy flow at a specific time. Also, EH optimization models fall into the two following categories [8]:

1) Structural optimization finds the ideal configuration and arrangement of an EH developed on unique energy requirements and relevant objectives' OFs.
2) Operational optimization involves finding the optimal power dispatch within a single EH or the optimal power flow within a network of interlinked EH, given a specific system structure.

7.2.3 MES Modeling Process and Challenges

As discussed earlier, energy system modeling encompasses various sectors, including energy infrastructure, technology, RES, and consumer behavior. Modeling these components accurately is crucial for accounting for the uncertain characteristics of RES and the actions of market players. However, developing precise MES models can be a daunting task, and it involves several challenges, including overall features such as model formulation and data, as well as specific aspects related to time and space [41].

- **Overall features:** One of the main challenges is to define the system boundaries and select appropriate modeling tools and techniques. Since MES models involve multiple domains such as electricity, gas, and heat, it is necessary to take a holistic approach to modeling to capture the interconnections between these domains accurately. Although energy system models are commonly used to investigate the technological and economic impacts, they are seldom utilized to explore the effects of factors such as human behavior, socio-political dynamics, or non-financial challenges regarding technology adoption. These factors can be important in the development of energy systems, and it is imperative to incorporate them in the modeling process. Energy systems consist of multiple interrelated fields, such as power generation, transmission and distribution, end-use consumption, and energy storage. With the increasing integration of distributed RES, the increasing interconnectedness among the distinct energy carriers will continue to rise, adding to the intricate nature of the system and increasing maintenance costs [43].

- **Mathematical formulation:** Mathematical formulation is a challenging aspect of MES modeling, as it involves determining the appropriate level of detail and accuracy for the models. The models must include the key interactions and dependencies among several energy systems, but not be so complex that they are computationally infeasible to solve. As a system becomes more interconnected, the complexity of the mathematical problem increases. One potential strategy involves breaking down energy flows to derive linear models, but this could result in significant inaccuracies due to the nonlinear nature of power flow equations for electricity and hydraulic networks. Optimizing nonlinear constraints requires additional optimality conditions, making the mathematical task even more challenging to solve. MES models need to be complex enough to capture the essential interactions, but not so complex that they become computationally infeasible to solve. A balance must be struck between model accuracy and computational efficiency. To address these challenges, various modeling techniques, such as mixed-integer linear programming, dynamic optimization, and agent-based modeling have been proposed in the literature. Each technique has its strengths and weaknesses, and the selection of the appropriate modeling technique depends on the specific problem being addressed.

The development of a model for operational planning or scenario exploration purposes often involves multi-criteria examination, which is investigated within the framework of MES. The process of multi-criteria decision-making involves several steps, including identification, characterization, and formulation of the problem, establishing the objectives, and mathematical modeling. This approach can be applied to MES planning problems, where a decision-maker designs a comprehensive and forward-thinking analysis of a specific system, incorporating diverse criteria, with a forward-looking and enduring perspective. This holistic approach captures strategies from the demand and supply sides of the power system, heating/cooling system, fuel systems, transport sector, and their interactions. To identify an adequate decision situation, a group of decision variables, alternatives, and criteria must be established [44]. The available options denote potential resolutions to the problem and can be determined by evaluating the investments and policies under examination. Each alternative includes a group of decision variables, such as modifications in the demand, installed capacity of RES and other MES technologies, use of storage systems, and sustainable mobility options. Multiple metrics can be utilized to enhance decision-making efficacy in addressing various challenges within MES. The process of selecting criteria forms a fundamental step in the decision-making process, and they must be sufficiently intelligible for each stakeholder, viewed as a valuable instrument for assessing and contrasting alternative actions, and must be a coherent family of criteria that is exhaustive, consistent, and non-redundant. Other desirable properties include intelligibility and operationality. Each criterion is associated with an attribute, which serves as a yardstick for assessing the criteria. Attributes are commonly used to describe the features, qualities, or performance parameters of alternatives, and are quantifiable measures represented by metrics or indicators that gauge the extent of fulfillment for a particular criterion. Criteria and their corresponding attributes/metrics that are typically used in the context of MES may include [44]:

- *Economics*, which may involve characteristics and measures such as annual costs, revenues, profits, net present values, and payback times [45].
- *Energy consumption/savings*, considering primary energy use and savings compared to a reference case, broken down by type of energy vector or overall [46].
- *Emissions/emission savings*, which may encompass global emissions and emission savings of GHG/CO_2, as well as local emissions/pollution such as CO, NO_x, etc. [47].
- *Integration of RES and low-carbon technologies*, using indicators such as the quantity of RES integrated into a given system, renewable energy curtailment, and hosting capacity of the distribution network, among other pertinent metrics.
- *Security of supply, reliability, and resilience*, evaluated using performance measures such as anticipated energy shortfall, frequency and duration of service interruptions, and so on.
- *Comfort level*, evaluated through indicators like average indoor temperature [40] and projected energy reduction by building management systems, among others.

- **Data collection and management:** The development of accurate and robust models for MES requires large amounts of data from various sources, including the following data:
  - *Weather data*, including historical and current weather data, such as temperature, wind speed, solar irradiation, and precipitation, which are essential for modeling RES like solar and wind.
  - *Energy consumption data*, including data on the consumption of energy by different sectors, such as residential, commercial, and industrial, as well as data on energy imports and exports.
  - *Infrastructure data*, incorporating data on the physical infrastructure of the energy system, such as the location, capacity, and efficiency of power plants, transmission and distribution lines, storage facilities, and other energy-related infrastructure.
  - *Economic data*, containing data on energy prices, tariffs, subsidies, and taxes.
  - *Demographic data*, including data on population size, growth, and distribution, as well as data on household and commercial characteristics, such as building size and age, which can help to model energy demand.
  - *Environmental data*, incorporation of data on environmental factors such as air quality, water availability, and land use, which can have an impact on energy production and consumption.
  - *Social data*, such as demographic data, data on energy use patterns and behaviors, and data on energy access and affordability.

  The availability and quality of data can pose significant challenges, particularly for emerging technologies and systems that lack historical data. Therefore, data quality and availability are crucial for developing accurate and robust models for MES. Uncertainty is inherent in all models, and to address it, two types of uncertainty are discussed in the excerpt: epistemic and aleatory. Epistemic uncertainty arises when the modeler thinks that uncertainty can be lowered through improved data and models, whereas aleatory uncertainty arises from randomness and cannot be reduced by better data or models [48, 49]. Formal methods, such as the Monte Carlo method, can be used to address aleatory uncertainty. This method involves iterating through input data to analyze the variations in the inputs and outputs of a model. For accurate and robust results, time and space modeling are crucial for MES models. However, acquiring sufficiently fine-resolution data for RES can be challenging. Traditional optimization models, that consider energy-based approaches, may not entirely encompass the resolution obstacles. Furthermore, assigning deviation ranges to input data and parameters is preferable, although challenges such as data quality, availability, and compliance with data protection laws may hinder access to necessary input information, which could be limited to aggregated levels. Unfortunately, omit comprehensive explanations of the approaches utilized to handle uncertainties pertaining to their input data, making it difficult or impossible to determine the uncertainty. Therefore, researchers must take measures to address uncertainties in their models by using formal methods and acquiring high-quality data from various sources to ensure that the MES models are accurate and robust [50].

- **Spatial and temporal requirement:** Energy supply may not always align with demand in terms of location and timing, which can result in an energy imbalance. This can be mitigated through spatial balancing via the grid or temporal balancing through energy storage, allowing for energy to be released at a later time when demand is higher [51, 52]. Nevertheless, highly detailed models with fine spatial and temporal resolution can demand excessive computational resources that may not be practical within a reasonable time frame. On the other hand, using a coarse resolution can reduce computational complexity; but, it can result in less accurate outcomes, as it averages out the extreme points that are crucial for designing the system [53]. MES models require high-resolution data, particularly for RES, to accurately represent the spatial and temporal variations in energy supply and demand. However, acquiring sufficiently fine-resolution data for RES can be challenging. Spatial resolution refers to the level of detail in geographic information. For example, solar irradiance varies with latitude, longitude, and altitude. Therefore, it is necessary to have high-resolution data on solar irradiance to accurately represent the potential energy yield of a solar power plant. However, as mentioned earlier, obtaining such data can be difficult and expensive, particularly in remote or inaccessible areas. Furthermore, the data may not be available at a small-enough scale to capture all the variations in the energy supply. It is essential to balance the level of detail required for accuracy with the computational cost of the model. Furthermore, models that do not fully incorporate the fluctuations in supply and demand may overstate the portion of demand fulfilled by renewable energies. Therefore, it is crucial to consider the full variability of energy supply and demand in MES modeling to accurately represent the performance of the system. In summary, space and time resolution is a significant challenge in MES modeling, requiring high-resolution data and balancing the level of detail required for accuracy with the computational cost of the model. It is essential to account for the entire spectrum of variation in energy supply and demand to accurately represent the performance of the system.

7.3 Integration of EVs in MES

Integrating EVs into MES involves incorporating these vehicles into existing energy systems that may include power, gas, and heat networks. This integration is important because changes in electricity load resulting from the adoption of EVs can affect the operation of energy conversion systems such as combined heat and power (CHP) systems, gas-fired generators, boilers, and others. To balance the supply and demand of the energy, EVs can be used as energy storage systems in MES [18, 19]. Some studies have explored the integration of EVs and renewable sources with combined power and gas networks to increase the share of renewable sources. Gas-fired generators have been shown to enhance the interdependency between natural gas and electricity networks. In addition, some studies have integrated residential charging stations with an integrated power and gas network, including hydrogen storage, CHP, gas-fired units, non-gas-fired units, and renewable sources. However, some studies have focused only on charging EVs and did not consider discharging them [54–59] have investigated EV charging in integrated power distribution and transportation networks, microgrids with buildings that have space heating, cooling, and electrical loads, and the presence of EVs in integrated power and district heat systems. The objective of integrating EVs into MES is to achieve better energy efficiency, reduce emissions, and create a more sustainable energy future. To successfully integrate EVs into MES, it is important to consider various technical aspects such as the capacity and power rating of charging stations, the control and coordination of charging schedules, and the impact of EV charging on the distribution network. By addressing these technical considerations, EVs can play an important role in balancing supply and demand in MES, enhancing the use of RES, and creating a more sustainable energy future.

7.3.1 Integration of EV with RES

Integrating EVs with RES is a critical component of the development of MES. By integrating EVs with RES, the overall energy system can be optimized for maximum efficiency and minimal environmental impact. RES like solar, wind, and hydroelectric power can provide a substantial amount of energy to power EVs, and by utilizing them, the carbon footprint of transportation can be reduced. EVs can serve as energy storage systems, allowing excess renewable energy generated during periods of low demand to be stored in EV batteries and used later when demand is high, which can address the issue of intermittency in RES. Figure 7.3 illustrates the overall framework that outlines the various RES, EH, and EV chargers. In different parts of the city, EV charging stations that are powered by EH will be strategically placed. These areas could include residential sites, workplaces, schools, hospitals, and other notable places where vehicles may be parked for extended periods. While superchargers exist for EVs that can charge a battery within 30 minutes, giving a range of 270 km, standard chargers (Level 2) typically provide an additional 18 km/hour of charging time [60, 61]. Hence, Level-2 charging stations are suitable for such locations. To establish this infrastructure, multiple charging points will be installed throughout the city in

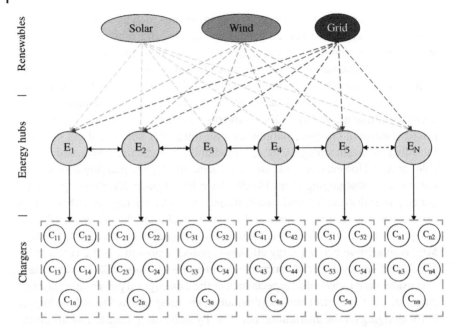

Figure 7.3 Overall framework for EV charging and energy infrastructure. *Source:* [60] / MDPI / CC BY 40.

locations where vehicles are likely to be parked for extended periods. These charging points will be energized by EH, which enables the integration of RES. In Figure 7.3, E indicates an EH at a specific site, like a rooftop, while C_{ij} among the green rectangle symbolizes each charger that connects to the EH. Moreover, apart from providing power to charge EVs, the energy generated by these hubs can also be utilized to meet some of the energy needs of the surrounding area [60].

Researchers have explored various scenarios, such as the optimal integration of EV charging stations without RES, the optimal integration of EV charging stations and RES with a controlled charging procedure, and the optimal integration of EV charging stations and RES with a controlled charging and discharging scheme. The most promising results were achieved with the optimal integration of EV charging stations and RES with a controlled charging and discharging strategy. This approach significantly reduced voltage deviation, energy not supplied, and total cost compared to the base case, with reductions of 73.97%, 50.43%, and 29.55%, respectively [62]. Research has primarily focused on the potential contribution of EVs to facilitate the integration of RES in the power system [63–65]. This research topic is centered on the smart grid paradigm, which uses communication, control technology, power-electronics technologies, and storage technologies to balance production and consumption at all levels. EVs can act as a controllable load or storage, charging or discharging part of their battery capacity back to the grid, according to the vehicle-to-grid (V2G) concept [64]. When integrating EVs with RES in MES, several technical tips must be considered. First, capacity planning should be considered to ensure that RES and the charging infrastructure can handle the load requirements of EVs. Second, implementing smart charging strategies that consider the availability of RES and the changing needs of EVs can help balance the supply and demand of energy and prevent the overloading of the power grid. Third, V2G technology should be considered, which allows EVs to charge from the grid and discharge energy back into the grid when needed, balancing the grid during peak demand, and supporting the integration of RES. Fourth, energy storage systems, such as batteries or hydrogen fuel cells, should be integrated to store excess renewable energy and use it to charge EVs, when needed. Finally, advanced system control and management tools should be implemented to monitor and optimize the operation of the MES, ensuring energy efficiency, reducing emissions, and enhancing the reliability and stability of the system. By considering these technical tips, EVs can be effectively integrated with RES in MES, providing more efficient and sustainable energy solutions while reducing carbon emissions.

7.3.1.1 Integration of EV with Wind Energy

The integration of EVs with wind power is advantageous for providing ancillary services due to their ability to respond quickly to changes in the grid's needs. Ancillary services are support services provided by power systems to ensure the security, stability, and reliability of the electric grid. These services include frequency regulation, voltage regulation, and reserve capacity, among others. EV owners may face a higher upfront cost than those, who opt for internal combustion engine (ICE) cars. However, Ref. [66] highlights how EVs can offer economic benefits in the long term. The study examined the potential for EVs to provide ancillary services in the western Danish power system, with a particular focus on the management of secondary reserves and load frequency control (LFC).

The authors used simulation models to demonstrate how EVs can efficiently regulate power imbalances arising from wind-power variability, reducing the reliance on conventional power plants. This could lead to lower costs for EV owners and a more sustainable energy system overall. In a similar vein, Ref. [67] introduces a concept whereby large clusters of EVs and household appliances could be leveraged to offer secondary reserves and LFC services in the power system. Simulation data suggests that EV batteries undergo extensive energy excursions, frequently transitioning from empty to full charge levels. The study highlights the potential of aggregating distributed energy resources for enhancing system stability and efficiency, while also underscoring the importance of battery management strategies for maintaining EV performance and longevity. The authors in Ref. [68] present an optimized operation model for EVs in the presence of wind turbines and dispatchable generation units. They evaluate the impact of coordinated and non-coordinated EV operations on the stochastic generation of wind turbines, using distribution functions to model wind turbine variability across various scenarios. In one scenario, they assumed that there are some EVs connected to the grid, but they are not coordinated in the operation with other units. In this case, it reveals that the absence of coordination resulted in a rise in the dispatchable unit generation, increased costs, and ultimately decreased profits for the system operator. In another scenario, the authors suppose that EVs are charged and discharged in coordination with other system units, utilizing coordinated scheduling to optimize energy utilization. This approach enables EVs to charge during periods of high wind turbine generation or low-power demand, while stored energy is discharged during peak hours when demand is high, or generation is low. Coordinated charge and discharge operations can enhance system efficiency and reduce total costs, with uncoordinated operations leading to increased power system stress and reduced profitability. The study emphasizes the importance of effective coordination strategies for maximizing energy utilization and achieving financial returns in the renewable energy sector. In Ref. [69], the authors examine the viability of coordinated EV load management for enhancing system frequency stability in microgrids with wind power, leveraging a droop control approach. Their findings indicate that coordinated EV load management can enable even greater wind power penetration levels. Based on this literature, however, the potential of EVs and power systems as controllable loads in simulation environments is highlighted, but some unresolved issues such as EV control requirements and element responses during coordination represent significant barriers to practical implementation and the need for continued research into effective coordination strategies for maximizing the potential of RES is underscored [70].

7.3.1.2 Integration of EV with Solar Energy

The research landscape regarding the integration of solar power and EVs is considerably more dynamic compared to that of wind power and EVs. Importantly, solar power can be harnessed at both medium and low voltage levels, offering greater flexibility within the power system. Moreover, this approach presents opportunities for integrating PV generation with EVs, further enhancing the potential of both technologies. These findings highlight the importance of considering a range of these RES in optimizing the functioning of EVs and the power network as a whole. Nevertheless, the concurrent incorporation of EVs and PVs into the power grid brings forth several challenges due to the uncertain power generation, dynamic loading conditions, and charging requirements.

In such a complicated and dynamic environment, appropriate scheduling and control schemes are required to ensure the stability and reliability of the electric system. Among various contributions suggesting environmentally conscious charging surveyed in Ref. [71], Bessa and Matos emphasize the importance of optimizing the cost-effectiveness of EV charging during the irradiation period. On the contrary, another valuable application is identified in Ref. [72], where the concept of charging EVs throughout the daytime at parking spots, located in workplaces, has been introduced. This approach enables EVs to be recharged completely during working hours, allowing for the implementation of the solar-to-vehicle approach. Additionally, research has shown that the energy generated in each parking area is crucial for generating sufficient electricity to meet the transportation requirements of EV operators. These contributions shed light on the potential of optimizing EV charging practices to enhance cost-effectiveness and efficiency while reducing carbon emissions. In Ref. [73], to determine the most effective configuration for meeting daily charging demands, a new solar-based grid-tied charging station (SGTCS) is proposed. The proposed method is evaluated through a techno-economic assessment that considers the technical, economic, and environmental impacts of a solar-powered charging station. The model is built to account for losses, and the results demonstrate a reduction in energy costs from $.200/kWh to $.016/kWh, as well as a decrease in grid load of 254,030 kWh/year. It also can support up to 7.7 charging sessions daily, using only 13% of the power supplied. The leftover 87% can be sold back to the grid, providing a significant source of revenue. The authors in Ref. [74] propose an SGTCS that maximizes the exploitation of solar energy by optimizing the charging of EVs through an advanced scheduling scheme. That study shows that the proposed grid-connected PV-based highway charging station provides electricity at a cost per unit of 0.05161 $/kWh, much less than the grid's cost of 0.20 $/kWh. It also concludes that the PV generates 64.5% of the power, creating 1,752,305 kWh/year. Consequently, only 34.5% of the electricity is drawn from the grid to fulfill the load requirement.

7.3.2 Integration of EV with Power Grids

The growing number of electric transportation options presents new challenges to electric energy systems, but EVs also provide significant advantages to power networks, as depicted in Figure 7.4. Through the continued development of EV networks, regulatory frameworks, and innovative technologies, the power grid can become more resilient, sustainable, and reliable, meeting the energy needs of society while reducing the environmental impact of transportation. In this regard, successful incorporation of EVs into the power grid requires the development of EV networks, which can contribute to mitigating harmonic distortions, providing reactive power, reducing peak demand, and more. To fully harness the potential of EVs, regulatory bodies specializing in EV aggregators are necessary. They aggregate EVs based on their owners' preferences to maximize business

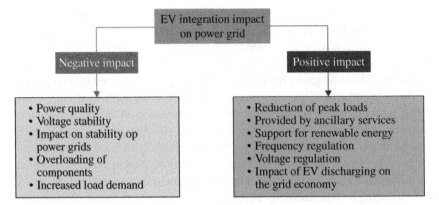

Figure 7.4 Advantages and disadvantages of integrating EVs into the power grid. *Source:* [68] / Springer Nature.

opportunities in the electricity sector. While EVs alone may not significantly impact the electricity industry, when coupled with energy storage systems like accumulators, they can significantly transform it.

As global climate change intensifies, integrated electrical energy systems, including cooling, heating, and electricity, are becoming increasingly important [18]. The integration of these systems enables multiple components to fulfill diverse requirements concurrently, making the incorporation of EVs essential. The fluctuation in electrical load leads to changes in energy conversion system performance, such as electric motors and boilers, further highlighting the importance of EV adoption in integrated energy systems. Moreover, as mentioned earlier, EVs can act as an energy storage system, helping to stabilize supply and demand. Additionally, Fattori et al. [56] examines the impact of EV penetration in electricity and heat-integrated energy systems on a regional level. To maximize the benefits of EV integration, numerous technical and logistical challenges must be addressed. One significant challenge is the potential increase in harmonic distortions due to the rapid charging of EVs. Another challenge is the difficulty in managing the high variability and uncertainty of EV charging demand, which can lead to an unstable power grid. To address these challenges, several academic investigations have examined the optimal scheduling of EV charging, leveraging machine learning techniques to predict EV charging demand, and the use of V2G technology to enable bidirectional power flow between EVs and the power grid.

Drawing on the above explanation, EV charging can be carried out using the central power grid or the distributed renewable/non-renewable energy resources. In the following, we review the role of EV smart charging into the distribution systems, microgrids, home and buildings, and VPP.

7.3.2.1 EV and Distribution Systems

EVs have the potential to function as a means of energy storage and supply the required energy to both the power grid and consumers, allowing them to function as distributed/decentralized energy units. Under such circumstances, EVs can contribute to the reliability, stability, and flexibility of the electricity distribution network, whenever required. The ability of EVs to store and supply energy during peak energy demand periods can have significant benefits for the electricity distribution system. During peak hours, electricity demand can exceed the available supply, leading to strain on the distribution infrastructure and potentially resulting in blackouts or brownouts. By using the stored energy in EVs to supplement the grid's energy supply during these times, the strain on the distribution infrastructure can be alleviated. This can help prevent costly upgrades to the infrastructure and improve the overall reliability of the electricity system. Additionally, EVs can provide backup power to critical infrastructure during emergencies, further enhancing the system's resiliency. Furthermore, the interconnectivity of EVs and the distribution network can also support the integration of RES in MES, such as solar and wind power. These renewable sources are unpredictable, meaning that their energy production fluctuates with weather conditions and time of day. However, by storing excess energy from renewable sources in EVs, the supply and demand of electricity on the grid can be balanced. This can reduce the need for fossil fuel-powered backup generators and increase the share of renewable energy in the electricity mix, contributing to a more sustainable energy system.

Reference [75] provides a thorough review of the modeling methods for grid-connected EV-PV charging systems. It outlines an overall structure for designing these systems, with a specific emphasis on incorporating smart charging algorithms. The article also provides a comprehensive analysis of the modeling techniques used to handle uncertainties associated with linking the EV and PV to the grid, such as EV demand, electrical load, and PV generation. In Ref. [76], the authors investigate the incorporation of parking depots with a large number of EVs into distribution networks. A solution for addressing the issue of dynamic power imbalance caused by the integration of EVs and PV systems is proposed in a recent study referred to as Ref. [77]. The study introduces a dynamic power balance system that models EV charging loads and PV generation in the distribution network, allowing for investigation and mitigation of imbalanced power flow among adjacent feeders. The aggregated EV charging loads are implemented in Monte-Carlo simulation, considering several factors that may affect the load profile. The potential role of battery-exchange stations in improving the reliability level of distribution systems was studied. The authors of Ref. [78] present a novel evaluation methodology that employs an analytical approach to provide fresh insights into the reliability assessment of V2G strategies. This approach effectively models the stochastic nature of vehicle owners' driving behaviors and fluctuations in system load, shedding new light on the reliability evaluation of V2G programs.

7.3.2.2 EV and Microgrids

To reduce reliance on fossil fuels, energy managers are incorporating greater amounts of RES into their production mix. This has led to the development of microgrids, which consist of intermittent microgenerators as a source of energy [32, 79]. Microgrids offer support for the smart charging of EVs, which can improve the reliability of the energy system while optimizing energy consumption and improving the economic aspects of energy management. By integrating EV charging with the microgrid, EV charging can be coordinated to take advantage of times when energy is most abundant and cheapest. In addition to their use in supporting EV charging, microgrids also offer a range of benefits for energy management. They can be designed to operate independently of the main power grid, providing backup power during emergencies or blackouts. Additionally, the incorporation of energy storage systems, such as batteries, can help to balance the supply and demand of energy on the grid, making the microgrid more resilient to disruptions. Moreover, microgrids can operate in a more sustainable and environmentally friendly manner by utilizing RES, such as solar and wind power. This can reduce the reliance on fossil fuels and contribute to a more sustainable energy system. Overall, by incorporating RES, energy storage systems, and coordination with EV charging, microgrids can provide a reliable, sustainable, and cost-effective source of energy.

Extensive research is conducted by researchers in Ref. [80] to explore the incentive schemes offered to EV owners for their participation in microgrid demand-response mechanisms. The authors in Ref. [81] design hybrid island systems that integrate EV parking and energy storage to decrease construction and operational expenses, while also considering various uncertainties. The research reveals that including EV parking in the design results in a considerable reduction in the costs associated with energy storage installation. Ref. [82] suggests incorporating solar PV microgrids as a means to meet the energy requirements of EV charging in Rwanda, an East African community nation. Using the Hybrid Optimization of Multiple Energy Resources (HOMER) Grid software [83], the financial implications of a scheduled EV charging station integrated with a grid-connected solar PV microgrid and storage are analyzed. It shows that the suggested method can reduce the overall electricity cost by 139.7%. In Ref. [84], an incentivized energy trading approach is introduced to study the interaction between EVs and critical load in microgrids. EV mobility and battery degradation are studied to ensure they do not deter EV participation. Bidder satisfaction is also introduced, which allows EV owners to enforce their energy trading conditions. An advanced stochastic optimization model, as detailed in Ref. [85], is devised to facilitate the control entity in effectively managing generation and storage through precise control of charging behavior of plug-in EVs. This study proposes a new strategy for reducing reliability costs to achieve the lowest total cost. In this regard, the V2G tool is used to reduce the overall system cost. The presented energy management model for the microgrid in the grid-connected mode takes into consideration the uncertainty of output power of WTs and PVs, as well as the EVs' charging and discharging.

7.3.2.3 EVs and Homes/Buildings

In the near future, a significant amount of EVs is anticipated to be integrated with residential or other buildings, enabling them to charge via either the central grid or decentralized energy systems. Through strategic charging and discharging at suitable times, EVs have the potential to decrease peak demand and lower the electricity costs of smart buildings. By working in harmony with the energy system, EVs can contribute to the total effectiveness of smart buildings and networks, maximizing their performance.

The authors in Ref. [86] present a multi-objective techno-economic optimization framework in pursuit of the maximum profitability of interconnected buildings with bilateral agreements and coordinated EV charging. The framework proposes that primary and secondary buildings agree on a minimum contracted load and contract tariff based on the varying load shape of the secondary building(s). The study found that a fleet of EVs can be optimally charged through a combination of grid, solar PV, or energy storage systems at the parking station located in the primary building. This fleet of EVs can then sell electricity to local loads in the primary and secondary buildings, storage, and the grid, depending on their power state, and arrival and departure times. The research results also show that the primary building can experience up to 62% daily revenue after accounting for the deployment costs of solar, storage, and charging stations. Moreover, the secondary buildings without solar, storage, and charging facilities can earn up to 20% cost savings based on the nature of their bilateral agreements with the primary building. Reference [87] describes a comprehensive study of integrating EVs with homes that have wind turbines, energy storage, and combined heat and power generators to optimize energy consumption. The authors conducted extensive research to explore the potential benefits of combining these technologies to enhance energy efficiency and reduce costs.

The authors in Ref. [88] present a two-level distributed deep reinforcement learning (DRL) algorithm that focuses on minimizing electricity costs by optimizing the scheduling of energy consumption for two controllable home appliances (an air conditioner and a washing machine), as well as managing an ESS and an EV charging/discharging. This algorithm ensures that the satisfaction of the consumer and the operational features of the appliances are preserved, while effectively reducing the overall electricity costs associated with these devices. The comparative case studies under different weather and driving patterns of the EV with different initial states of energy confirm that the proposed approach can successfully minimize the cost of electricity within the consumer's preference. Hou et al. [89] present a comprehensive approach for scheduling various physical equipments involved in home energy management systems (HEMS) that center on the preferences of users. Additionally, it introduces a specially developed charging/discharging approach for both ESS and EVs while taking into account their upfront costs to incorporate them into HEMS. This strategy provides greater flexibility and economic benefits and ensures that battery life is prolonged.

7.3.2.4 EV and EH

As was mentioned in Section 7.2, in the modeling of MES, an EH serves as a central facility that efficiently manages energy supply and demand by integrating various energy sources and carriers. This is achieved using advanced optimization algorithms and control strategies, which help to balance the flow of energy across different sectors, including residential, commercial, industrial, and transportation. EH can incorporate different energy sources, including renewable sources like solar, wind, and geothermal, as well as non-renewable sources like fossil fuels and nuclear power. The integration of EV into EH requires careful energy consumption management plans to maintain the demand balance. This integration can potentially upset the equilibrium in such systems, emphasizing the importance of developing optimal strategies for managing the energy consumption of EV within EH.

In Ref. [90], the energy management of an EH integrated with an EV is investigated. The study establishes optimal operational costs by taking into account the thermal loads and flexible power requirements. It also involves extensive research, and the results highlight the importance of developing strategies that ensure the efficient use of energy resources in EH while maintaining optimal operational costs. In Ref. [91], an optimal load dispatch form for an EH to reduce the total costs of the EH, such as CO_2 emission costs is addressed. The considered EH includes a heat storage unit, CHP unit, PV arrays, gas boiler, wind turbine, and EV. The proposed optimization algorithm models EV uncertainty via Monte Carlo simulation and takes into account the electric and thermal demand-response (DR) schemes extensively. It shows that the EV coordinated charge/discharge mode can reduce the total costs by 12% compared to the uncoordinated charge mode. The results also showed that by implementing the DR methods, the total cost can be reduced by about 5%. In Ref. [92], research is carried out on decreasing the purchase cost and tax of building an EH with EVs.

7.3.2.5 EV and Virtual Power Plants

A VPP is a system of distributed and moderately sized power generation facilities, power consumers, and flexible storage systems that are aggregated and managed as a single entity to provide grid services, such as balancing supply and demand and providing ancillary services. The resources that make up a VPP can include a variety of DERs, such as solar panels, wind turbines, energy storage systems (such as batteries), and demand-response programs. These resources are connected to the grid and can be controlled through a centralized platform that enables the VPP operator to monitor and optimize their performance in real time.

The key advantage of a VPP is its ability to provide a flexible and reliable source of power that can help to balance the grid and support the integration of intermittent RES. For example, a VPP can store excess energy produced by photovoltaic systems during daylight hours and release it during peak demand periods in the evening, reducing the need for expensive and polluting peaking power plants. In addition to providing grid services, VPPs can also offer benefits to energy consumers. EVs can play a dual role in a VPP as both means of large-scale energy storage and consumer of energy for transportation. Along with DER, EV integration into VPP contributes to both sides. Integrating EVs with VPPs contributes to stabilizing electricity demand by compensating for instantaneous increases in electricity consumption within a building, thus reducing the additional energy drawn from the grid. Treating EVs as custom power devices connected to the grid helps mitigate issues arising from the grid-to-consumer interface. Moreover, balancing electricity demand in residential areas with the use of EVs enhances grid flexibility and reduces imbalances [93]. Leveraging EVs as virtual power plants involves supplying electrical energy from EVs to the grid during optimal times, offering multiple benefits in addition to charging the energy units within the system [94].

In Ref. [95], the authors provide a comprehensive analysis of the different ways EVs can be integrated into VPPs, including the arrangement of the system, interface designs, marketing strategies, and future outlooks. The authors examine the integration concepts of vehicle-grid integration (VGI) in stand-alone, grid-connected, transitional, and grid-supported operations. They also discuss VGI topologies in terms of the types of energy generation/storage units used in EVs, as well as the single-stage, two-stage, and hybrid-multi-stage systems employed, and the types and parameters of grid connections. Marra et al. [96] examines the impact of EVs on the frequency response of VPPs, as well as their potential to improve power storage capabilities. An all-encompassing scheme for stochastic optimization is presented in Ref. [97] for the coordinated operation of a fleet of EVs with a wind power producer (WPP) in a three-settlement pool-based market. An aggregator acquires sufficient energy for the EVs by examining their daily driving patterns and efficiently manages the stored energy to offset fluctuations in the WPP generation.

7.3.3 EV Charging/Discharging Strategies

EV fleets can offer a large storage capacity for electricity. Nevertheless, the predominant effective charging behavior will be influenced by the energy mix of the power grid. The integration of EVs will differ between regions where solar-based generation is more prevalent compared to regions where wind power is more dominant. Adopting smart charging techniques can enable EVs to serve as a flexible resource, reducing the requirement for investment in carbon-intensive fossil-fuel power plants. Implementing this approach promptly, it can play a significant role in balancing renewable energy systems. Several factors can be considered when analyzing the effect of EVs on the power grid. These include the level of EV adoption, charging strategies, characteristics of EV batteries, charging locations, charging patterns, charging times, the state of charge (SoC) of the battery, the charging patterns of EV fleets, driving patterns of EVs, driving distances, tariff schemes, and demand-response techniques. By examining these factors, a comprehensive assessment can be made of the impacts that EVs have on the power network. Uncoordinated EV charging can result in the equivalent of a large electric load, leading to higher power-system peak loads and distribution grid congestion issues [98–100].

Figure 7.5 demonstrates the consequences of uncoordinated EV charging on generation, transmission, and distribution systems. In Ref. [101] a thorough analysis of the implications of uncontrolled EV charging, and highlights of the advantages of implementing intelligent charging methods and V2G is presented. This

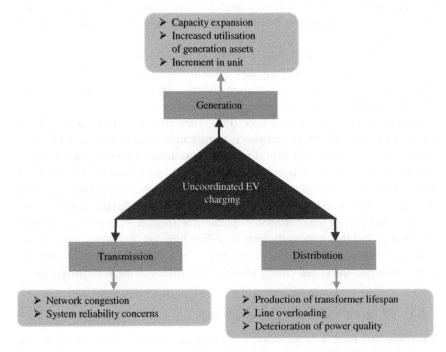

Figure 7.5 Consequences of uncoordinated EV charging on grid. *Source:* Adapted from [101].

study reveals that unmanaged charging of EVs leads to various problems such as increased peak load, high power losses, voltage violations, voltage unbalance, reduction in transformer life span, and harmonic distortion. On the other hand, it is found that smart charging can address these issues while also providing a range of economic, social, and environmental benefits. The study emphasizes the crucial role of smart charging in achieving zero net carbon emissions by the transportation sector. Additionally, the research investigates the necessary policy, regulatory frameworks, and standards required to promote the adoption of smart charging and V2G amid EV owners and charging infrastructure designers. Such publications provide valuable insights into these issues.

Smart charging is a flexible and adaptable approach for charging EVs that takes into consideration both the needs of vehicle users and the conditions of the power system. By enabling a degree of management over the EV charging procedure, smart charging helps to integrate EVs into the power system while meeting mobility needs. This approach includes various pricing and technical charging options, such as time-of-use pricing that encourages consumers to charge their vehicles during off-peak periods. However, as EV penetration levels increase, more advanced smart-charging mechanisms will be required to deliver close-to-real-time balancing and ancillary services. The primary types of advanced smart charging include V1G, V2G, V2H, and V2B, as explained in Figure 7.6. These approaches offer various possibilities to enhance the flexibility of power systems and facilitate the integration of variable renewable energy (VRE), such as wind and solar PV.

Smart charging has the potential to give flexibility both at the system and local levels, as illustrated in Figure 7.7. At the system level, smart charging can help balance the wholesale market by controlling EV charging patterns to flatten peak demand, fill load valleys, and enable the dynamic adjustment of EV charging levels to facilitate real-time grid balancing. V2G charging can also enable EVs to inject electricity back into the grid, providing ancillary services to transmission system

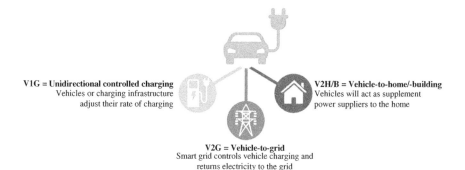

V1G = Unidirectional controlled charging
Vehicles or charging infrastructure
adjust their rate of charging

V2H/B = Vehicle-to-home/-building
Vehicles will act as supplement
power suppliers to the home

V2G = Vehicle-to-grid
Smart grid controls vehicle charging and
returns electricity to the grid

Figure 7.6 Advanced smart charging mechanisms. *Source:* [102] / IRENA.

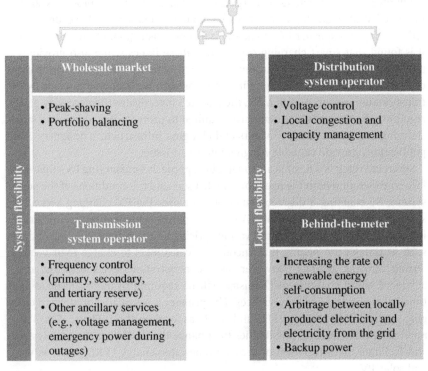

Figure 7.7 Diverse flexibility capabilities of EVs. *Source:* [102] / IRENA.

operators. Smart charging can assist distribution system operators in managing congestion and can also help customers manage their energy consumption, increasing their rates of renewable power self-consumption. Overall, smart charging represents an important tool for enhancing grid flexibility and enabling the integration of RES [102].

7.3.3.1 Vehicle-to-Everything (V2X)

Based on the previous explanation, V2X refers to the interaction between EVs, other entities, and the infrastructure of a smart city. It can be in the form of vehicle-to-vehicle (V2V), vehicle-to-infrastructure (V2I), vehicle-to-home (V2H), vehicle-to-grid (V2G), vehicle-to-network (V2N), vehicle-to-building (V2B) or vehicle-to-load (V2L). Regarding the interaction of EVs with the power grid, which is the focus of this report, V2G, V2H, and V2L are considered. In this context, EVs are considered mobile energy storage systems that can locally support the power grid to balance supply and demand. Like battery storage systems, these technologies can contribute to the grid by load leveling, voltage and frequency regulation, and peak load shaving. However, several technical and economic challenges should be solved to achieve the benefit. The viability of V2G is impacted by several factors, including the availability of EVs, which relies on the acceptance of this emerging technology by vehicle owners, driving habits, willingness to engage in V2G schemes, infrastructure readiness (e.g., availability of chargers), technical limitations (e.g., battery degradation), market preparedness, and adherence to technical standards. V2G can provide valuable support to the grid for voltage and frequency ancillary services, better demand-side management, compensating for variable RES, and reducing the infrastructure upgrade cost compared to the G2V-only cases. Indeed, the V2G technology is a flow of energy, information, and money between the EV owners, aggregators, and the power grid to make a stable balance between the demand and supply. Despite these opportunities, implementing V2G has clear consequences and costs for EV owners and grid operators, such as battery degradation, the necessity of sophisticated communication among EVs, and grid infrastructure upgrades. There are also social, political, cultural, and technical obstacles [103]. Failure to optimally integrate V2G may result in overloaded transformers, decreased system efficiency, and voltage and frequency perturbations resulting in disruptive harmonics in the grid. V2G may cause a reduction in power quality, if not well managed. It can cause voltage compliance issues, harmonics, and overloading of transformers. In the past, utility companies have typically employed devices such as voltage regulators, capacitor banks, and transformer taps to enhance power consistency. "Flexibility" is an important factor in EV smart charging that can be an alternative to alleviate power-quality issues. A US study shows that average personal vehicles are engaged in travel only 4–5% of the time during the idle hours spent in home garages or parking lots throughout the day [104, 105].

V2G-enabled EVs can support the grid by frequency/voltage services, support for intermittent solar/wind power, reactive power support, load balancing, valley filling, and peak shaving. In order to use V2G for ancillary services, we need to have a sufficient number of V2G-enabled vehicles simultaneously connected to the power network, which can be the case for shopping-center parking with EV chargers. Therefore, these locations are the first candidates for V2G trials. In addition, for V2G technology to be effective, the EV battery must possess adequate capacity to meet travel requirements and allow for the injection of surplus energy into the power network.

7.3.3.2 Smart Bidirectional Charging

In general, aggregators can control the charging/discharging of EVs in different control architectures as follows:

- **Centralized control approach:** In an MES, the central controller would need to balance the supply and demand of electricity from all sources, including the changing needs of EVs. The central controller collects information from EVs about its power requirements and communicates with each aggregator to ensure that the appropriate amount of power is being supplied to meet the changing needs of all EVs. Centralized charging is a type of EV charging management system that uses a central controller to determine the timing and charging rate of multiple EVs. Under the centralized model, the optimization of EV-charge scheduling is executed centrally at the aggregator level after accumulating information about the power requirements of the EVs. The EVs can provide information such as maximum battery capacity, state of charge (SOC), and charge rate to the aggregator, which then contracts with the independent system operator (ISO), based on the aggregated power requirements [106]. The ISO receives the agreements from several aggregators and decides the efficient and optimal power share for each aggregator within their agreed boundary, taking into account other loads, generation capacity, and grid constraints. Each aggregator then implements an optimization routine to schedule the EV charging to meet the anticipated energy requirements of the EVs.

Centralized charging has several advantages, including improved efficiency of the charging process, reduced stress on the power grid during peak demand periods, and increased use of RES. However, it requires a high level of coordination and communication between the central controller and the EVs, as well as a robust and flexible charging infrastructure capable of supporting multiple EVs charging simultaneously. It should be noted that the size of the optimization problem will increase with the number of EVs in an area, and control architectures have been developed to address this issue by separating the loads associated with EVs based on their geographic location, with localized controllers overseeing the distributed power management for different categories. This differs from the central controller, which only controls the group's demand [107, 108]. In other words, centralized approaches may lose their effectiveness and become impractical as the number of grid-connected EVs grows. Scalability poses a significant challenge for centralized approaches, particularly as the number of connected EVs and the planning time horizon increase, resulting in a larger optimization problem. This can potentially make a centralized approach computationally intractable and lead to long implementation times. Moreover, as the quantity of variables to control and limitations for each EV grow, the complexity of the problem increases. Additionally, the centralized approach requires complete charging requirements and technical specification information from EV users, which can create practical difficulties such as communication bottlenecks, bandwidth limitations, and costly infrastructure expansion to handle the increased data from the growing number of EVs [109].

- **Decentralized control approach:** In the decentralized charging method, each EV comes equipped with its own computing capabilities and communicates its net energy needs to an aggregator. The aggregator then utilizes this information to determine an optimal charging schedule [110]. Decentralized control offers numerous benefits, including scalability, enabling the scheduling process to accommodate a large number of EVs [106]. Despite its advantages, decentralized charging control encounters difficulties due to inadequate information at the level of each EV. This limitation can lead to suboptimal charging schedules. In contrast to centralized systems, decentralized systems assign individual decision-making roles to each EV, enabling them to solve their own small-scale problems independently. However, this approach does not guarantee optimal charging schedules, particularly when complete information is unavailable at the level of each EV. Despite the challenges, decentralized solutions are gaining popularity because of their scalability and feasibility for implementation in the field. In an MES, there are two types of decentralized control architectures [109]. The first type, T1, features a center-free design, with each EV locally computing and adjusting its schedule by exchanging information with other EVs until a global equilibrium is reached. However, this approach comes with a significant communication overhead, especially with many EVs. In contrast, the second type, T2, involves an indirect aggregator that gathers specific information and transmits coordination signals to all EVs. This design significantly reduces communication overheads and can handle a larger number of EVs in the scheduling process.

Charging schemes that operate in a decentralized manner can demonstrate greater resilience in the face of network failures, especially if the controllers are programmed to function even when the centralized communication system fails. In an MES, it is crucial to consider the integration of RES and the impact of EV charging on the overall system, including other loads and grid constraints. Decentralized control architectures should also be designed to match the grid rules and the electricity tariff mechanisms to ensure proper integration of EV loads with the overall system.

- **Hierarchical control approach:** The hierarchical control method for EV charging in MESs is a strategy that involves dividing the charging process into two levels of control. The first level is the local control level, where individual EVs make charging decisions based on their own needs and constraints. The second level is the global control level, where an aggregator or central controller manages the charging process by coordinating the charging decisions of individual EVs. At the local control level, each EV uses its own decision-making capability to determine when to charge, how much to charge, and for how long. This decision is based on factors such as the battery state of charge, charging station availability, and user's preference. The local controller communicates with the global controller to provide information on the EV's charging needs and constraints. At the global control level, the aggregator uses the information from individual EVs to create an optimal charging schedule that takes into account the availability of energy sources, such as renewable energy, and the constraints of the grid. The global controller also ensures that the charging process is coordinated to avoid overloading the grid and causing blackouts. Hierarchical control methods offer several advantages over other control strategies. In contrast to centralized systems, hierarchical systems distribute control and computational burden among multiple aggregators using a tree-like communication topology. This approach minimizes the requirement for network-wide communication, as each aggregator manages a group of EVs and influences the decisions of other aggregators. The hierarchical architecture combines the benefits of centralized and decentralized architectures in a unique way, providing a balance between the two [109]. Moreover, they allow for a greater level of flexibility and responsiveness to changing conditions in the system. They can also improve the overall efficiency of the charging process by optimizing the use of available energy sources and reducing the risk of overloading the grid. Additionally, hierarchical control methods are well suited for large-scale deployments of EVs, where the number of individual EVs and charging stations is high. However, hierarchical control methods also have some drawbacks. They can be complex to implement and require significant computational resources to manage the charging process. The communication between the local and global controllers can also introduce delays and increase the overall cost of the charging process. Despite these challenges, hierarchical control methods remain a promising approach for managing the charging of EVs in MES.

- **Static approach:** The static scheduling model for EV charging in MES is a method of charging EVs that treats the EVs as stationary loads, ignoring their mobility characteristics [111]. This means that the EVs are assumed to be stationary, and the charging schedule is determined without considering their mobility patterns, such as when they will be driven and how far they will travel. The advantage of the static scheduling model is its simplicity, which makes it easy to formulate and solve the problem of charging multiple EVs in an MES. However, this model lacks realism as it does not consider the dynamic nature of EV mobility, such as changes in driving patterns or unexpected charging needs. Therefore, it may not provide an accurate representation of the actual charging needs of EVs in a real-world setting. Despite its limitations, the static scheduling model can be useful for investigating the impact of other parameters on the power grid, such as the effect of different charging rates or the use of RES for charging. It can also be used as a baseline for comparing the performance of other charging models that consider the mobility of EVs.

- **Mobility aware approach:** The mobility aware scheduling model is an advanced method of scheduling EV charging that considers various aspects of mobility and uncertainty associated with EVs. This approach considers factors such as the arrival and departure times of EVs at charging stations, the trip history of EVs, and the possibility of unplanned departures. This method adds a realistic aspect to the charging scheduling problem, making it more suitable for real-world scenarios. By modeling the spatiotemporal behavioral traits of EVs, the mobility aware scheduling model provides a more accurate representation of the charging load on the grid. This modeling technique can predict the expected arrival times of EVs at charging stations and the impact of their charging requests on the grid load. Such information can be used to optimize the charging schedules, reduce the impact on the grid, and ensure efficient charging demand management.

The utilization of intelligent V2G technology presents a robust opportunity for demand response, resulting in decreased peak loads and improved asset utilization within the network (i.e., relieving the network congestion and reducing network upgrade necessity). However, customer engagement in V2G programs depends on how the smartness of these algorithms guarantees benefits to EV owners, considering battery degradation. For example, aggregators can consider limitations for battery availability to address the owner's usage requirements for their EV. As a part of its Powerloop offering, Octopus offers an application that allows customers to input their charging schedule and receive cashback only if they fulfill a minimum number of V2G sessions each month. Additionally, the application enables customers to override V2G sessions if they choose to do so. The agreement between Octopus and its customers guarantees a minimum battery state of charge of 30% [112]. The inclusion of an override option is crucial as it gives EV owners a sense of complete control, even if they choose not to use it. However, the app-based model of EV chargers assumes a single user, which may not be ideal for households with multiple members sharing a car. To address this limitation, some chargers, like the Indra V2G charger used in Ovo Energy's program, feature a boost mode that can be activated through the app or directly via the charger's boost button, overcoming this issue [113]. Figure 7.8 showcases REVOLVE, the foresight optimization model used in the Sciurus trial, capable of simulating the charging and discharging behavior of a large number of EVs at half-hourly intervals over a year. This model optimizes the charging and discharging behavior of individual EVs based on minimum cost considerations, using import and export tariffs available to the EVs. It covers an entire year by optimizing weekly blocks one at a time. While algorithms and programs to address owner requirements primarily fall under retail or distribution concerns, the structure of the system that enables V2G services to the transmission grid, involving EV owners, DNSPs, and aggregators, should be regulated [115].

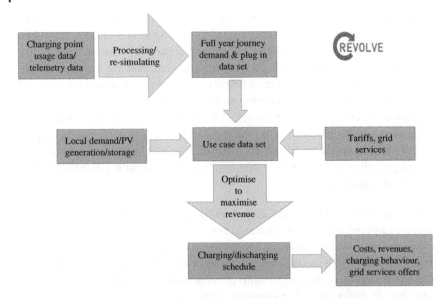

Figure 7.8 REVOLVE EV charging/discharging optimization model diagram.
Source: Adapted from [114].

7.4 Conclusion

This chapter presents an overview of various aspects of MES and its integration with EVs. MES is a system of integrating several energy networks, such as natural gas, transportation, and electricity distribution networks, to promote efficient energy use. EVs are a component of MESs and their integration into the MESs is important because changes in electricity load resulting from the adoption of EVs can affect the operation of energy conversion systems such as CHP systems, gas-fired generators, boilers, and others. To balance the supply and demand of energy, EVs can be used as energy storage systems in MES as they provide an opportunity for sustainable transport and decarbonization of the transport sector. The increased use of EVs for charging and routing will have a significant impact on the energy management of EH and stations. This chapter reviews original research works related to modeling, management, and intelligent controls of MESs integrating EV routing and charging. In light of the knowledge acquired through this review, the challenges as well as positive and negative effects of integrating EV within MES can be summarized as follows:

Challenges of integrating EVs within MES:
- *Charging infrastructure:* Establishing an extensive and reliable charging infrastructure is a key challenge. This includes deploying various types of charging stations, such as slow chargers, fast chargers, and ultra-fast chargers, at

strategic locations to meet the charging needs of different types of EVs. Building and maintaining the charging infrastructure can be capital-intensive and require coordination with multiple stakeholders, including utilities, local authorities, and charging equipment manufacturers.

- *Grid integration:* The integration of EVs with the electric grid requires careful planning and coordination to manage the increased demand for electricity. The charging of a large number of EVs simultaneously can cause load peaks, leading to grid instability, voltage fluctuations, and potential grid overloading. Grid upgrades may be necessary to handle the increased demand, and advanced grid management techniques, such as demand response, load shifting, and smart grid technologies, may need to be employed to ensure smooth integration.

- *Energy management and optimization:* Effective energy management and optimization strategies are crucial for integrating EVs within MESs. This includes scheduling the charging and discharging of EVs to align with renewable energy generation, grid demand, and other energy sources. Advanced energy management systems, including predictive analytics, machine learning algorithms, and optimization algorithms, may be required to optimize energy flows, storage, and distribution among various energy systems within the MES.

- *Interoperability and standardization:* Interoperability and standardization of EV-charging equipment, communication protocols, and data formats are critical for seamless integration. Lack of standardization can lead to compatibility issues, hinder interoperability, and limit the effectiveness of integrating EVs with other energy systems. The adoption of common standards, such as the Open Charge Point Interface and the Open Charge Point Protocol, can facilitate interoperability and simplify integration.

- *Cost and economic viability:* The cost of integrating EVs within MESs can be a significant challenge. It includes the initial costs of setting up charging stations, grid upgrades, and energy management systems, as well as ongoing operational and maintenance costs. Economic viability needs to be carefully evaluated, considering factors such as capital costs, operational costs, revenue-generation potential, and return on investment to ensure the sustainability of the integrated system.

Positive Effects of integrating EVs within MESs:

- *Renewable energy integration:* EVs can act as energy storage devices, enabling better integration of intermittent RES like solar and wind power. EVs can charge during times of excess renewable energy generation, store the energy in their batteries, and discharge it during periods of high demand or low renewable energy availability, helping to balance the grid and increase the utilization of renewable energy. This can contribute to reduced reliance on fossil fuels and lower greenhouse gas emissions.

- *Demand-side management:* EVs can be used for demand-side management by scheduling their charging/discharging to align with grid demand and other energy systems. This can help optimize energy consumption, reduce peak loads, and lower electricity costs by charging during off-peak hours or when renewable energy generation is high, thereby reducing strain on the grid. Advanced energy management techniques, such as V2G technology, can enable bidirectional power flow between EVs and the grid, allowing EVs to serve as a flexible load or a source of energy to support grid stability.
- *Reduced emissions:* EVs are known for their zero tailpipe emissions, which can aid in the reduction of air pollution and the mitigation of greenhouse gas emissions. When integrated into MESs powered by RES, EVs can contribute to improved air quality and mitigated climate change impacts.
- *Increased energy efficiency:* MESs can optimize energy use by integrating EVs into the overall energy management strategy. For example, EVs can be charged during periods of excess renewable energy generation, thereby utilizing surplus energy that would otherwise be wasted. Additionally, EVs can recover and store energy during regenerative braking, which can increase the overall energy efficiency of the transportation system and reduce energy waste.
- *Diversification of energy sources:* Integrating EVs within MESs can diversify energy sources and reduce dependence on fossil fuels. EVs can be charged from various energy sources, including renewable energy, grid electricity, and on-site generation such as solar panels, allowing for a more diverse and distributed energy supply. This can enhance energy security and resilience, reducing the reliance on a single energy source and providing flexibility in the energy supply chain.

Negative Effects of integrating EVs within MESs:

- *Grid overloading and instability:* The increased demand for electricity from charging multiple EVs simultaneously can strain the electric grid and lead to overloading and instability. This can cause voltage fluctuations, grid congestion, and potential disruptions in energy supply. Grid upgrades and advanced grid management techniques may be required to address these challenges and ensure grid stability.
- *Complex energy management and control:* Integrating EVs within MESs requires sophisticated energy management and control strategies. Coordinating the charging and discharging of EVs, aligning with renewable energy generation and grid demand, optimizing energy flows, and maintaining energy balance among various energy systems can be complex and challenging. Advanced energy management systems and expertise in optimizing

multi-energy systems may be required, which can increase operational complexities and costs.

- *Cost and return on investment:* The initial setup costs of charging infrastructure, grid upgrades, and energy management systems can be significant, and the return on investment may not be immediate. It may take time to recover the costs through revenue generation from charging services or other mechanisms. The economic viability of integrating EVs within MESs needs to be carefully evaluated to ensure sustainability and financial feasibility.
- *Interoperability and standardization challenges:* Lack of interoperability and standardization among different EV charging equipment, communication protocols, and data formats can hinder seamless integration. Incompatibility issues can result in limitations in interoperability, data exchange, and system efficiency. Adoption of common standards and protocols across different stakeholders and technologies may be necessary to overcome these challenges.
- *Technology and infrastructure limitations:* The current technology and infrastructure for EVs, including charging stations, battery capacity, and charging speed, are still evolving and may have limitations. This includes challenges such as limited charging station availability in certain areas, long charging times for some types of EVs, and limited battery capacity for long-distance travel. These limitations can impact the convenience and practicality of using EVs within MESs.

Based on the information provided in this chapter, some potential recommendations related to modeling, management, and intelligent controls of MESs integrating EV routing and charging can be concluded as follows:

- Develop more comprehensive and accurate models for MESs that integrate EV routing and charging. These models should consider various factors such as energy demand, energy supply, energy storage, and the behavior of EVs, as well as the interactions between these factors.
- Design and implement effective management strategies for MESs that can handle the improved utilization of charging patterns of EVs. These strategies should aim to balance the energy supply and demand of the entire system, minimize energy waste, and ensure the reliable and efficient operation of the system.
- Use intelligent control techniques such as artificial intelligence and machine learning to optimize the performance of MESs. These techniques can help to predict energy demand and supply, optimize energy generation and storage, and improve the overall efficiency and reliability of the system.
- Develop and test new technologies for energy conversion, storage, and distribution that can better integrate with EVs. This could include the development of new battery technologies, energy storage systems that can be integrated with

EVs, and smart charging systems that can better coordinate the charging of multiple EVs.

- Conduct more research on the social, economic, and environmental impacts of MESs integrating EV routing and charging. This research should focus on the potential benefits and challenges of these systems, as well as the trade-offs between different design and operation strategies. This information can help policymakers and stakeholders make informed decisions about the development and deployment of these systems.

References

1 World Bank (2023). https://www.worldbank.org/en/topic/urbandevelopment/ overview#:~:text=Today%2C%20some%2056%25%20of%20the,people%20will% 20live%20in%20cities. (accessed 11 Nov 2023).

2 United Nations Environment Programme. Global initiative for resource efficiency in cities. www.unep.org/pdf/GI-REC_4pager.pdf (accessed October 2014).

3 European Environment Agency (2013). Overview of the European energy system. www.eea.europa.eu/data-and-maps/indicators/overview-of-the-european-energy-system/assessment (accessed October 2014).

4 Van Beuzekom, I., Gibescu, M., and Slootweg, J.G. (2015). A review of multi-energy system planning and optimization tools for sustainable urban development." *2015 IEEE Eindhoven PowerTech*, 1–7. Eindhoven, Netherlands.

5 Wang, D., Zhi, Y.Q., Jia, H.J. et al. (2019). Optimal scheduling strategy of district integrated heat and power system with wind power and multiple energy stations considering thermal inertia of buildings under different heating regulation modes. *Applied Energy* 240: 341–358.

6 Xie, S., Hu, Z., Wang, J., and Chen, Y. (2020). The optimal planning of smart multi-energy systems incorporating transportation, natural gas and active distribution networks. *Applied Energy* 269: 115006.

7 Beigvand, S.D., Abdi, H., and La Scala, M. (2017). A general model for energy hub economic dispatch. *Applied Energy* 190: 1090–1111.

8 Onen, P.S., Mokryani, G., Zubo, R.H.A. et al. (2022). Planning of multi-vector energy systems with high penetration of renewable energy source: a comprehensive review. *Energies* 15 (15): 5717.

9 Hussain, B. and Thirkill, A. (2018). *Multi-Energy Vector Integration Innovation Opportunities – Preliminary assessment of innovation opportunities for SMEs.* Catapult Energy Systems: Birmingham, UK.

10 O'Malley, M., Kroposki, B., Hannegan, B. et al. (2016). *Energy Systems Integration. Defining and Describing the Value Proposition.* Golden, CO: National Renewable Energy Lab. (NREL).

11 Zarif, M., Khaleghi, S., Javidi, M.H. et al. (2015). Assessment of electricity price uncertainty impact on the operation of multi-carrier energy systems. *IET Generation, Transmission and Distribution* 9 (16): 2586–2592.

12 Ata, M., Erenoğlu, A.K., Şengör, İ. et al. (2019). Optimal operation of a multi-energy system considering renewable energy sources stochasticity and impacts of electric vehicles. *Energy* 186: 115841.

13 Reynolds, J., Ahmad, M.W., Rezgui, Y. et al. (2018). Holistic modelling techniques for the operational optimisation of multi-vector energy systems. *Energy and Buildings* 169 (15): 397–416.

14 Huang, W., Zhang, N., Yang, J. et al. (2019). Optimal configuration planning of multi-energy systems considering distributed renewable energy. *IEEE Transactions on Smart Grid* 10 (2): 1452–1464.

15 Son, Y.G., Oh, B.C., Acquah, M.A. et al. (2021). Multi energy system with an associated energy hub: a review. *IEEE Access* 9: 127753–127766.

16 Huang, W., Du, E., Capuder, T. et al. (2021). Reliability and vulnerability assessment of multi-energy systems: an energy hub based method. *IEEE Transactions on Power Systems* 36 (5): 3948–3959.

17 Lan, Y., Guan, X., and Wu, J. (2016). Rollout strategies for real-time multi-energy scheduling in microgrid with storage system. *IET Generation, Transmission and Distribution* 10: 688–696.

18 Sadeghian, O., Oshnoei, A., Mohammadi-Ivatloo, B. et al. (2022). Concept, definition, enabling technologies, and challenges of energy integration in whole energy systems to create integrated energy systems. In: *Whole Energy Systems: Bridging the Gap via Vector-Coupling Technologies* (ed. V. Vahidinasab and B. Mohammadi-Ivatloo), 1–21. Springer International Publishing.

19 Sadeghian, O., Oshnoei, A., Mohammadi-Ivatloo, B. et al. (2022). A comprehensive review on electric vehicles smart charging: solutions, strategies, technologies, and challenges. *Journal of Energy Storage* 54: 105241.

20 Wang, S., Luo, F., Dong, Z.Y. et al. (2019). Joint planning of active distribution networks considering renewable power uncertainty. *International Journal of Electrical Power & Energy Systems* 110: 696–704.

21 Wang, S., Dong, Z.Y., Chen, C. et al. (2020). Expansion planning of active distribution networks with multiple distributed energy resources and EV sharing system. *IEEE Transactions on Smart Grid* 11 (1): 602–611.

22 Luo, L., Gu, W., Wu, Z. et al. (2019). Joint planning of distributed generation and electric vehicle charging stations considering real-time charging navigation. *Applied Energy* 242: 1274–1284.

23 Wei, W., Mei, S., Wu, L. et al. (2017). Optimal traffic-power flow in urban electrified transportation networks. *IEEE Transactions on Smart Grid* 8 (1): 84–95.

24 Xie, S., Hu, Z., Wang, J. et al. (2019). Scenario-based comprehensive expansion planning model for a coupled transportation and active distribution system. *Applied Energy* 255: 113782.

25 Matanov, N. and Zahov, A. (2020). Developments and challenges for electric vehicle charging infrastructure. *12th Electrical Engineering Faculty Conference (BulEF)*, 1–5 (9 September 2020). IEEE.

26 Valipour, E., Nourollahi, R., Taghizad-Tavana, K. et al. (2022). Risk assessment of industrial energy hubs and peer-to-peer heat and power transaction in the presence of electric vehicles. *Energies* 15 (23): 8920.

27 Weis, A., Jaramillo, P., and Michalek, J. (2014). Estimating the potential of controlled plug-in hybrid electric vehicle charging to reduce operational and capacity expansion costs for electric power systems with high wind penetration. *Applied Energy* 115: 190–204.

28 Ma, T., Wu, J., Hao, L. et al. (2018). The optimal structure planning and energy management strategies of smart multi energy systems. *Energy* 160: 122–141.

29 Sheikhi, A., Rayati, M., Bahrami, S. et al. (2015). Integrated demand side management game in smart energy hubs. *IEEE Transactions on Smart Grid* 6 (2): 675–683.

30 Obi, M., Slay, T., Bass, R. et al. (2020). Distributed energy resource aggregation using customer-owned equipment: a review of literature and standards. *Energy Reports* 6: 2358–2369.

31 Braun, M. and Strauss, P. (2008). A review of aggregation concepts of controllable distributed energy units in electrical power systems. *International Journal of Distributed Energy Resources* 4 (4): 297–319.

32 Guerrero, J.M., Vasquez, J.C., Matas, J. et al. (2010). Hierarchical control of droop-controlled ac and dc microgrids – a general approach toward standardization. *IEEE Transactions on Industrial Electronics* 58 (1): 158–172.

33 Naughton, J., Wang, H., Riaz, S. et al. (2020). Optimization of multi-energy virtual power plants for providing multiple market and local network services. *Electric Power Systems Research* 189: 106775.

34 Mishra, S., Bordin, C., Leinakse, M. et al. (2021). Virtual power plants and integrated energy system: current status and future prospects. In: *Handbook of Smart Energy Systems* (ed. M. Fathi, E. Zio, and P.M. Pardalos). Cham: Springer.

35 Liu, T., Zhang, D., Dai, H. et al. (2019). Intelligent modeling and optimization for smart energy hub. *IEEE Transactions on Industrial Electronics* 66 (12): 9898–9908.

36 Geidl, M., Koeppel, G., Favre-Perrod, P., et al. (2007). The energy hub – a powerful concept for future energy systems. *Third Annual Carnegie Mellon Conference on the Electricity Industry*, Pittsburgh, PA (13–14 March 2007).

37 Soroudi, A. (2017). *Power System Optimization Modeling in GAMS*, vol. 78. Berlin/Heidelberg, Germany: Springer.

38 Geidl, M. and Andersson, G. (2005). A modeling and optimization approach for multiple energy carrier power flow. *Proceedings of the 2005 IEEE Russia Power Tech*, 1–7 (27–30 June 2005). St. Petersburg, Russia.

39 Geidl, M., Koeppel, G., Favre-Perrod, P. et al. (2006). Energy hubs for the future. *IEEE Power and Energy Magazine* 5 (1): 24–30.

40 Good, N., Karangelos, E., Navarro-Espinosa, A. et al. (2015). Optimization under uncertainty of thermal storage-based flexible demand response with quantification of residential users' discomfort. *IEEE Transactions on Smart Grid* 6 (5): 2333–2342.

41 Kriechbaum, L., Scheiber, G., Kienberger, T. et al. (2018). Grid-based multi-energy systems – modelling, assessment, open source modelling frameworks and challenges. *Energy, Sustainability and Society* 8 (35).

42 Bahmani, R., Karimi, H., Jadid, S. et al. (2021). Cooperative energy management of multi-energy hub systems considering demand response programs and ice storage. *International Journal of Electrical Power & Energy Systems* 130: 106904.

43 Pfenninger, S., Hawkes, A., Keirstead, J. et al. (2014). Energy systems modeling for twenty-first century energy challenges. *Renewable and Sustainable Energy Reviews* 33: 74–86.

44 Mancarella, P., Andersson, G., Peças-Lopes, J.A., et al. (2016). Modelling of integrated multi-energy systems: drivers, requirements, and opportunities. *Power Systems Computation Conference (PSCC)*, Genoa, Italy (20–24 June 2016).

45 Biezma, M.V. and San Cristobal, J.R. (2006). Investment criteria for the selection of cogeneration plants – a state of the art review. *Applied Thermal Engineering* 26 (5–6): 583–588.

46 Chicco, G. and Mancarella, P. (2007). Trigeneration primary energy saving evaluation for energy planning and policy development. *Energy Policy* 35 (12): 6132–6144.

47 Mancarella, P. and Chicco, G. (2009). Global and local emission impact assessment of distributed cogeneration systems with partial-loadmodels. *Applied Energy* 86 (10): 2096–2106.

48 Blaud, P.C., Haurant, P., Claveau, F. et al. (2020). Modelling and control of multi-energy systems through multi-prosumer node and economic model predictive control. *International Journal of Electrical Power & Energy Systems* 118: 105778.

49 Fodstad, M., del Granado, P.C., Hellemo, L. et al. (2022). Next frontiers in energy system modelling: a review on challenges and the state of the art. *Renewable and Sustainable Energy Reviews* 160: 112246.

50 Keirstead, J., Jennings, M., and Sivakumar, A. (2012). A review of urban energy system models: approaches, challenges and opportunities. *Renewable and Sustainable Energy Reviews* 16 (6): 3847–3866.

51 Dranka, G.G., Ferreira, P., and Vaz, A.I.F. (2021). Integrating supply and demand-side management in renewable-based energy systems. *Energy* 232: 120978.

52 Seljom, P., Kvalbein, L., Hellemo, L. et al. (2021). Stochastic modelling of variable renewables in long-term energy models: dataset, scenario generation & quality of results. *Energy* 236: 121415.

53 Ludig, S., Haller, M., Schmid, E. et al. (2011). Fluctuating renewables in a long-term climate change mitigation strategy. *Energy* 36 (11): 6674–6685.

54 Nikoobakht, A., Aghaei, J., Shafie-khah, M. et al. (2020). Co-operation of electricity and natural gas systems including electric vehicles and variable renewable energy sources based on a continuous-time model approach. *Energy* 200: 117484.

55 AlHajri, I., Ahmadian, A., and Elkamel, A. (2021). Stochastic day-ahead unit commitment scheduling of integrated electricity and gas networks with hydrogen energy storage (HES), plug-in electric vehicles (PEVs) and renewable energies. *Sustainable Cities and Society* 67: 102736.

56 Fattori, F., Tagliabue, L., Cassetti, G., and Motta, M. (2019). Enhancing power system flexibility through district heating – potential role in the Italian decarbonisation. *IEEE International Conference on Environment and Electrical Engineering and IEEE Industrial and Commercial Power Systems Europe (EEEIC/I&CPS Europe)*, Genova, Italy (11–14 June 2019).

57 Unterluggauer, T., Rich, J., Andersen, P.B., and Hashemi, S. (2022). Electric vehicle charging infrastructure planning for integrated transportation and power distribution networks: a review. *eTransportation* 12: 100163.

58 Arias, A., Sanchez, J., and Granada, M. (2018). Integrated planning of electric vehicles routing and charging stations location considering transportation networks and power distribution systems. *International Journal of Industrial Engineering Computations* 9 (4): 535–550.

59 Geng, L., Lu, Z., He, L. et al. (2019). Smart charging management system for electric vehicles in coupled transportation and power distribution systems. *Energy* 189: 116275.

60 Taqvi, S.T., Almansoori, A., Maroufmashat, A., and Elkamel, A. (2022). Utilizing rooftop renewable energy potential for electric vehicle charging infrastructure using multi-energy hub approach. *Energies* 15 (24): 9572.

61 Shahan, Z. (2015). Electric car charging 101-types of charging, charging networks, apps, & more! EV Obsession. https://evobsession.com/electric-car-charging-101-types-of-charging-apps-more. (accessed 11 November 2023).

62 Asaad, A., Ali, A., Mahmoud, K. et al. (2023). Multi-objective optimal planning of EV charging stations and renewable energy resources for smart microgrids. *Energy Science & Engineering* 11 (3): 1202–1218.

63 Das, H.S., Nurunnabi, M., Salem, M. et al. (2022). Utilization of electric vehicle grid integration system for power grid ancillary services. *Energies* 15 (22): 8623.

64 Mojumder, M.R.H., Ahmed Antara, F., Hasanuzzaman, M. et al. (2022). Electric vehicle-to-grid (V2G) technologies: impact on the power grid and battery. *Sustainability* 14 (21): 13856.

65 Sevdari, K., Calearo, L., Andersen, P.B., and Marinelli, M. (2022). Ancillary services and electric vehicles: an overview from charging clusters and chargers technology perspectives. *Renewable and Sustainable Energy Reviews* 167: 112666.

66 Pillai, J.R. and Bak-Jensen, B. (2011). Integration of vehicle-to-grid in the western Danish power system. *IEEE Transactions on Sustainable Energy* 2 (1): 12–19.

67 Galus, M.D., Koch, S., and Andersson, G. (2011). Provision of load frequency control by PHEVs, controllable loads, and a cogeneration unit. *IEEE Transactions on Industrial Electronics* 58 (10): 4568–4582.

68 Shafiekhani, M. and Zangeneh, A. (2020). Integration of electric vehicles and wind energy in power systems. In: *Electric Vehicles in Energy Systems* (ed. A. Ahmadian, B. Mohammadi-ivatloo, and A. Elkamel). Cham: Springer.

69 Lopes, J.P., Polenz, S.A., Moreira, C.L., and Cherkaoui, R. (2010). Identification of control and management strategies for LV unbalanced microgrids with plugged-in electric vehicles. *Electric Power Systems Research* 80 (1): 898–906.

70 Longo, M., Foiadelli, F., and Yaïci, W. et al. (2018). Electric vehicles integrated with renewable energy sources for sustainable mobility. New Trends in Electrical Vehicle Powertrains, January. IntechOpen: Ch. 10.

71 Bessa, R.J. and Matos, M.A. (2011). Economic and technical management of an aggregation agent for electric vehicles: a literature survey. *International Transactions on Electrical Energy Systems* 22: 334–350.

72 Birnie, D. (2009). Solar-to-vehicle (S2V) systems for powering commuters of the future. *Journal of Power Sources* 186 (2): 539–542.

73 Shafiq, A., Iqbal, S., Rehman, A.U. et al. (2023). Integration of solar based charging station in power distribution network and charging scheduling of EVs. *Frontiers in Energy Research* 11: 1086793.

74 Ullah, Z., Wang, S., Wu, G. et al. (2023). Optimal scheduling and techno-economic analysis of electric vehicles by implementing solar-based grid-tied charging station. *Energy* 267: 126560.

75 Mohammad, A., Zamora, R., and Lie, T.T. (2020). Integration of electric vehicles in the distribution network: a review of PV based electric vehicle modelling. *Energies* 13 (17): 4541.

76 Chen, T., Pourbabak, H., Liang, Z., and Su, W. (2017). An integrated evoucher mechanism for flexible loads in real-time retail electricity market. *IEEE Access* 5: 2101–2110.

77 Su, J., Lie, T.T., Zamora, R. et al. (2020). Integration of electric vehicles in distribution network considering dynamic power imbalance issue. *IEEE Transactions on Industry Applications* 56 (5): 5913–5923.

78 Farzin, H., Moeini-Aghtaie, M., and Fotuhi-Firuzabad, M. (2016). Reliability studies of distribution systems integrated with electric vehicles under battery-exchange mode. *IEEE Transactions on Power Delivery* 31 (6): 2473–2482.

79 Ramachandran, B., Srivastava, S.K., and Cartes, D.A. (2013). Intelligent power management in micro grids with EV penetration. *Expert Systems with Applications* 40 (16): 6631–6640.

80 Thomas, D., Deblecker, O., and Ioakimidis, C.S. (2018). Optimal operation of an energy management system for a grid-connected smart building considering photovoltaics' uncertainty and stochastic electric vehicles' driving schedule. *Applied Energy* 210: 1188–1206.

81 Yang, Z., Ghadamyari, M., and Khorramdel, H. (2021). Robust multi-objective optimal design of islanded hybrid system with renewable and diesel sources/stationary and mobile energy storage systems. *Renewable and Sustainable Energy Reviews* 148: 111295.

82 Bimenyimana, S., Wang, C., Nduwamungu, A. et al. (2021). Integration of microgrids and electric vehicle technologies in the national grid as the key enabler to the sustainable development for Rwanda. *International Journal of Photoenergy* 2021: 9928551.

83 Homer energy. https://homerenergy.my.site.com/supportcenter/s/

84 Umoren, I.A., Shakir, M.Z., and Ahmadi, H. (2023). VCG-based auction for incentivized energy trading in electric vehicle enabled microgrids. *IEEE Access* 11: 21117–21126.

85 Hai, T. and Zhou, J. (2023). Optimal planning and design of integrated energy systems in a microgrid incorporating electric vehicles and fuel cell system. *Journal of Power Sources* 561: 232694.

86 Ahsan, S.M. and Khan, H.A. (2022). Optimized power dispatch for smart building(s) and electric vehicles with V2X operation. *Energy Reports* 8: 10849–10867.

87 Zhang, D., Shah, N., and Papageorgiou, L.G. (2013). Efficient energy consumption and operation management in a smart building with microgrid. *Energy Conversion and Management* 74: 209–222.

88 Lee, S. and Choi, D.H. (2020). Energy management of smart home with home appliances, energy storage system and electric vehicle: a hierarchical deep reinforcement learning approach. *Sensors* 20 (7): 2157.

89 Hou, X., Wang, J., and Huang, T. (2019). Smart home energy management optimization method considering energy storage and electric vehicle. *IEEE Access* 7: 144010–144020.

90 Alhelou, H.H., Siano, P., and Tipaldi, M. (2020). Primary frequency response improvement in interconnected power systems using electric vehicle virtual power plants. *World Electric Vehicle Journal* 11 (2): 40.

91 Li, R. and SaeidNahaei, S. (2022). Optimal operation of energy hubs integrated with electric vehicles, load management, combined heat and power unit and renewable energy sources. *Journal of Energy Storage* 48: 103822.

92 Qi, F., Wen, F., Liu, X., and Salam, M.A. (2017). A residential energy hub model with a concentrating solar power plant and EVs. *Energies* 10: 1159.

93 Stawska, A., Romero, N., de Weerdt, M., and Verzijlbergh, R. (2021). Demand response: For congestion management or for grid balancing? *Energy Policy* 148: 111920.

94 Naval, N. and Jose, M.Y. (2021). Virtual power plant models and electricity markets – a review. *Renewable and Sustainable Energy Reviews* 149: 111393.

95 Inci, M., Savrun, M.M., and Çelik, Ö. (2022). Integrating electric vehicles as virtual power plants: a comprehensive review on vehicle-to-grid (V2G) concepts, interface topologies, marketing and future prospects. *Journal of Energy Storage* 55: 105579.

96 Marra, F., Sacchetti, D., Pedersen, A.B., et al. (2012). Implementation of an electric vehicle test bed controlled by a virtual power plant for contributing to regulating power reserves. *IEEE Power and Energy Society General Meeting*, 1–7 (22–26 July 2012), San Diego, CA, USA.

97 Abbasi, M.H., Taki, M., Rajabi, A. et al. (2019). Coordinated operation of electric vehicle charging and wind power generation as a virtual power plant: a multi-stage risk constrained approach. *Applied Energy* 239: 1294–1307.

98 Clement-Nyns, K., Haesen, E., and Driesen, J. (2009). The impact of charging plug-in hybrid electric vehicles on a residential distribution grid. *IEEE Transactions on Power Systems* 25 (1): 371–380.

99 Akhavan-Rezai, E., Shaaban, M.F., El-Saadany, E.F., and Zidan, A. (2012). Uncoordinated charging impacts of electric vehicles on electric distribution grids: normal and fast charging comparison. *IEEE Power and Energy Society General Meeting*, 1–7 (2012), San Diego, CA, USA.

100 EV Integration Project – Milestone 10: Recommendations for EV Integration UoM-ENA-C4NET-EV_Integration_M10 30th September 2022.

101 Tirunagari, S., Gu, M., and Meegahapola, L. (2022). Reaping the benefits of smart electric vehicle charging and vehicle-to-grid technologies: regulatory, policy and technical aspects. *IEEE Access* 10: 114657–114672.

102 IRENA (2019). *Innovation Outlook: Smart Charging for Electric Vehicles*. Abu Dhabi: International Renewable Energy Agency.

103 Ravi, S.S. and Aziz, M. (2022). Utilization of electric vehicles for vehicle-to-grid services: progress and perspectives. *Energies* 15 (2): 589.

104 Foundation for traffic safety, New American Driving Survey, April 2021.

105 Vatandoust, B., Ahmadian, A., Golkar, M.A. et al. (2018). Risk-averse optimal bidding of electric vehicles and energy storage aggregator in day-ahead frequency regulation market. *IEEE Transactions on Power Systems* 34 (3): 2036–2047.

106 Singh, P.P., Wen, F., Palu, I. et al. (2023). Electric vehicles charging infrastructure demand and deployment: challenges and solutions. *Energies* 16 (1): 7.

107 Taghizad-Tavana, K., Alizadeh, A.A., Ghanbari-Ghalehjoughi, M., and Nojavan, S. (2023). A comprehensive review of electric vehicles in energy systems: integration with renewable energy sources, charging levels, different types, and standards. *Energies* 16 (2): 630.

108 Esmaili, M. and Goldoust, A. (2015). Multi-objective optimal charging of plug-in electric vehicles in unbalanced distribution networks. *International Journal of Electrical Power & Energy Systems* 73: 644–652.

109 Nimalsiri, N.I., Mediwaththe, C.P., and Ratnam, E.L. (2020). A survey of algorithms for distributed charging control of electric vehicles in smart grid. *IEEE Transactions on Intelligent Transportation Systems* 21 (11): 4497–4515.

110 Moeini-Aghtaie, M., Abbaspour, A., Fotuhi-Firuzabad, M., and Dehghanian, P. (2013). PHEVs centralized/decentralized charging control mechanisms: requirements and impacts. *North American Power Symposium (NAPS)*, 1–6 (22–24 September 2013). Manhattan, KS, USA.

111 Yao, W., Zhao, J., Wen, F. et al. (2014). A multi-objective collaborative planning strategy for integrated power distribution and electric vehicle charging systems. *IEEE Transactions on Power Systems* 29: 1811–1821.

112 Octopus Energy. Powerloop. https://www.octopusev.com/powerloop. (accessed 28 November 2023)

113 Energy OVO (2020). Vehicle-to-Grid Trial: Building better grid for everyone | OVO Energy (https://www.ovoenergy.com/electric-cars/vehicle-to-grid-charger) (accessed 11 November 2023).

114 Cenex (2019). Understanding the True Value of V2G Report. https://www.cenex.co.uk/app/uploads/2019/10/True-Value-of-V2G-Report.pdf (accessed 11 November 2023).

115 Amamra, S. and Marco, J. (2019). Vehicle-to-grid aggregator to support power grid and reduce electric vehicle charging cost. *IEEE Access* 7: 178528–178538.

8

Self-Driving Vehicle Systems in Intelligent Transportation Networks

Yigit Cagatay Kuyu

R&D department, Karsan Otomotiv Sanayi ve Tic, Bursa, Turkey

8.1 Introduction

Self-driving vehicles consist of intelligent sub-systems that are capable of driving themselves entirely or in part for reducing pollution and traffic problems along with safety purposes. While numerous technological breakthroughs have made significant progress in the development of more capable vehicles, the technology is not yet perfect and the complete elimination of human intervention remains a challenge. Furthermore, public reservations and psychological barriers associated with the use of driverless vehicles raise concerns regarding trust in personal safety and security [1].

In the near future, there is a strong possibility that fully autonomous vehicles (AVs) will become the dominant component of intelligent transportation networks, replacing human drivers. This has sparked intense competition among major car manufacturers and technology companies, as they strive to be the first to introduce fully self-driving vehicles to the market. However, experts suggest that the transition to fully autonomous technology and its widespread adoption may not occur until around 2030 due to the aforementioned limitations [2]. Intelligent vehicles have a set of complicated systems that use data coming from sensors to identify the environment and generate appropriate behaviors for subsequent actions [3]. The Society of Automotive Engineers (SAE) has classified AVs based on their level of intelligence from zero to five. Basically, level 0 means no automation, while level 5 refers to full automation. The detailed functionality of the SAE levels is given in Table 8.1.

Interconnected Modern Multi-Energy Networks and Intelligent Transportation Systems: Towards a Green Economy and Sustainable Development, First Edition. Edited by Mohammadreza Daneshvar, Behnam Mohammadi-Ivatloo, Amjad Anvari-Moghaddam, and Reza Razzaghi.
Published 2024 by John Wiley & Sons, Inc.

Table 8.1 Level of driving automation proposed by SAE.

Level	Description	Definition
0	No automation driving	Vehicle is fully controlled by the driver in all conditions.
1	Driver assistance	Vehicle features a single automated system.
2	Partial automation	Vehicle can assist in performing basic tasks.
3	Conditional automation	Vehicle can control specific actions.
4	High automation	Vehicle can drive itself for limited conditions.
5	Full automation	Vehicle can drive itself totally.

Source: Adapted from Ref. [4].

The acceleration of computing methodologies has reshaped mobility and enabled the development of promising technologies and practical applications for self-driving vehicles. These key methodologies, which support making vehicles move from level 0 to level 5, primarily involve sensing, perception, and planning/control, which will be elaborated upon in the following sections.

The motivation behind studying autonomous and intelligent vehicle systems arises from the potential to address pressing challenges in modern transportation, including pollution reduction, traffic congestion alleviation, energy consumption minimization, and safety enhancement. With the rapid advancement of vehicle technology and the increasing complexity of these systems, it becomes crucial to comprehend the underlying components and their integration into intelligent transportation networks. This work aims to provide a comprehensive review of the state of the art in AV research, with a specific focus on key areas such as sensing technologies, perception techniques (e.g., object detection and tracking, lane tracking, traffic sign detection, vehicle and pedestrian detection), simultaneous localization and mapping (SLAM), and planning and control strategies [5–7]. By examining the latest advancements in these areas, this work seeks to establish a solid foundation for further research and innovation, ultimately contributing to the development of AVs and their seamless integration into future transportation systems.

8.2 Brief History

Self-driving vehicle systems with rapid developments in sensor and computing technologies are an active research area aimed at making the components work in harmony for safe driving. However, autonomous-based vehicles and driving technologies are not a new field; they date back to two decades. One of the first self-sufficient attempts, called Navlab, appeared at Carnegie Mellon University in the 1980s, combining image processing, neural networks, and control [8]. The European EUREKA project "Prometheus," launched in 1986, has demonstrated the usability of fully autonomous driving on the highway near Paris [9]. By the mid-1990s, the ARGO Project was announced with the aim of developing an active safety system for fully AV technology [10]. By the early 2000s, the Defense Advanced Research Projects Agency (DARPA) issued a challenge that accelerated the research for autonomous driving. In the first year, no vehicle finished the race. A year later, Stanford's racing team won the challenge [11]. In 2010, Google announced the prototype of a car with self-driving capability. The company has also revealed that the developed car has logged 500,000 miles in autonomous mode with no crashes [12]. In 2014, Tesla, the electric car company, integrated autopilot with advanced driver-assistance system features into Model S, which makes it one of the most advanced autonomous cars in the market [13]. When it comes to the 2020s, the increasing focus is not only on autonomous personal cars but also on AVs in public transportation. HONDA, the well-known Japanese car company, introduced a limited number of Legend Hybrid EX models with hands-free driving functioning [14]. Mercedes integrated the conditional self-driving system into the S-Class series that is currently legally used in Germany [15]. EasyMile, a commercial autonomous shuttle bus manufacturer, introduced the first-ever Level 4 service model on public roads [16]. KARSAN automotive, one of the leading electric bus manufacturers, launched the commercial autonomous bus called "Autonomous e-Atak," which is capable of following the traffic signs while driving on urban lines without the need for driver control [17]. A concise overview of the advancements in autonomous vehicles is presented in Figure 8.1.

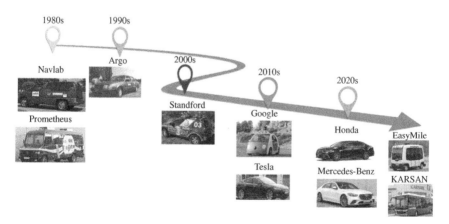

Figure 8.1 A brief summary of the developments of autonomous vehicles. *Source:* Argo - Alberto Broggi, Massimo Bertozzi, Alessandra Fascioli/Department of Information Engineering/University of Parma; Navlab - Carnegie Mellon University; Prometheus - Mercedes Benz; Google - Flickr; Tesla - nakhon100/Wikimedia Common; Honda - Honda Motor Co; EasyMile - Ulkl/Ulf Klingström/Wikimedia Commons.

8.3 Literature Review

The rapid advancement of autonomous and intelligent vehicle systems has led to a wealth of research spanning various domains. The increasing complexity of these systems necessitates a deeper understanding of their components and the challenges associated with their integration into existing transportation networks. This extended literature review delves into key areas of research, highlighting their importance in the development and deployment of AVs [18].

Sensing and perception form the foundation of AV systems. A wide range of sensors, including cameras, lidar, and radar, have been investigated and implemented to provide a comprehensive understanding of the vehicle's surroundings [19]. Lidar-based object tracking systems have been proposed for AVs, emphasizing the importance of accurate and reliable perception [20]. Cooperative maneuvering in mixed urban traffic has been studied, highlighting the role of vehicle-to-vehicle (V2V) communication and the combination of various sensor data for safe and efficient driving [21]. Additionally, the fusion of data from multiple sensors, such as cameras, lidar, and radar, has been explored to improve the robustness and reliability of AV perception systems [22, 23]. Furthermore, semantic scene understanding, which involves recognizing and interpreting the context of objects and scenes detected by sensors, has been explored to support more informed decision-making by AVs [24]. A probabilistic planner for AVs has been developed, allowing for robust navigation in complex, dynamic environments [25]. Machine learning techniques, including deep learning and reinforcement learning (RL), have been applied to develop adaptive control algorithms that enable AVs to handle various driving scenarios more effectively [26]. Model predictive control (MPC) has been employed for trajectory planning and obstacle avoidance, providing a framework that considers the vehicle's dynamics and constraints while optimizing its trajectory [27]. In addition, game theory has been applied to analyze interactions between AVs and other road users, enabling the development of decision-making algorithms that account for the strategic behavior of multiple agents in the driving environment [28]. Swarm intelligence, inspired by the collective behavior of social insects, has been explored as a means to optimize the coordination and decision-making processes of connected AVs in large-scale transportation systems [29].

8.4 Advantages and Challenges

One of the key advantages of self-driving vehicles lies in their capacity to enhance safety on the roads. Given that human error represents a predominant factor contributing to accidents, AVs can address this concern through the utilization of advanced sensors, algorithms, and control systems, enabling them to make precise and prompt decisions surpassing those made by human drivers [30]. Despite the numerous advantages of AVs, the deployment of these vehicles also faces several challenges that need to be addressed before their full potential can be realized.

One of the primary advantages of self-driving vehicles is the potential for increased safety on the roads. Human error is a leading cause of accidents, and AVs can mitigate this risk by utilizing advanced sensors, algorithms, and control systems to make more accurate and timely decisions than human drivers [31]. Additionally, AVs have the potential to contribute to the reduction of greenhouse gas emissions and air pollution. Through optimized routing, reduced traffic congestion, and facilitating smoother traffic flow, AVs can help to minimize environmental impact [32]. Moreover, integrating electric vehicles (EVs) into the AV fleet further enhances the environmental benefits, as EVs produce zero tailpipe emissions [33, 34]. Furthermore, AVs have the promising capability to enhance mobility for individuals facing transportation challenges, such as the elderly, disabled, or those without access to personal transportation [35]. By providing reliable and accessible transportation options, AVs can improve the quality of life for these individuals and promote social inclusion.

In addition to the advantages, the development of fully AVs necessitates the use of sophisticated hardware and software systems that can reliably operate in complex and dynamic environments. Achieving high levels of reliability and performance for these systems remains a significant technical challenge [36]. Furthermore, the successful deployment of AVs requires the establishment of comprehensive legal, regulatory, and ethical frameworks, developed through collaboration among policymakers, industry stakeholders, researchers, and consumer advocacy groups [37, 38]. These frameworks should address various challenges unique to self-driving vehicles, including liability, insurance, data privacy, cybersecurity, and ethical decision-making. Striking a balance between fostering innovation and ensuring public safety is essential for the effective integration of AVs into society.

8.5 Sensing

AVs are equipped with several fundamental sensors to perceive their surroundings. The quality of these sensors plays a critical role in implementing Advanced Driver Assistance Systems (ADAS) by providing input data for interpretation [39]. Given that each sensor has its own strengths and limitations, the use of a multisensor configuration enables AVs to mitigate undesired weaknesses in the system while driving. Certain sensors are essential for increasing the autonomy level of vehicles from one to five, with frequently utilized sensors including GNSS, lidar, camera, and radar [40].

GNSS, or Global Navigation Satellite System, serves as the AV's navigation system by providing precise location positioning anywhere on Earth. It is a vital component that assists in SLAM, improving the estimation of the vehicle's position within the surrounding environment. GNSS is also commonly employed in ADAS to determine the global location of the vehicle [41].

Lidar, which stands for Light Detection and Ranging, is a robust component for autonomous driving primarily utilized for object detection, mapping, and ranging. By scanning the environment, lidar can detect objects and generate a detailed 3D point cloud map of the surroundings. This is achieved by emitting pulsed laser waves that are reflected by objects, providing information about static and dynamic environments. However, it is important to note that lidar can be relatively expensive and sensitive to harsh weather conditions [42, 43].

Camera serves as a visual sensor that captures the surrounding environment, providing a visual representation of the world in color. It is capable of detecting obstacles, objects, traffic signs, and road markings. Many AV applications employ multiple cameras along with image processing and artificial intelligence (AI) algorithms to classify detected objects statistically. To enhance the vehicle's perspective, cameras are typically positioned at the front, back, and wing mirrors of the vehicle. While cameras have a lower cost compared to lidar, they are susceptible to changes in illumination and cannot provide distance information [43].

Radar, or Radio Detection and Ranging, emits radio waves that are reflected back from obstacles, allowing for the measurement of size and distances. This enables the acquisition of collision avoidance, distance, and velocity information. Radar can detect objects with high precision, depending on the speed at which the waves propagate. One advantage of radar is its ability to provide high resolution even in adverse weather conditions with limited visibility. However, radar cannot precisely determine the shape of objects like cameras can.

The use of multiple sensors directly impacts the sensing capabilities of AVs and is crucial for safety-related tasks. Sensor fusion is a framework that involves the cooperative integration of two or more diverse sensors to leverage their respective advantages, as illustrated in Figure 8.2. Since delivering all the necessary information for AVs with a single sensor type is challenging, an appropriate sensor fusion framework becomes essential to achieve higher levels of autonomy, particularly in perception, and gain a comprehensive understanding of the surrounding environment [44]. Consequently, there is an increased demand for an appropriate framework that facilitates a higher level of autonomy, particularly in perception, to gain a comprehensive understanding of the surrounding environment. This multi-sensor approach not only enhances the accuracy and reliability of the collected data but also provides redundancy, ensuring system operation even if one sensor fails or encounters limitations [45]. Integrating diverse sensor modalities, such as lidar, camera, and radar, offers a more holistic representation of the environment, enabling AVs to better comprehend and respond to dynamic situations. Ultimately, the fusion of data from multiple sensors is crucial for developing robust and reliable AV systems capable of navigating complex environments safely and efficiently.

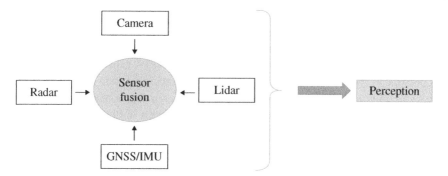

Figure 8.2 Representative sensor fusion diagram.

8.6 Perception

After sensing, the data received from sensors are fed into the perception stage for processing in order to gather useful and consistent information about the surrounding environment. The whole process is called perception, which is one of the main modules of the AV systems. It is essential to note that perception plays a key role in the decision-making of AVs [46]. This section will review the fundamental aspects of detection and tracking that are frequently faced in many applications.

8.6.1 Object Detection and Tracking

Each perception module has to accomplish a particular task to ensure the efficiency and stability of the overall AV system for road safety. However, an intelligent system can sometimes be manipulated, resulting in potential damage and collision due to missed detections and wrong tracking results.

Semantic Segmentation, a vital component of perception in AVs, goes beyond object detection and tracking to provide a more detailed understanding of the environment. It involves assigning semantic labels to individual pixels or regions in an image, enabling the vehicle to differentiate between different objects and their respective classes, such as pedestrians, vehicles, and obstacles [47]. By leveraging advanced computer vision techniques, such as convolutional neural networks (CNNs), semantic segmentation enhances the perception capabilities of self-driving vehicles by providing a fine-grained analysis of the scene [48]. This additional level of understanding enables more precise decision-making, leading to improved safety and performance. Integrating semantic segmentation with other perception modules, such as object detection and tracking, can further enhance the overall perception capabilities of AVs.

Lane Tracking is essential for AV perception and should be robust to a variety of driving environments. It is the process that detects lines on the roads and continually monitors the position of the vehicle within a lane for safe driving. Many vehicles have implemented this feature over the decades. One of the first implementations was started by Stanford Research Institute's Artificial Intelligence Center in the early 1980s [49]. In 2006, the DARPA urban challenge made it possible to operate many vehicles with the lane tracking system [11]. Currently, this system is adapted to advanced vehicles with a high degree of safety. It should be noted that the figure of the lane tracking can be seen in Figure 8.3.

Traffic Sign Detection is a useful application for ADAS, which recognizes traffic signs on the road. It is important to highlight that successful sign detection can contribute to safer and easier driving. However, it can sometimes be challenging to detect signs accurately due to the presence of other objects and light reflections. This system plays a crucial role in alerting the driver in case a sign is missed. The detection of traffic signs on roads has been extensively studied by researchers and engineers, and its performance relies on various computer vision and AI techniques. The performance of this system capable of reading and classifying road signs depends on the success of several computer vision and AI techniques [50]. A vast majority of existing approaches have been proposed to perform the system in many kinds of adverse situations [51]. Some of the applications of traffic sign detection can be seen in Figure 8.4.

Vehicle/Pedestrian Detection is a critical technology for object detection in a traffic scene and many of the techniques are based on deep learning. You Only Look Once (YOLO), MultiBox Detector (SSD), and Region-based Convolutional Neural Networks (R-CNN) are among the most recognized algorithms in the literature [52–55]. Detecting vehicles and pedestrians can be challenging due to their unpredictable behaviors in traffic. The system

Figure 8.3 Lane tracking.

Figure 8.4 Traffic sign detection.

should be capable of recognizing them regardless of their pose or shape. Additionally, providing timely visual or voice warnings can help prevent many dangerous situations. The figure of the aforementioned system can be seen in Figure 8.5.

Figure 8.5 Vehicle/Pedestrian detection.

8.6.2 Simultaneous Localization and Mapping

SLAM is a method that is able to explore the environment without any prior knowledge and localize itself within it while moving. The goal of SLAM is to effectively perform localization and mapping simultaneously by utilizing sensors [56]. Accurate localization not only provides the absolute position of AV, but also determines how far the vehicle is from its surrounding objects [57]. Mapping is essential for AV to visualize an unknown environment, as well as assist in building the path of AV for navigation [58]. The basic representation of SLAM is given in Figure 8.6.

SLAM technology can be divided into two categories, visual SLAM and lidar SLAM, according to sensor types, such as camera and Lidar [56]. SLAM, depending on visual information, is called visual SLAM (vSLAM) [59]. The technical difficulty of vSLAM can be higher than that of lidar SLAM due to the limited field of view, but it can be less expensive than that of lidar-based systems. The early development stage of vSLAM using a monocular camera dates back to the 2000s, which is called the feature-based approach. Besides, a direct approach is proposed to deal with a whole image directly for textureless or featureless environments [60]. MonoSLAM, a filter-based algorithm, can be an example of feature-based approaches developed in 2003 [61]. The algorithm is utilized by an extended Kalman filter to use camera motion and 3D structure of an unknown environment simultaneously. Parallel tracking and mapping (PTAM) has been proposed and provides less computational cost by using different threads on the CPU in parallel [62]. Dense tracking and mapping (DTAM) is the direct approach based on the dense 3D surface model for camera tracking via whole image registration [63].

Lidar-based SLAM technology is one of the popular methods in robotics and AVs. It is arguably more sophisticated but less complex than classical visual

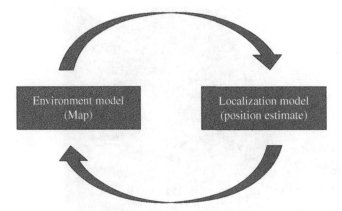

Figure 8.6 Schematic representation of SLAM.

SLAMs [64]. One of the key issues in this technology is the point cloud registration process. Regarding this problem, the standard iterative closest point (ICP) algorithm is generally used for aligning two surfaces of scans [65]. The extended ICP algorithm is also proposed to estimate the continuous-time trajectory for SLAM from actuated lidar [66]. More and more algorithms have been developed to solve the drawbacks of existing ones [67, 68].

8.7 Planning and Control

The planning and Control mechanism needs feedback from environmental representation supported by mapping and localization. The architecture of the process basically consists of route planning, behavioral planning, and motion planning.

Route planning refers to the process of determining the road network from the current location to the destination that the vehicle will reach. It is important due to the direct effect on the decisions of the vehicle in finding the optimal path. It can be classified into two strategies: global and local route planning [69]. Global route planning is the first step of the navigation task, which is mainly aimed at finding the most optimal and safe route for the vehicle to reach its destination. Local route planning mainly addresses to navigate the vehicle safely by determining a precise path at tactical level, such as obstacle avoidance. This strategy considers driving conditions, such as obstacles and road boundaries, as well as instantaneous factors, such as vehicle movement, dynamic objects, and generates candidate paths ensuring driving safety [70]. There are some popular methods for route planning, such as Dijkstra [71] and A* [72] approaches. Dijkstra is an algorithm that allows us to find the shortest path from a starting point to a target point on a map. A* combines the principles of Dijkstra's algorithm with the heuristic function to provide a faster solution in reaching the shortest path between an initial point and a final point on a map. The comparative survey of the routing algorithms can also be found in Ref. [73]. An example of the route planning can be found in Figure 8.7.

Behavioral planning, which is essentially responsible for decision-making, focuses on determining driving actions that the vehicle should take along the route planned in the previous layer according to the information of traffic, static, and dynamic objects provided by the perception layer [74]. This layer can also make a decision among the candidate paths; thus, it can work with local route planning in parallel. Mainly, the proposed approaches in behavior planning include motion prediction and decision-making [75].

Regarding motion prediction, the improved event data recorder has been proposed for driving safety. The driver-behavior models have been developed based on Gaussian mixture models (GMMs) to determine automatically safe and risky

Figure 8.7 An example of route planning.

driving behavior from the recorded driving data. The proposed system is able to give the driver a suitable response/warning under some circumstances [76]. Quantum Markov decision process (QMDP) with Monte Carlo tree search (MCTS) has also been offered to avoid risky behavior. The outcomes demonstrate how input uncertainty can be propagated to estimate uncertainty in the output of candidate maneuvers [77]. Driving behavior has been realized by constructing state-action mapping rules in which fuzzy logic partition sets and fuzzy driving rules have been benefited, and are then embedded into a neural network structure that models two types of driving behavior: car-following behavior and evasive behavior [78]. AI techniques have been commonly used in the literature to predict accidents by building models [79]. Artificial neural network (ANN) is one of the examples that was used for the detection of possible accidents. The number of data collected from complex traffic and geometrical characteristics including 52,447 crash cases has been used to build the ANN model by selecting 25 independent variables. The results show that the multilayer perceptron models obtain better outcomes [80]. In another study, 77,800 complete accident data have been used for the development of ANN models that benefited from the sigmoid activation function. According to the results, the most important parameter that influences the number of accidents on highways is the degree of vertical curvature [81]. Temporal difference (TD) learning is a technique that is frequently used in RL to predict the future reward [82]. Several frameworks have been proposed to detect unsafe driving behavior by using TD learning in RL. An approach to learn a danger level function to be used as an alert to the users in advance of dangerous situations has been proposed. TD learning has been used to obtain the approximation by propagating the penalty/reward observable under some constraints [83]. Interested readers can also find RL-based methods in behavior planning architecture in Ref. [84].

Regarding decision-making, a solution to real-time decision-making has been proposed. The proposed method is divided into two parts: the former is for safety-critical tasks, while the latter addresses non-safety-critical driving. A multiple criteria decision-making model is built for the second decision-making part and the performance of the model is compared with the existing solutions. The results illustrate that the developed model presents several benefits [85]. A technique has been presented for the planning of future trajectories regarding dynamic obstacles for AVs. The proposed technique utilizes an optimization algorithm to simultaneously optimize multiple continuous contingency paths [86]. An evolutionary-based trajectory planning technique has been proposed for AVs in traffic. The presented structure is able to navigate the vehicle in several scenarios including a large number of obstacles in a predefined map for correct decision-making [**87**]. RL-based methods have also been used for the decision-making process in AVs. Curriculum learning has been presented to facilitate the learning process of the model by training on easy tasks and continuously increasing the complexity of the tasks introduced to the learning agent [88]. Reverse curriculum generation for RL has been proposed without any prior knowledge. This technique generates a curriculum of starting positions that adapts to the agent's performance for difficult simulated navigation problems [89]. In another study, building the model for a partially observable Markov decision process (POMDP) allows the algorithm to consider uncertainties of the decision-making process. It can draw the robust characteristics for perception inputs [90]. The aforementioned RL-based algorithms have been ineffective in continuous tasks, such as the control of the acceleration, braking, and steering actions. To overcome this, the study has presented a framework to implement continuous actions in which an agent is trained exclusively with deep RL [91]. The hierarchical RL structure has also been proposed including a heuristic-based rules-enumeration policy and applied to the decision-making part of AVs. The presented structure is able to generate continuous actions by using the hybrid RL and heuristic techniques [92]. The diagram of the AV software stack framework can be seen in Figure 8.8.

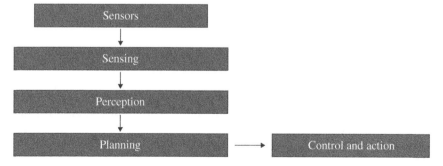

Figure 8.8 AV framework diagram.

8.8 Conclusion

This chapter has provided a comprehensive overview of autonomous and intelligent vehicle systems addressing the pressing challenges in modern transportation. The review of the state of the art in AV research has highlighted the historical development and current advancements in the field, focusing on key areas such as sensing technologies, perception techniques, and planning and control strategies.

The hierarchical arrangement of self-driving vehicle sub-modules and the levels of automation ranging from level 0 to level 5 have been discussed, emphasizing the importance of successful integration and collaboration among these modules for safe and efficient AV operation.

Building upon the advancements discussed in the preceding sections, further research and innovation in the field of AVs are encouraged. The successful development and deployment of AVs rely on continuous improvements in sensing, perception, and decision-making capabilities, as well as addressing the challenges associated with their implementation.

The advantages of AVs, including improved safety, reduced traffic congestion, and enhanced overall efficiency of road networks, have been highlighted. These vehicles have the potential to revolutionize the transportation sector and address critical issues such as pollution reduction, traffic congestion alleviation, energy conservation, and enhanced safety.

As research and development in AVs progress, it is essential to consider the interdisciplinary nature of this field and foster collaborations among researchers, engineers, policymakers, and stakeholders. This collective effort will contribute to the realization of a future where autonomous and intelligent vehicle systems play a pivotal role in shaping a sustainable and efficient transportation ecosystem.

In summary, the advancements in autonomous and intelligent vehicle systems discussed in this chapter provide a strong foundation for future endeavors, aiming to harness the potential of these technologies and usher in a new era of transportation.

Acknowledgment

The author thanks the KARSAN Company for supporting this study under the R&D Department.

References

1 Yuen, K.F., Chua, G., Wang, X. et al. (2020). Understanding public acceptance of autonomous vehicles using the theory of planned behaviour. *International Journal of Environmental Research and Public Health* 17: 4419.

2 Gao, P., Kaas, H.-W., Mohr, D., and Wee, D. (2016). *Automotive Revolution–Perspective Towards 2030: How the Convergence of Disruptive Technology-Driven Trends Could Transform the Auto Industry*. USA: Wiley.

3 Vishnukumar, H.J., Butting, E.B., Müller, C., and Sax, E. (2017). Artificial intelligence core for lab and real-world test and validation for adas and autonomous vehicles: AI for efficient and quality test and validation. *In* 2017 *Intelligent Systems Conference*, London, UK, 714–721.

4 Sae Standards. https://www.sae.org/news/2019/01/sae-updates-j3016-automated-driving-graphic (accessed 2 January 2022).

5 Zhou, Q., Shen, Z., Yong, B. et al. (2022). *Theories and Practices of Self-Driving Vehicles*. Elsevier.

6 Fossen, T., Pettersen, K.Y., and Nijmeijer, H. (2017). *Sensing and Control for Autonomous Vehicles*. Springer International.

7 Stavens, D.M. (2011). *Learning to Drive: Perception for Autonomous Cars*. Stanford University.

8 Thorpe, C.E. (1990). *Vision and Navigation: The Carnegie Mellon Navlab*. USA: Kluwer Academic Publishers.

9 Franke, U., Gavrila, D., Gorzig, S. et al. (1998). Autonomous driving goes downtown. *IEEE Intelligent Systems and Their Applications* 13: 40–48.

10 Broggi, A., Bertozzi, M., and Fascioli, A. (1999). Argo and the millemiglia in automatico tour. *IEEE Intelligent Systems and their Applications* 14: 55–64.

11 Thrun, S. (2006). Winning the DARPA grand challenge: a robot race through the mojave desert. *In 21st IEEE/ACM International Conference on Automated Software Engineering*, Tokyo, Japan, 11.

12 Lutin, J.M., Kornhauser, A.L., and Masce, E.L.L. (2013). The revolutionary development of self-driving vehicles and implications for the transportation engineering profession. *ITE Journal* 83: 28.

13 Zhang, B. and Musk, E. (2016). In 2 years your tesla will be able to drive from New York to la and find you. *Yahoo Finance* 10.

14 Honda legend hybrid ex. https://global.honda/newsroom/news/2021/4210304eng-legend.html (accessed 2 January 2022).

15 Mercedes-benz. https://www.mercedes-benz.com/en/innovation/autonomous (accessed 2 January 2022).

16 Easymile. https://easymile.com/about-us (accessed 2 January 2022).

17 Karsan. https://www.karsan.com/en/autonomous-bus (accessed 2 January 2022).

18 Dresner, K. and Stone, P. (2008). A multiagent approach to autonomous intersection management. *Journal of Artificial Intelligence Research* 31: 591–656.

19 Zhang, Y., Carballo, A., Yang, H., and Takeda, K. (2023). Perception and sensing for autonomous vehicles under adverse weather conditions: A survey. *ISPRS Journal of Photogrammetry and Remote Sensing* 196: 146–177.

20 Ziegler, J., Bender, P., Schreiber, M. et al. (2014). Making bertha drive – an autonomous journey on a historic route. *IEEE Intelligent Transportation Systems Magazine* 6: 8–20.

21 Milanes, V., Shladover, S.E., Spring, J. et al. (2014). Cooperative adaptive cruise control in real traffic situations. *IEEE Transactions on Intelligent Transportation Systems* 15: 296–305.

22 Cho, K. and Wang, J. (2015). Autonomous ground vehicles – sensing, planning, and control. *IEEE Control Systems Magazine* 35: 78–97.

23 Yurtsever, E., Lambert, J., Carballo, A., and Takeda, K. (2020). A survey of autonomous driving: common practices and emerging technologies. *IEEE Access* 8: 58443–58469.

24 Valada, A., Vertens, J., Dhall, A., and Burgard, W. (2018). Adapnet: adaptive semantic segmentation in adversarial environments. *IEEE International Conference on Robotics and Automation*, Singapore (2017), 3369–3376.

25 Montemerlo, M., Becker, J., Bhat, S. et al. (2020). Junior: the Stanford entry in the urban challenge. *Journal of Field Robotics* 9: 569–597.

26 Pan, X., You, Y., Wang, Z., and Lu, C. (2017). Virtual to real reinforcement learning for autonomous driving. *arXiv* 1704.03952.

27 Wang, Y., Wang, R., and Yang, J. (2018). A model predictive control approach for improving traffic safety and efficiency of automated vehicles. *Transportation Research Part C: Emerging Technologies* 323-338: 96.

28 Fudenberg, D. and Tirole, J. (1991). *Game Theory*. Cambridge, MA: MIT Press.

29 Huang, H., Li, G., Wang, H., and Liu, H. (2017). Swarm intelligence for autonomous cooperative agents of connected vehicles. *IEEE Transactions on Vehicular Technology* 66: 9021–9033.

30 Borenstein, J., Bucher, J., and Herkert, J. (2022). *Autonomous Vehicle Ethics: The Trolley Problem and Beyond*. Oxford University Press.

31 Noy, I.Y., Shinar, D., and Horrey, W.J. (2018). Automated driving: safety blind spots. *Safety Science* 102: 68–78.

32 Conlon, J. and Lin, J. (2019). Greenhouse gas emission impact of autonomous vehicle introduction in an urban network. *Transportation Research Record* 2673: 142–152.

33 Liu, F., Zhao, F., Liu, Z., and Hao, H. (2019). An autonomous vehicle reduce greenhouse gas emissions? A country-level evaluation. *Energy Policy* 132: 462–473.

34 Massar, M., Reza, I., Rahman, S.M. et al. (2021). Impacts of autonomous vehicles on greenhouse gas emissions – positive or negative? *International Journal of Environmental Research and Public Health* 18: 5567.

35 Scott, J., Zhuravleva, N.A., Durana, P., and Cug, J. (2020). Public acceptance of autonomous vehicle technologies: attitudes, behaviors, and intentions of users. *Contemporary Readings in Law and Social Justice* 12: 23.

36 Burton, S., McDermid, J.A., Garnett, P., and Weaver, R. (2021). Safety, complexity, and automated driving: holistic perspectives on safety assurance. *Computer* 8: 22–32.

37 Boeglin, J. (2015). The costs of self-driving cars: reconciling freedom and privacy with tort liability in autonomous vehicle regulation. *Yale Journal of Law and Technology* 17: 171.

38 Carp, J.A. (2018). Autonomous vehicles: problems and principles for future regulation. *University of Pennsylvania* 4: 81.

39 Varghese, J.Z. and Boone, R.G. (2016). Overview of autonomous vehicle sensors and systems. *International Conference on Operations Excellence and Service Engineering*, Orlando, USA, 178–191.

40 Liu, S., Li, L., Tang, J. et al. (2020). Creating autonomous vehicle systems. *Synthesis Lectures on Computer Sciences* 8: i–216.

41 Joubert, N., Reid, T.G., and Noble, F. (2020). Developments in modern GNSS and its impact on autonomous vehicle architectures. 2020 *IEEE Intelligent Vehicles Symposium,* Las Vegas, USA, 2029–2036.

42 Kai, Z., Andrew, B., Sreevatsan, B., and Ronald, C. (2019). *Autonomous Vehicle Lidar: A Tutorial.* Independently Published.

43 Liu, S., Li, L., Tang, J. et al. (2020). Autonomous vehicles: problems and principles for future regulation. *Synthesis Lectures on Computer Science* 8: 216.

44 Burlet, J. and Dalla, F.M. (2012). Robust and efficient multi-object detection and tracking for vehicle perception systems using radar and camera sensor fusion. *IET and ITS Conference on Road Transport Information and Control*, London, UK, 1–6.

45 Yeong, D.J., Velasco-Hernandez, G., Barry, J., and Walsh, J. (2020). Sensor and sensor fusion technology in autonomous vehicles: a review. *Creating Autonomous Vehicle Systems* 21: 2140.

46 Van Brummelen, J., O'Brien, M., Gruyer, D., and Najjaran, H. (2020). Autonomous vehicle perception: the technology of today and tomorrow. *Transportation Research Part C Emerging Technologies* 2018: 384–406.

47 Holder, C.J. and Shafique, M. (2022). On efficient real-time semantic segmentation: a survey. *arXiv* 2206.08605.

48 Bhavadharshini, V., Mridula, S., Sakthipriya, B., and Gracewell, J.J. (2023). Semantic segmentation using convolutional neural networks. *International Conference on Computing Methodologies and Communication*, Erode, India, 481–488.

49 Tsugawa, S., Yatabe, T., Hirose, T., and Matsumoto, S. (1979). An automobile with artificial intelligence. *6th international joint conference on Artificial intelligence-Volume 2*, Tokyo, Japan, 893–895.

50 De La Escalera, A., Moreno, L.E., Salichs, M.A., and Armingol, J.M. (1997). Road traffic sign detection and classification. *IEEE Transactions on Industrial Electronics* 44: 848–859.

51 Fu, M.Y. and Huang, Y.S. (2010). A survey of traffic sign recognition. *International Conference on Wavelet Analysis and Pattern Recognition*, Qingdao, China, 119–124.

52 Bochkovskiy, A., Wang, C.Y., and Liao, H. (2020). Yolov4: optimal speed and accuracy of object detection. *arXiv* arXiv:2004.10934.

53 Liu, W., Anguelov, D., Erhan, D. et al. (2016). Single shot multibox detector. *The 14th European Conference on Computer Vision*, Amsterdam, Holland, 21–37.

54 Girshick, R., Donahue, J., Darrell, T., and Malik, J. (2014). Rich feature hierarchies for accurate object detection and semantic segmentation. *IEEE conference on computer vision and pattern recognition*, Columbus, USA, 580–587.

55 Kuyu, Y.C. and Ozekmekci, N. (2022). Grey wolf optimizer to the hyperparameters optimization of convolutional neural network with several activation functions. *IEEE International Symposium on Multidisciplinary Studies and Innovative Technologies (ISMSIT)*, Ankara, Turkey, 13–17.

56 Bresson, G., Alsayed, Z., Yu, L., and Glaser, S. (2017). Simultaneous localization and mapping: a survey of current trends in autonomous driving. *IEEE Transactions on Intelligent Vehicles* 2: 194–220.

57 Durrant-Whyte, H. and Bailey, T. (2006). Simultaneous localization and mapping: part I. *IEEE Robotics and Automation Magazine* 13: 99–108.

58 Chen, W., Zhou, C., Shang, G. et al. (2022). Slam overview: from single sensor to heterogeneous fusion. *Remote Sensing* 14: 6033.

59 Billinghurst, M., Clark, A., and Lee, G. (2015). Found trends human-computer interact. *A Survey of Augmented Reality*, 8:73–272, 2015.

60 Taketomi, T., Uchiyama, H., and Ikeda, S. (2017). Visual slam algorithms: a survey from 2010 to 2016. *IPSJ Transactions on Computer Vision and Applications* 9: 1–11.

61 Davison, A.J (2003). Real-time simultaneous localisation and mapping with a single camera. *International Conference on Computer Vision*, Nice, France, 1403–1410.

62 Klein, G. and Murray, D.W. (2007). Parallel tracking and mapping for small AR workspaces. *International Symposium on Mixed and Augmented Reality*, Nara, Japan, 225–234.

63 Newcombe, R.A., Lovegrove, S.J., and Davison, A.J. (2011). DTAM: dense tracking and mapping in real-time. *International Conference on Computer Vision*, Barcelona, Spain, 2320–2327.

64 Taheri, H. and Xia, Z.C. (2021). Slam; definition and evolution. *Engineering Applications of Artificial Intelligence* 97: 104032.

65 Censi, A. (2011). An ICP variant using a point-to-line metric. *IEEE International Conference on Robotics and Automation*, Shanghai, China, 19–25.

66 Alismail, H., Baker, L.D., and Browning, B. (2014). Continuous trajectory estimation for 3d slam from actuated lidar. *IEEE International Conference on Robotics and Automation*, Hong Kong, China, 6096–6101.

67 Jost, T. and Hugli, H. (2002). A multi-resolution scheme ICP algorithm for fast shape registration. *IEEE International Symposium on 3D Data Processing Visualization and Transmission*, Padua, Italy, 19–21 June 2002.

68 Men, H., Biruk, G., and Kishore, P. (2011). Color point cloud registration with 4d ICP algorithm. *IEEE International Conference on Robotics and Automation*, Shanghai, China.

69 Zhuang, H., Dong, K., Qi, Y. et al. (2021). Multi-destination path planning method research of mobile robots based on goal of passing through the fewest obstacles. *Applied Sciences* 11: 7378.

70 Benelmir, R., Bitam, S., and Mellouk, A. (2020). An efficient autonomous vehicle navigation scheme based on lidar sensor in vehicular network. *IEEE Conference on Local Computer Network*, Sydney, Australia, 349–352.

71 Dijkstra, E.W. (1959). A mobile automaton: an application of artificial intelligence techniques. *Numerische Mathematik* 1: 269–271.

72 Nilsson, N.J. (1969). A mobile automaton: an application of artificial intelligence technique. Technical Report, DTIC Document.

73 Bast, H., Delling, D., Goldberg, A. et al. (2016). Route planning in transportation networks. *Algorithm Engineering* 19–80.

74 Huang, Y., Ding, H., Zhang, Y. et al. (2019). A motion planning and tracking framework for autonomous vehicles based on artificial potential field elaborated resistance network approach. *IEEE Transactions on Industrial Electronics* 67: 1376–1386.

75 Sharma, O., Sahoo, N.C., and Puhan, N.B. (2021). Recent advances in motion and behavior planning techniques for software architecture of autonomous vehicles: a state-of-the- art survey. *Engineering Applications of Artificial Intelligence* 101: 104211.

76 Takeda, K., Miyajima, C., Suzuki, T. et al. (2011). Improving driving behavior by allowing drivers to browse their own recorded driving data. *IEEE Conference on Intelligent Transportation Systems*, Washington, USA, 44–49.

77 Naghshvar, M., Sadek, A.K., and Wiggers, A.J. (2018). Risk-averse behavior planning for autonomous driving under uncertainty. *arXiv* 1812.01254.

78 Chong, L., Abbas, M.M., Flintsch, A.M., and Higgs, B. (2013). A rule-based neural network approach to model driver naturalistic behavior in traffic. *Transportation Research Part C: Emerging Technologies* 32: 207–223.

79 Halim, Z., Kalsoom, R., Bashir, S., and Abbas, G. (2016). Artificial intelligence techniques for driving safety and vehicle crash prediction. *Artificial Intelligence Review* 46: 351–387.

80 Rezaie, M.F., Afandizadeh, S., and Ziyadi, M. (2011). Prediction of accident severity using artificial neural networks. *International Journal of Civil Engineering* 9: 41–48.

81 Yasin Çodur, M. and Tortum, A. (2015). An artificial neural network model for highway accident prediction: a case study of Erzurum, Turkey. *Promet – Traffic & Transportation* 27: 217–225.

82 O'Doherty, J.P., Dayan, P., Friston, C.H., and K., and R. J. Dolan. (2003). Temporal difference models and reward-related learning in the human brain. *Neuron* 38: 329–337.

83 Ning, H., Xu, W., Zhou, Y. et al. (2008). Temporal difference learning to detect unsafe system states. *IEEE International Conference on Pattern Recognition*, Florida, USA, 1–4.

84 Zhu, Z. and Zhao, H. (2021). A survey of deep RL and IL for autonomous driving policy learning. *IEEE Transactions on Intelligent Transportation Systems* 23: 14043–14065.

85 Furda, A. and Vlacic, L. (2011). Enabling safe autonomous driving in real-world city traffic using multiple criteria decision making. *IEEE Intelligent Transportation Systems Magazine* 3: 4–17.

86 Hardy, J. and Campbell, M. (2013). Contingency planning over probabilistic obstacle predictions for autonomous road vehicles. *IEEE Transactions on Robotics* 29: 913–929.

87 Kala, R. and Warwick, K. (2014). Heuristic based evolution for the coordination of autonomous vehicles in the absence of speed lanes. *Applied Soft Computing* 19: 387–402.

88 Bengio, S., Vinyals, O., Jaitly, N., and Shazeer, N. (2015). Scheduled sampling for sequence prediction with recurrent neural networks. *Advances in Neural Information Processing Systems* 28.

89 Florensa, C., Held, D., Wulfmeier, M. et al. (2017). Reverse curriculum generation for reinforcement learning. *In Conference on Robot Learning*, California, USA, 482–495.

90 George, E.M. (1982). State of the art – a survey of partially observable markov decision processes: theory, models, and algorithms. *Management Science* 28: 1–16.

91 Hausknecht, M. and Stone, P. (2015). Deep reinforcement learning in parameterized action space. *arXiv* 1511.04143.

92 Qiao, Z. Reinforcement learning for behavior planning of autonomous vehicles in urban scenarios. Thesis. Carnegie Mellon University.

9

Energy Storage Technologies and Control Systems for Electric Vehicles

Mariem Ahmed Baba[1], Mohamed Naoui[2], and Mohamed Cherkaoui[1]

[1] Engineering for Smart and Sustainable Systems Research Center, Mohammadia School of Engineers, Mohammed V University in Rabat, Rabat, Morocco
[2] Research Unit of Energy Processes Environment and Electrical Systems, National Engineering School of Gabes, University of Gabés, Gabés, Tunisia

Acronyms

AFC	Alkaline fuel cell
BLDC	Brushless DC motor
CO_2	Carbon dioxide
DMFC	Direct methanol fuel cell
EMF	Electromotive force
EVs	Electric vehicles
FC	Fuel cell
Li-ion	Lithium-ion
MCFC	Molten carbonate fuel cell
NiCd	Nickel-cadmium
Ni-MH	Nickel-metal-hydride
PAFC	Phosphoric acid fuel cell
Pb-acid	Lead-acid
PEMFC	Proton exchange membrane fuel cell
PI	Proportional-integral controller
PID	Proportional-integral-derivative controller
PMSM	Permanent magnet synchronous motors
SOFC	Solid oxide fuel cell

Interconnected Modern Multi-Energy Networks and Intelligent Transportation Systems: Towards a Green Economy and Sustainable Development, First Edition. Edited by Mohammadreza Daneshvar, Behnam Mohammadi-Ivatloo, Amjad Anvari-Moghaddam, and Reza Razzaghi.
© 2024 The Institute of Electrical and Electronics Engineers, Inc.
Published 2024 by John Wiley & Sons, Inc.

9.1 Introduction

The current transportation sector is based on conventional vehicles, increasing the danger of destroying the environment through global warming. The latest studies focus on finding an alternative solution that eliminates fossil fuel use in vehicles. Electric vehicles are considered the most promising solution in the automotive industry. There are several EV technologies, including hybrid electric vehicles (HEVs), battery electric vehicles (BEVs), plug-in hybrid electric vehicles (PHEVs), and fuel cell electric vehicles (FCEVs) [1].

These EVs generally are taken as a means of storage of the battery pack and the ultracapacitor with a possible collaboration of fuel cells [2]. The operating system of EVs relies on the fact that the wheels of the vehicle are driven by the rotational power provided by the electric motor. FCEVs may become more prevalent in the transportation sector if investments in refueling stations continue to increase (Odabaş, İ. H., Akca 2020). Fuel cells can be classified into several categories according to different criteria, such as their chemical properties and operating temperature [3].

The automotive industry is making a great effort to find a compatible solution to keep pace with market demands with minimal emissions of climate-destroying substances. Among the most promising solutions, in this case, are pure BEV, FCEV, and HEV [4]. Indeed, BEVs have a particular system that eliminates using a traditional internal combustion engine and instead requires an additional (external) source to recharge the battery. Regarding FCEV, it can play the role of a battery or an internal combustion engine, depending on the power system. It produces electrical energy through a chemical process that resembles the battery system. At the same time, it can operate as an internal combustion engine in the case of the adoption of hydrogen as fuel [5]. HEVs rely on several storage sources to provide sufficient electricity for electric propulsion, whether by integrating batteries, fuel cells, or internal combustion engines [6]. FCEV and FCHEV vehicles generally rely on hybridization between fuel cells, rechargeable batteries, and ultracapacitors because the existence of the latter two ensures sufficient energy to provide the necessary power. The storage system is the most influential factor in the efficiency and range of hybrid vehicles, as it enables energy to be stored and quickly released as needed [7]. In the case of South Korea, for example, the number of FCEVs in 2017 increased five times more compared to 2015 [8]. The brushless motor has become popular in the automotive industry due to its high-performance and well-developed control system. There are many technologies to control this motor, of which this chapter presents three widespread control modes: PI controller, PID controller, and fuzzy logic controller. The work presented in this chapter provides a concrete study of energy storage systems used in electric vehicles (fuel cells and batteries). Several car manufacturers, such as Toyota and Honda, have adopted the BLDC engine due to its high efficiency, hence the importance of this work, which also focuses on modeling and controlling a BLDC engine designed for electric vehicles.

The first part of this chapter examines a study on the different types of fuel cells, highlighting their role in the development of electric cars. Furthermore, the second section presents the new battery technologies used in electric vehicles and their corresponding characteristics. In the end, the section examines the different control strategies of the brushless motor applied to EVs.

9.2 Fuel Cell

The fuel cell consists mainly of three components: the anode, the cathode, and the electrolyte. The anode's "oxidation" phase occurs, where hydrogen is oxidized into protons and electrons. On the other hand, at the cathode, the "reduction" phase takes place where oxygen is reduced to produce heat and water [9, 10].

$$½ O_2 + 2H^+ + 2e^- \rightarrow H_2O + \text{Electricity} + \text{Heat}$$

The operating principle of this battery is based on hydrogen reacting with oxygen in the air. In the internal combustion engine system, for example, energy is released in the form of heat. In contrast, the fuel cell is an electrochemical reaction that provides electrical power and heat [2, 11]. In fact, these batteries sometimes play the role of a lifesaver; for example, in remote and rural areas, it is a significant source of energy, especially in the unavailability of public networks or when the costs of wiring and power supply are frequent.

In Ref. [12], a comparison between fuel cell (FC) and various power generation technologies in Table 9.1 shows a high efficiency compared to other sources

Table 9.1 Comparison between different means of energy production.

	Reciprocating engine: diesel	Turbine generator	Photovoltaic	Wind turbine	Fuel cells
Capacity range	500 kW–50 MW	500 kW–5 MW	1 kW–1 MW	10 kW–1 MW	200 kW–2 MW
Efficiency	35%	29–42%	6–19%	25%	40–85%
Capital cost ($/kW)	200–350	450–870	6600	1000	1500–3000
O & M cost ($/kW)	0.005–0.015	0.005–0.0065	0.001–0.004	0.01	0.0019–0.0153

(including diesel, turbogenerator, and wind turbine). Therefore, using hydrogen as a reagent offers better skills and, above all, a cleaner environment.

9.2.1 Types of Fuel Cells

There are currently several types of FC, of which the main ones are six categories [13]: proton exchange membrane fuel cell (PEMFC), phosphoric acid fuel cell (PAFC), alkaline fuel cell (AFC), molten carbonate fuel cell (MCFC), solid oxide fuel cell (SOFC), and direct methanol fuel cell (DMFC). These cells are generally classified according to the fuel type, the degree of operating temperature, and the chemical characteristics of the electrolyte [14]. In addition, the most widely used FCs today are PEMFCs and solid oxide fuel cells (SOFCs) [15].

9.2.1.1 Proton Exchange Membrane Fuel Cell

PEM is characterized by the exploitation of a solid polymer electrolyte which is considered as an ideal conductor and its role of insulation for electrons. The operating temperature of PEMs is typically 70°C–80°C, although this margin changes in the case of the automotive sector, where this temperature varies between 120°C and 150°C [16]. Furthermore, PEMFCs are considered the most promising fuel cell technology due to their high competence, including fast start-up technique, ability to perform many thermal cycles, low-operating temperature, and hydrogen conversion efficiency. Indeed, the most common use of PEMs is in the fields of transport and portable power generation. The application of PEMs in the transport sector (cars, buses, etc.) has increased in recent decades due to the high and unstable price of fossil fuels, the unavailability of fuel to meet needs, and the global consensus on the need to find an alternative to fossil fuels in the automotive sector to reduce CO_2 emissions [16, 17]. Actually, the introduction of these types of fuel cells in vehicles (FCV) has been incredibly successful in recent times, including the 2017 Toyota Mirai car, which has a power density of 3.1 kW/L and a fuel cell stack that reaches 114 kW, and Hyundai's Hyundai Tucson model [18].

Indeed, fuel cell buses present a perfect solution that continues to grow in the market as this type of car has safe and convenient hydrogen storage compared to other vehicles. The challenges facing PEMS remain an obstacle to its adaptation in other fields. Among these disadvantages are the low efficiency, which varies between 40% and 45%, and the high cost of the platinum catalyst used in constructing PEMs. Furthermore, a study in Ref. [19] concluded that the significant challenges of PEMs are the high cost for most applications, the low sustainability, and the non-existence of a hydrogen infrastructure. Therefore, Figure 9.1 represents the descriptive diagram of the PEM fuel cell.

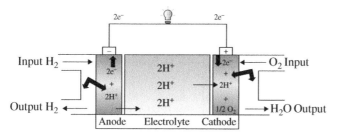

Figure 9.1 Proton exchange membrane fuel cell (PEMFC).

9.2.1.2 Phosphoric Acid Fuel Cell (PAFC)

Over the past decades, there has been a lot of research into developing PAFCs in various fields, such as electrochemical systems and the transport sector. The PAFC is characterized by a liquid electrolyte (phosphoric acid) compared to the DMFC and PEM. In the case of PAFCs, liquid phosphoric acid acts as the electrolyte, while hydrogen is considered the fuel [20]. It is necessary to mention that the operating temperature of PAFCs is higher than that of alkaline fuel cells and proton exchange membrane (PEM) fuel cells, as PAFCs have an operating temperature that varies between 150°C and 220°C. The principle of operation is characterized by the movement of electrons and protons in succession to produce electrical current and heat. These types of fuel cells are mainly used in the stationery and transport sectors.

Moreover, PAFCs offer the option of cogeneration through the effect of their high-operating temperature to produce electrical energy and heat simultaneously through the combustion of a specific fuel. On the other hand, PAFCs have much the same disadvantages as PEMs, which essentially boil down to the fuel cell's high cost due to using a platinum catalyst [20, 21]. However, compared with AFCs, hydrogen vapor impurities do not influence PAFCs. These fuel cells have an efficiency of about 40%–50%, whereas the efficiency of cogeneration can be as high as 85% [22]. Figure 9.2 shows the schematic of the PAFC battery with its main components.

9.2.1.3 Alkaline Fuel Cell

Alkaline fuel cells have been exploited in the field of space missions by the National Aeronautics and Space Administration (NASA) as the first type of fuel cell used in this sector [23]. They are characterized by an electrolyte, an aqueous solution of potassium hydroxide (KOH) as shown in Figure 9.3. The AFC system is characterized by ions being negatively charged toward the cathode, releasing water. In contrast, at the cathode, electrons returning from the anode react with oxygen and water to form new hydroxyl ions [24]. Therefore, fuel cells are reduced

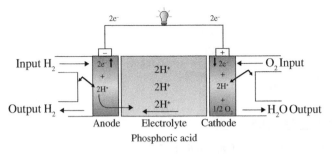

Figure 9.2 Phosphoric acid fuel cell (PAFC).

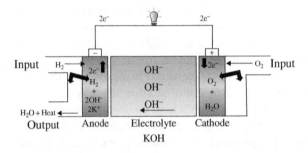

Figure 9.3 Alkaline fuel cell (AFC).

to pure hydrogen and oxygen elements to avoid CO_2 absorption, which impedes the conductivity of the electrolyte.

The temperature of these types of batteries is between 60°C and 120°C with an electrical efficiency that reaches 70% [25]. To keep the AFCs working correctly and to avoid cell destruction, it is recommended not to exceed the operating temperature of 70°C [26]. These types of FCs are designed to be used for low-power onboard applications in the range (of 1–100 kW). The choice of this type of fuel cell comes down to its several advantages, including good efficiency of 60%–70%, fast start-up, no greenhouse gas emissions, and high-power density. However, the problem of carbon dioxide sensitivity limits the use of AFCs despite all these advantages.

9.2.1.4 Molten Carbonate Fuel Cell

Molten carbonate fuel cells (MCFC) are fuel cells known for their ability to operate under high temperatures that can reach the range of 600°C–650°C and are characterized by an electrolyte in the form of a molten alkali carbonate [27]. Figure 9.4 shows the general structure of the molten carbonate fuel cell. MCFCs have a high

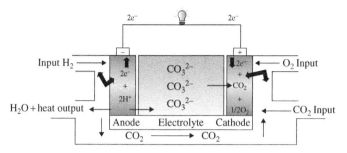

Figure 9.4   Molten carbonate fuel cell (MCFC).

efficiency ranging from 50% to 60%) with a low-power density compared to other fuel cells ranging from 100 to 150 mW/cm². It works because hydrogen combines with carbonate ions (CO_2^{2-}) in a redox reaction, whereby electrons are released to produce water and carbon dioxide. In a reduction reaction at the cathode, oxygen combines with carbon dioxide and incoming electrons to release carbonate ions [28]. Despite the disadvantages of MCFC, including low-power density, system complexity, and sulfur sensitivity, on the other hand, it has many advantages, including high efficiency, high internal reforming capacity, and the absence of metal catalyst in its structure. As a result, the use of MCFCs remains limited in large installations, making them more available in power plants and industrial cogeneration. According to MCFC studies, the significant challenge for improving the commercialization of conventional hydrogen-fueled and direct internal reforming MCFCs is the stability of the electrodes [29].

9.2.1.5   Solid Oxide Fuel Cell

Solid oxide fuel cells (SOFCs) are characterized by their high operating temperatures of up to 1000°C with a high efficiency of 65%. The electrolyte used in this type of fuel cell is yttrium-stabilized zirconium, which is considered to be a ceramic-based solid material. Two types of SOFCs are defined according to the mobile ion the proton-conducting SOFCs and oxygen ion-conducting SOFCs [30]. The SOFC is considered one of the most promising fuel cells despite the obstacles that prevent its commercialization, including mainly its high operating temperature [31], which is why recent research focuses on reducing this temperature. To solve this problem, the principal solution proposed is the improvement of anode SOFCs by manufacturing cells with a very thin electrolyte to reduce the temperature to about 700°C–800°C [32].

On the other hand, the SOFC-gas turbine hybrid system can be considered an efficient method for industrial power supplies and in the automotive sector due to the high competence that this system offers. Moreover, SOFCs have several

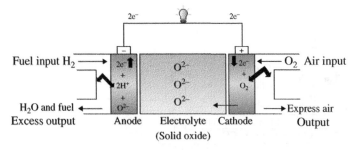

Figure 9.5 Solid-oxide fuel cell (SOFC).

advantages, such as high efficiency, internal reforming (IR) capability, and possibility to use non-conventional fuels such as biomass. Figure 9.5 shows the detailed system diagram of the solid-oxide fuel cell.

9.2.1.6 Direct Methanol Fuel Cell

The DMFC is an FC that uses methanol as fuel. Its architecture is similar to the PEM system, with the difference of using methanol instead of hydrogen. The DMFC system does not need a fuel processor, making its structure more superficial and efficient. Its operating principle generally consists of a membrane-electrode combination squeezed between two bipolar plates to pass the fuel through distribution channels [33]. Figure 9.6 shows the general concept of this fuel cell type. There are three types of DMFC depending on the choice of fuel and oxidant supply mode: direct passive methanol fuel cell, direct active methanol fuel cell, and direct semi-passive fuel cell. The DMFC is characterized by its low-operating temperature of up to 60°C with a power density of less than $100\,\mathrm{mW/cm^2}$. Its field of use includes mainly small electric vehicles and portable applications.

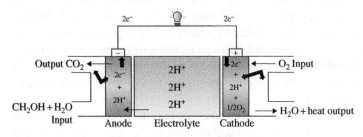

Figure 9.6 Direct methanol fuel cell (DMFC).

9.3 Battery Technologies for Electric Vehicles

The battery is defined as a device that ensures energy storage to deliver electrical energy. The battery has an operating system that serves to convert chemical energy directly into electrical energy through several chemical reactions [34]. It is divided into two categories: primary batteries and secondary batteries. Primary batteries are characterized by the provision of energy for a single time during discharge, while secondary batteries permanently offer storage energy during the process of charge and discharge during the whole life of the battery.

The characteristics of a battery are generally defined by several criteria indicating its performance which are:

- Energy density
- Cost in the market
- The number of charging cycles
- The discharge processes
- The influence on the environment
- The temperature ranges
- The memory effects

This section will focus on secondary batteries as they are used in electric or hybrid vehicles. Many batteries are used in EVs, generally based on lead acid, nickel, lithium metal, silver, and sodium-sulfur. The following battery technologies that are used in EVs will be described respectively: lead-acid (Pb-acid), nickel-cadmium (NiCd), nickel-metal-hydride (Ni-MH), and lithium-ion (Li-ion). Table 9.2 shows the characteristics of the different battery hybrid cars, their electrification systems, costs, and reduction of CO_2 emissions in each case.

9.3.1 Lead-Acid Batteries

Lead-acid batteries represent the oldest category of rechargeable batteries, which appeared in 1859; therefore, its use was widespread in gasoline-powered vehicles, but over time it has become widely used in electric cars [36]. This battery was first used in 1981, introduced in a three-wheeled electric vehicle with a maximum speed of 12 km/h. Therefore, in 1899, another use of this battery in the automotive sector was implemented by powering an electric car designed by Camille Jenatzy with a maximum speed of 100 km/h [37]. Although this type of battery is considered the most widespread battery technology in the automotive sector, its use is mainly based on powering a vehicle's starter motor. Many car manufacturers have adopted lead-acid batteries, including the GM EV1 manufactured by General Motors and the Toyota RAV4 EV. Lead-acid batteries are available in the market in all sizes and with different voltage ratings and are characterized by their ability

Table 9.2 Characteristics of the different battery hybrid cars, their electrification systems, costs, and CO_2 emission minimization.

	Micro-HEV start/stop and regenerative braking	Mild HEV + launch assist	Full HEV + power assist and limited e-drive	Plug-in HEV + extended e-drive
Preferred battery	EFB, AGM	Li-ion, LAB + Li-ion	Ni-MH, Li-ion	Li-ion
System voltage (V)	14(−48)	48–150	>200	>200
Battery power (kW)	2–10	7–20	>20	>20
Usable battery (additional to SLI) (kWh) Launch assist (kW)	0–0.25	0.25–1	0.7–2.5	4–10
Launch assist (kW)	0	<15	>15	>60
E-drive range (km)	0	0	−2	−32
OEM on-cost	€150–700	€1600–3000	€3000–5000	€6000–10,000
CO_2 benefit (%)	4–7	8–12	15–20	20+
Cost (€) to achieve each 1% reduction in CO_2 emissions	35–100	200–250	200–250	300–500

Source: Ref. [35]/with permission of Elsevier.

to provide high surge currents as needed due to their low internal impedance [38]. The efficiency of these batteries is because they offer high surge currents and a low cost that reaches about (150 US$/kWh) despite the low ratio between energy/ weight and between energy/volume. In addition, there are several advantages of the lead battery that have contributed to its use in electric vehicles, including the fact that they have a high cell voltage (2 V), a h specified f, and ic power that varies between 200 and 300 W/kg) [39]. However, they have several disadvantages, including low specific energy which varies between 30 and 45 Wh/kg) and a short life span between 400 and 600 cycles. Despite the success of lead-acid batteries during the last century, their use has begun to decline because of lead, which is harmful to the environment and makes it necessary to rely on other batteries.

This is explained by the fact that human health is affected when living near lead battery plants, as it was found in a survey in China that the blood of residents in Guangdong province reached up to 100 μg/L due to the waste emitted from a lead-acid battery plant [40]. There are several types of lead-acid batteries, the main ones being: SLI batteries, deep-cycle batteries, stationary batteries, and VRLA batteries [41]. Indeed, SLI batteries are commonly used in the automotive sector, they ensure in turn, the power supply of the starter to make the engine run. Therefore, a generator will charge these batteries during the engine's operation.

9.3.2 Nickel-Cadmium Battery (NiCd)

Waldemar Jungner invented the nickel-cadmium (NiCd) battery in 1899, which is considered to be the first battery to use an alkaline electrolyte in its system. It is a rechargeable battery with nickel and cadmium electrodes carefully placed in a potassium hydroxide solution [42]. Indeed,this battery is equipped with a second-ary cell with a margin of between 1.2 and 1.4 V with a density of energy double that of the batteries with lead. It also has a long service life of up to 1000 charges. During the system operation of the battery in normal conditions, the ampere-hour yield reaches 80%, while the watt-hour yield reaches 65%. It has the advantage over lead-acid batteries in terms of robustness and energy density, but its high price has been one of the obstacles to its diffusion. NiCd batteries also suffer from the "memory effect" in case they are exposed to continuous cycles of charge and discharge at the same level, which can suffer a sudden drop in voltage as if it had been discharged [43]. Cadmium is considered a toxic metal for the body and the environment, which requires special care in its use and recycling process. Cadmium can be produced from other elements such as copper, lead, and zinc. Therefore, Ni-Cd batteries are currently rarely used because of the toxicity of cadmium and are rather replaced by other less polluting batteries, such as nickel-metal hydride (NiMH) batteries.

9.3.3 Nickel-Metal-Hydride (Ni-MH)

The Ni-MH battery was invented in 1980 by Stanford R. Ovshinsky; it is characterized by an alloy electrode that replaces the cadmium electrode and which ensures in its turn the storage of hydrogen [44]. The Ni-MH battery is composed of a positive $Ni(OH)_2$ electrode, a negative metal hydride (MH) electrode, and a potassium hydroxide solution as electrolyte (KOH). In recent years, this type of battery has been used in hybrid vehicles such as the Toyota Prius and Honda Civic. Unlike cadmium metal, NiMH batteries are environmentally friendly. The Toyota Mirai hydrogen car is also equipped with a Ni-MH battery pack that weighs 29 kg and offers the accumulation of 1.6 kWh of energy with an output voltage that reaches 245 V [45]. The operating system of these batteries in electric vehicles ensures the recovery of energy during deceleration and braking phases to restore it when the vehicle moves at a low speed. Ni-MH has a very high-energy density compared to lead, and nickel-cadmium batteries, its chemical structure and simple mechanism offer high-power capacity and long life. In addition, Ni-MH is considered a robust technology that ensures high tolerance during the over-charge and over-discharge process.

9.3.4 Lithium-ion (Li-ion)

This technology was first commercialized in 1991 [46], and it is currently considered among the most dominant rechargeable batteries due to its simple architecture, high charge capacity, and long life under normal conditions. Li-ion is generally made up of a lithium element introduced as a cathode and graphite used as the anode. The electrolyte ensures the ionic transfer of lithium between the cathode and the anode by forming an ion channel. Li-ion batteries have high efficiency and high-energy density (due to the high electrochemical potential of lithium) ranging from 90 to 190 Wh/kg and low maintenance, which make them commonly used in portable applications and electric vehicles [47]. Li-ion batteries have been adopted by several car manufacturers such as General Motors in the Chevy-Volt model, Tesla has also used Li-ion in their Roadster model, and Mitsubishi in their iMiEV model. Despite the advantages of the Li-ion battery, it has some disadvantages, mainly its high cost, which varied between $900/kWh and $1300/kWh in 2012. However, this disadvantage is no longer seriously considered because the cost of Li-ion has clearly decreased to a range between 200 and 700 US$/kWh [48]. On the other hand, it also has a low safety margin because it has low resistance against temperatures exceeding 70°C–100°C. This phenomenon of thermal risk leads to thermal runaways, which can cause an explosion or an accident in electric vehicles. Lithium-ion batteries are found in the market under several categories that are mainly distinguished by the cathode, as it is the main source of all active lithium ions in a lithium-ion battery system. Below are the different types of lithium-ion batteries most commonly used in EVs.

9.3.4.1 Lithium Cobalt Oxide (LiCoO$_2$, LCO)

This type of material was used in the construction of the first Li-Ion commercialized by the Sony group in 1991 [49]. The LCO is considered the most used material thanks to its characteristics with specific energy between 150 and 190 Wh/kg and a nominal voltage equal to 3.6 V [50]. Its system consists essentially of a cathode made of cobalt oxide and an anode made of carbon graphite. During the discharge process, Li-ion ions move from the anode to the cathode, and the process is reversed during the charge. This type of battery has an average operating temperature, a limited specific power, and a service life that varies between 800 and 1000 depending on the charge, discharge, and temperature conditions.

9.3.4.2 Lithium Manganese Oxide (LiMn$_2$O$_4$, LMO/Spinel)

LiMn$_2$O$_4$ are cathode materials that offer fast charging and high current discharge due to the low internal resistance of the cells. This battery is characterized by a three-dimensional (3D) spinel architecture that ensures a more efficient ion flow at the electrode [51]. As a result, LiMn$_2$O$_4$ has several advantages, including its high-voltage rating of over 3 V, and high-energy density of 300 Wh/kg [52].

9.3.4.3 Lithium Iron Phosphate (LiFePO$_4$, LFP)

The lithium iron phosphate battery has the advantage of reliability due to the use of phosphate material as a cathode material, high safety, and low cost. The introduction of phosphate in the cathode system of this battery has guaranteed a strong resistance against high temperatures, and its olivine structure also ensures excellent stability in the overcharge process. LFPs offer an average voltage of 3.40 V, a specific capacity reaching 160 mAh/g [53], and a considered high life cycle that varies between 1000 and 2000. In addition, it has a specific energy that varies between 90 and 120 Wh/kg and an operating temperature of up to 300°C [54]. LFPs have been adopted by several automotive companies such as Tesla, BYD, and General Motors.

9.4 Overview of Brushless Motor

Permanent magnet synchronous motors can be classified into two main groups, the AC permanent magnet synchronous motor (PI) is characterized by a sinusoidal-shaped back EMF. The second is the brushless DC motor (BLDC) which has a trapezoidal-shaped back EMF [55]. Brushless motors are considered a new technology for conventional DC motors with an electronic commutation principle. This type of motor eliminates the need for a commutator and brushes, its system contains a rotating permanent magnet on the rotor and stationary magnets on the motor housing. The BLDC motor is widely used in industrial sectors such as aerospace and medical equipment, but also in the automotive sector such as Honda, Peugeot, and Toyota in its hybrid model named Toyota Prius. This motor offers a high power-to-weight ratio and efficiency compared to other machines, easy maintenance, and long life under normal conditions. However, this lifetime may be limited in the case of operation at high intensities. In addition, the problem of high cost frequently arises because the BLDC system requires control electronics, increasing the total approved cost [56]. On the other hand, BLDC motors can be used for energy storage and conversion. BLDC motors can contribute to the energy storage process. For example, if the BLDC motor functions as a generator during periods of excess energy, it can store the excess energy in a battery.

Before accessing the control of a brushless motor, it is first necessary to specify the modeling followed to ensure the mathematical set of corresponding equations. The following part deals with the global modeling of a brushless motor.

9.4.1 Mathematical Modeling of BLDC Motor

Modeling a BLDC motor requires the representation of all electrical and mechanical equations.

9.4.1.1 Electric Model of BLDC

Before accessing the control part, the mathematical model of the studied motor must be chosen. The following relations express basic voltage equations of the armature winding belonging to the BLDC:

$$
\begin{cases}
V_a = RI_a + L\left(\dfrac{di_a}{dt}\right) + e_a \\[2mm]
V_b = RI_b + L\left(\dfrac{di_b}{dt}\right) + e_b \\[2mm]
V_c = RI_c + L\left(\dfrac{di_c}{dt}\right) + e_c
\end{cases}
\tag{9.1}
$$

The vector of voltages across the three phases is presented as follows:

$$
\begin{bmatrix} V_a \\ V_b \\ V_c \end{bmatrix} =
\begin{bmatrix} R & 0 & 0 \\ 0 & R & 0 \\ 0 & 0 & R \end{bmatrix}
\begin{bmatrix} i_a \\ i_b \\ i_c \end{bmatrix} +
\frac{d}{dt}
\begin{bmatrix} L_{aa} & L_{ab} & L_{ac} \\ L_{ba} & L_{bb} & L_{bc} \\ L_{ca} & L_{cb} & L_{cc} \end{bmatrix}
\begin{bmatrix} i_a \\ i_b \\ i_c \end{bmatrix} +
\begin{bmatrix} e_a \\ e_b \\ e_c \end{bmatrix}
\tag{9.2}
$$

The expression of the corresponding electrical speed is written as follows:

$$
W = \frac{d\theta}{dt} = p\frac{d\theta_r}{dt} = pw_r
\tag{9.3}
$$

The set of electromotive forces of the three phases (a, b, c) can be expressed using the relationship between the speed of rotation and the EMF:

$$
\begin{cases}
e_a = k_e w_r f_a(\theta) \\
e_b = k_e w_r f_b(\theta) \\
e_c = k_e w_r f_c(\theta)
\end{cases}
\tag{9.4}
$$

The expression of the electromotive force depends on the rotor position and is characterized by an offset of 120°.

where k_e is the coefficient of the corresponding electromotive force:

$$\begin{cases} e_a = k_e w f(\theta_e) \\ e_b = k_e w f(\theta_e - 2\Pi/3) \\ e_c = k_e w f(\theta_e + 2\Pi/3) \end{cases} \tag{9.5}$$

The current equations are as follows:

$$\begin{cases} \dfrac{di_a}{dt} = \dfrac{1}{3Lm}(2V_{ab} + V_{bc} - 3Ri_a + \Omega PK(-2e_a + e_b + e_c)) \\[2mm] \dfrac{di_b}{dt} = \dfrac{1}{3Lm}(-V_{ab} + V_{bc} - 3Ri_b + \Omega PK(e_a - 2e_b + e_c)) \\[2mm] \dfrac{di_c}{dt} = -\left(\dfrac{di_a}{dt} + \dfrac{di_b}{dt}\right) \end{cases} \tag{9.6}$$

where *ea, eb,* and *ec* are the different electromotive forces of each phase, V_{ab} and V_{bc} are the voltages between phases, and Ω is the mechanical rotation speed.

The converter part was realized from the following equations:

$$\begin{cases} V_a = \dfrac{(S_1)V_d}{2} - \dfrac{(S_4)V_d}{2} \\[2mm] V_b = \dfrac{(S_3)V_d}{2} - \dfrac{(S_6)V_d}{2} \\[2mm] V_c = \dfrac{(S_5)V_d}{2} - \dfrac{(S_2)V_d}{2} \end{cases} \tag{9.7}$$

9.4.1.2 Mechanical Model of BLDC

The following form gives the mechanical equation of motion. This equation allows us to deduct the mathematical expression speed:

$$J\frac{d\Omega}{dt} = C_e - f\Omega - C_r \tag{9.8}$$

The expression of the electromagnetic torque can be established as follows:

$$C_e = \frac{P_e}{\Omega} \tag{9.9}$$

with

$$P_e = e_a i_a + e_b i_b + e_c i_c \tag{9.10}$$

P_e represents electromagnetic power.

Table 9.3 BLDC parameters.

Rated speed	3800 rpm
Stator inductance	58.9 μH
Supply voltage	12 V
Magnet excitation flux	0.0030225 Wb
Number of pairs of poles	4
E.m.f. constant (Ke)	0.05
Moment of inertia of rotating parts (J)	0.0000528 N.m.s²
Stator resistance	0.39 Ω

The electromagnetic torque:

$$C_e = P(e_a i_a + e_b i_c + e_c i_c) \tag{9.11}$$

P is the number of pairs of poles.

The BLDC motor parameters used in this chapter are listed above in Table 9.3

9.5 BLDC Motor Control Strategy for Electric Vehicles

Brushless motors are considered an alternative technology to conventional DC motors. For the control of BLDC in the case of electric vehicles, many control methods are applied, whether with the adoption of sensors or without sensors. There is a classical control for using proportional-integral (PI) controllers or proportional-integral derivative (PID) controllers. Other developed techniques are also implemented including fuzzy logic control, digital signal processor (DSP), adaptive neuro-fuzzy controller, etc. The following section discusses sensorless control methods [57].

9.5.1 PI Controller

This proportional-integral (PI) controller is used in various applications in the industrial sector. It ensures the delivery of control of a system using the difference between the set point and the output obtained. It is characterized by its simple architecture and its contribution to the realization of several control strategies, such as predictive control.

The proportional-integral (PI) controller also has a serial or parallel architecture, and its usual transfer function is represented as follows:

$$PI(S) = K_P . \frac{w_i}{S} . \left(1 + \frac{S}{w_i}\right) \tag{9.12}$$

where $Kp = K1$, $w_i = \dfrac{k_1}{k_2} = \dfrac{1}{T_i}$.

9.5.2 PID Controller

PID control is a name that symbolizes the three terms: P represents the proportional term, the letter I represents the integral term, and D represents the derivative term of the controller. The PID controller has proven its competence in the industrial field because it ensures the improvement of the stability, precision, and speed of the system it is applied to control the speed of a brushless motor, and the PID controller is a popular choice to ensure better results. PID controller can exist in three structures according to the combination of the three actions, they are series PID controller, parallel PID controller, and ideal PID controller. The architecture of the PID controller is shown in Figures 9.7–9.9.

The following transfer function describes the case of the parallel PID:

$$C(S) = K_p + \frac{K_I}{S} + K_d S \tag{9.13}$$

Reducing to the same denominator, we obtain

$$C(S) = \left(K_p S + K_I + K_d S^2\right)/S \tag{9.14}$$

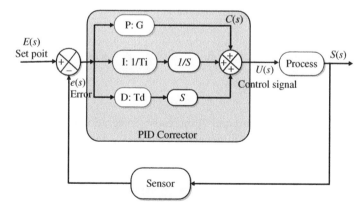

Figure 9.7 Block diagram of a PID controller (parallel form).

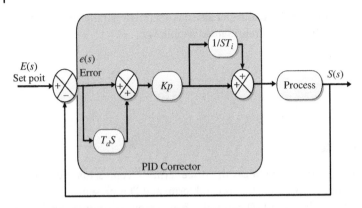

Figure 9.8 Block diagram of a PID controller (serial form).

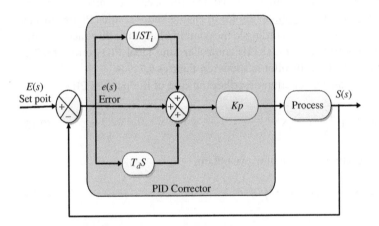

Figure 9.9 Block diagram of a PID controller (ideal form).

The following form obtains the transfer function of the PID series:

$$K(s) = K_p\left(1 + \frac{1}{T_i S}\right)(1 + T_d S) \tag{9.15}$$

where T_d represents the time constant and T_i represents integration time.

The PID equation for the ideal structure case is described as follows:

$$G(s) = \frac{1}{S + T_i} + S T_d \tag{9.16}$$

PID controllers are considered the most widely used industrial controllers in the last decades. To ensure the control of a PI or a PID, several methods have been proposed to fix in a fast way the different parameters of the controller. These are the Ziegler-Nichols method, the critical point method, and the index response method.

9.5.3 Fuzzy Logic Controller

Fuzzy logic control technology has proven its high capacity very quickly, becoming one of the most used control techniques in industrial applications. LF consists of expressing operational rules in linguistic forms instead of mathematical equations. The integration of LF has contributed to the simplicity of complex systems to be controlled. Comparative studies have shown the success of LF control compared to those providing conventional control algorithms. This has ensured the presence of this control mode in various electric motor control systems, including the brushless motor.

9.5.3.1 Fuzzification

This phase consists of transforming real quantities into fuzzy quantities, each input value is represented in linguistic form by the intermediary of membership functions, by allotting to each input its own membership function for a moment t. In fuzzy logic, the linguistic terms are generally represented in logical elements such as the expression of the rules (If.then).

9.5.3.2 Fuzzy Inference

Fuzzy inference is an operation that ensures the formulation of the mapping of a determined input to an output by the principle of fuzzy logic. The fuzzy inference technique has presented great success in several fields, such as regulating automated systems and electronic applications.

9.5.3.3 Defuzzification

This step consists of converting the internal fuzzy output variables into real variables so the system can use these variables. This operation can be done by different methods, the most famous of which are: the max criterion method, the height method, the centroid method, or the center of area method.

For the speed control of a brushless motor, it is necessary to specify the applied fuzzy inference rules. Table 9.4 represents the set of fuzzy inference rules adopted in this case, whose corresponding fuzzy membership functions are the following.

Table 9.4 Fuzzy logic controller rules set.

E/CE	NB	NM	NS	ZE	PS	PM	PB
NB	NB	NB	NB	NB	NM	NS	ZE
NM	NB	NB	NB	NM	NS	ZE	PS
NS	NB	NB	NB	NM	ZE	PS	PM
ZE	NB	NM	NS	ZE	PS	PM	PB
PS	NM	NS	ZE	PS	PM	PB	PB
PM	NS	ZE	PS	PM	PB	PB	PB
PB	ZE	PS	PM	PB	PB	PB	PB

NG: negative large, EZ: about zero, and PG: positive large.

9.6 Simulation Results

The simulation results are obtained from the realization of three different control systems of a brushless motor: PI, PID, and fuzzy logic control, as in Figures 9.10–9.12. Indeed, these figures illustrate the rotor speed of a BLDC motor and the currents when different controllers are used. The motor achieves a speed of 2000 rpm in 0.5 seconds with turbulence (resistive torque Cr) in time 0.3 seconds, then we change the speed to 1000 rpm in all circumstances, although with varying rise time, settling time, and stability. We note that the speed curve in PID and PI contains many oscillations contrary to the curve of fuzzy logic. However, the current curve in the three control cases is considered close with slight differences. For example, by observing the current curves in Figures 9.10 and 9.11, we can see that the PID has succeeded in reducing the disturbance that appeared in the case of the PI controller between the instants 0 and 0.1 seconds.

The results obtained from the simulation in Matlab proved that the fuzzy logic was the best among the three implemented controllers, as shown in the speed curve where the disturbances were reduced powerfully. Therefore, the PID controller showed its effectiveness before the PI controller as shown in the speed and current curves presented above. The PID controller has a shorter rise time than the PI and fuzzy controllers. The PI and PID controllers have the same and shorter settling times as the fuzzy controller. The PI overshoot percentage is 1.5% higher than the fuzzy controller, and the PID overshoot percentage is 0.7% higher than the fuzzy controller.

Table 9.5 compares various operating characteristics such as rising time, settling time, and overshoot percentage.

(a)

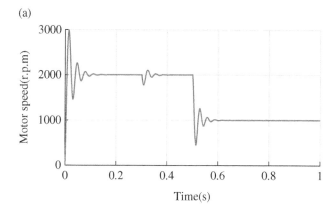

(b)

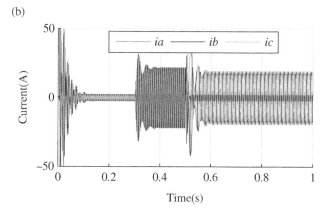

Figure 9.10 Speed and current for the PI control: (a) speed and (b) current.

9.7 Environnemental Impact of EVs

Although it is generally accepted that electric cars do not harm the environment, they can indirectly contribute to pollution. Indeed, producing electricity in these vehicles is accompanied by greenhouse gas emissions. The effect of greenhouse gas emissions caused by any type of EV always depends on the primary materials or other sources producing the electricity. For example, coal-fired EVs severely damage the environment with a global warming potential up to 27% higher than diesel and gasoline vehicles [58]. While HEVs and PHEVs have lower emissions than ICE vehicles [59]. In general, EVs are considered the best means of transportation to save the planet from climate pollution. Electric vehicles recharged from renewable energy sources are a non-polluting technology that deserves to be more widely available in the automotive market.

(a)

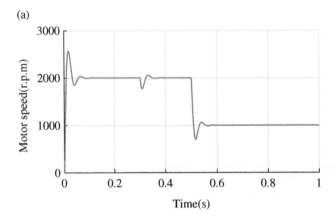

(b)

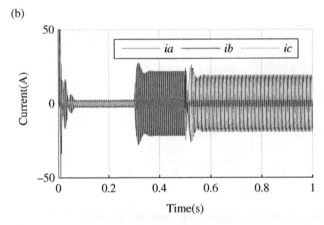

Figure 9.11 Speed and current for the PID control: (a) speed and (b) current.

9.8 EVs and Modern Technologies

Electric cars are known for their environmental friendliness, especially concerning air pollution, reducing damage to human health compared to the harmful emissions produced by conventional vehicles [60]. In addition, several innovative artificial intelligence technologies have been recently used to improve and optimize storage systems in electric vehicles as in Ref. [61] whose authors have used the fuzzy logic controller to ensure better battery management. Several modern strategies have been applied to maintain the safety of vehicles while driving. In Ref. [62], an Internet of things (IoT)-based system is used to organize vehicles' movement and specify their parking. A new branch of IOT called the Internet of vehicles

(a)

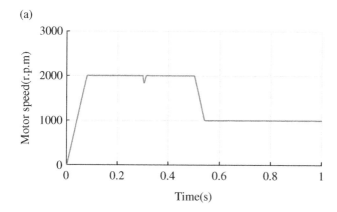

(b)

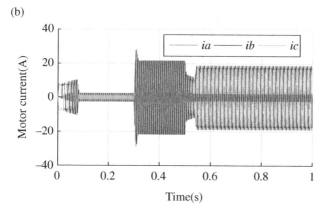

Figure 9.12 Speed and current for the fuzzy logic: (a) speed and (b) current.

Table 9.5 Speed response characteristics.

Controller	Rise time (seconds)	Settling time (seconds)	Overshoot percentage (%)
PI controller	0.03	0.06	1.5
PID controller	0.01	0.06	0.7
FUZZY controller	0.08	0.11	0

(IOV) has been developed to offer exceptional communication techniques and strong security measures. Among these techniques, the most widespread are vehicle-to-infrastructure (V2I), vehicle-to-pedestrian (V2P), vehicle-to-vehicle (V2V), vehicle-to-network (V2G), and vehicle-to-anything (V2X).

9.9 Challenges and Perspectives of EVs

Among the most worrying challenges in the construction of EVs is autonomy because using batteries reduces the profitability of these vehicles regarding driving time because of the need for this type of car to recharge every period. For example, a battery-powered vehicle is able to travel a distance of up to 250 km [63]. This capacity is considered an evolution in EVs, but it is insufficient to replace conventional vehicles. Moreover, with the development of electric vehicle technologies, the dream of replacing traditional vehicles with electric ones has become a reality. Therefore, a study presented in the United States [64] estimates that when conventional vehicles are replaced by battery-powered vehicles, greenhouse gas emissions can be reduced by about 25%. The same study also cited that when fuel-cell electric vehicles are used instead of conventional vehicles, emissions will be reduced by 40%.

9.10 Conclusion

This work has examined the main terms in electric vehicles: energy storage systems and electric motor control strategies for adopting the brushless motor. The study carried out in this work allows researchers to know the best energy storage technologies for EVs with the advantages and disadvantages of each method. On the other hand, within the framework of BLDC engine control, the results obtained during simulation present the effect of each of the three controllers used. First, a detailed study of the different types of fuel cells has been presented. In the second part, the different battery technologies used in EVs were discussed, and several brushless motor control strategies were proposed to make a comparison between PI, PID, and fuzzy logic control when simulating these systems with Matlab software. The simulation results clearly showed the efficiency of the fuzzy logic control compared to the other proposed control models.

Acknowledgments

The author is deeply grateful to the Organization of Women in Science for the Developing World (OWSD) and the Swedish International Development Cooperation Agency (SIDA) for offering the fellowship during her PhD studies.

References

1 Propfe, B., Redelbach, M., Santini, D.J., and Friedrich, H. (2012). Cost analysis of plug-in hybrid electric vehicles including maintenance & repair costs and resale values. *World Electric Vehicle Journal* 5 (4): 886–895.

2 İnci, M., Büyük, M., Demir, M.H., and İlbey, G. (2021). A review and research on fuel cell electric vehicles: topologies, power electronic converters, energy management methods, technical challenges, marketing, and future aspects. *Renewable and Sustainable Energy Reviews* 137: 110648. https://doi.org/10.1016/j.rser.2020.110648.

3 Miler, P. (2008). The environmental impact of using different supply voltages for HEVs and FCEVs. *IEEJ Transactions on Industry Applications* 128 (7): 880–884.

4 Eberle, U. and Von Helmolt, R. (2010). Sustainable transportation based on electric vehicle concepts: a brief overview. *Energy & Environmental Science* 3 (6): 689–699.

5 Das, V., Padmanaban, S., Venkitusamy, K. et al. (2017). Recent advances and challenges of fuel cell based power system architectures and control – a review. *Renewable and Sustainable Energy Reviews* 73: 10–18.

6 Campanari, S., Manzolini, G., and De la Iglesia, F.G. (2009). Energy analysis of electric vehicles using batteries or fuel cells through well-to-wheel driving cycle simulations. *Journal of Power Sources* 186 (2): 464–477.

7 Hannan, M.A., Hoque, M.M., Mohamed, A., and Ayob, A. (2017). Review of energy storage systems for electric vehicle applications: issues and challenges. *Renewable and Sustainable Energy Reviews* 69: 771–789.

8 Kim, J.H., Kim, H.J., and Yoo, S.H. (2019). Willingness to pay for fuel-cell electric vehicles in South Korea. *Energy* 174: 497–502.

9 O'hayre, R., Cha, S.W., Colella, W., and Prinz, F.B. (2016). *Fuel Cell Fundamentals*. Wiley.

10 Sazali, N., Wan Salleh, W.N., Jamaludin, A.S., and Mhd Razali, M.N. (2020). New perspectives on fuel cell technology: A brief review. *Membranes* 10(5): 99. https://doi.org/10.3390/membranes10050099.

11 Sharaf, O.Z. and Orhan, M.F. (2014). An overview of fuel cell technology: fundamentals and applications. *Renewable and Sustainable Energy Reviews* 32: 810–853.

12 Kirubakaran, A., Jain, S., and Nema, R.K. (2009). A review on fuel cell technologies and power electronic interface. *Renewable and Sustainable Energy Reviews* 13: 2430–2440.

13 Gasik, M. (2008). Materials for fuel cells. *Woodhead Publishing Series in Electronic and Optical Materials* https://doi.org/10.1533/9781845694838.1.

14 Inci, M. and Türksoy, Ö. (2019). Review of fuel cells to grid interface: configurations, technical challenges and trends. *Journal of Cleaner Production* https://doi.org/10.1016/j.jclepro.2018.12.281.

15 Costamagna, P., De Giorgi, A., Magistri, L. et al. (2019). A classification approach for model-based fault diagnosis in power generation systems based on solid oxide fuel cells. *IEEE Transactions on Energy Conversion* https://doi.org/10.1109/TEC.2015.2492938.

16 Wilkinson, D.P., Zhang, J., Hui, R. et al. (2009). *Proton Exchange Membrane Fuel Cells: Materials Properties and Performance*. CRC Press.

17 Albarbar, A. and Alrweq, M. (2018). Proton exchange membrane fuel cells. In: *Proton Exchange Membrane Fuel Cells*. Springer International Publishing https://doi.org/10.1007/978-3-319-70727-3_2.

18 Wang, Y., Diaz, D.F.R., Chen, K.S. et al. (2020). Materials, technological status, and fundamentals of PEM fuel cells – a review. *Materials Today* https://doi.org/10.1016/j.mattod.2019.06.005.

19 Barbir, F. and Yazici, S. (2008). Status and development of PEM fuel cell technology. *International Journal of Energy Research* 32 (5): https://doi.org/10.1002/er.1371.

20 Eapen, D.E., Suseendiran, S.R., and Rengaswamy, R. (2016). Phosphoric acid fuel cells. In: *Compendium of Hydrogen Energy*, 57–70. Woodhead Publishing.

21 Park, C., Jung, Y., Lim, K. et al. (2021). Analysis of a phosphoric acid fuel cell-based multi-energy hub system for heat, power, and hydrogen generation. *Applied Thermal Engineering* https://doi.org/10.1016/j.applthermaleng.2021.116715.

22 Song, C. (2002). Fuel processing for low-temperature and high-temperature fuel cells: challenges, and opportunities for sustainable development in the 21st century. *Catalysis Today* 77 (1–2): 17–49.

23 Guo, H., Liu, X., Zhao, J.F. et al. (2014). Experimental study of two-phase flow in a proton exchange membrane fuel cell in short-term microgravity condition. *Applied Energy* 136: 509–518.

24 Revankar, S.T. and Majumdar, P. (2014). *Fuel Cells: Principles, Design, and Analysis*. CRC Press.

25 Couture, G., Alaaeddine, A., Boschet, F., and Ameduri, B. (2011). Polymeric materials as anion-exchange membranes for alkaline fuel cells. *Progress in Polymer Science* 36 (11): 1521–1557.

26 Lin, B.Y., Kirk, D.W., and Thorpe, S.J. (2006). Performance of alkaline fuel cells: a possible future energy system? *Journal of Power Sources* 161 (1): 474–483. https://doi.org/10.1016/j.jpowsour.2006.03.052.

27 Mehmeti, A., Santoni, F., Della Pietra, M., and McPhail, S.J. (2016). Life cycle assessment of molten carbonate fuel cells: state of the art and strategies for the future. *Journal of Power Sources* 308: 97–108.

28 Adzic, R. and Marinkovic, N. (2020). Electrochemical energy conversion in fuel cells. In: *Platinum Monolayer Electrocatalysts*, 19–25. Cham: Springer.

29 Antolini, E. (2011). The stability of molten carbonate fuel cell electrodes: a review of recent improvements. *Applied Energy* 88 (12): 4274–4293. https://doi.org/10.1016/j.apenergy.2011.07.009.

30 Bhattacharyya, D. and Rengaswamy, R. (2009). A review of solid oxide fuel cell (SOFC) dynamic models. *Industrial & Engineering Chemistry Research.* https://doi. org/10.1021/ie801664j.

31 Zhu, B. (2009). Solid oxide fuel cell (SOFC) technical challenges and solutions from nano-aspects. *International Journal of Energy Research.* https://doi.org/10.1002/ er.1600.

32 Tarancón, A. (2009). Strategies for lowering solid oxide fuel cells operating temperature. *Energies* 2 (4): 1130–1150.

33 Aricò, A.S., Baglio, V., and Antonucci, V. (2009). Direct methanol fuel cells: history, status and perspectives. In: *Electrocatalysis of Direct Methanol Fuel Cells* (ed. H. Liu and J. Zhang). Wiley.

34 Cho, J., Jeong, S., and Kim, Y. (2015). Commercial and research battery technologies for electrical energy storage applications. *Progress in Energy and Combustion Science* https://doi.org/10.1016/j.pecs.2015.01.002.

35 Garche, J., Moseley, P.T., and Karden, E. (2015). Lead–acid batteries for hybrid electric vehicles and battery electric vehicles. In: *Advances in Battery Technologies for Electric Vehicles*, 75–101. Woodhead Publishing.

36 Han, J., Kim, D., and Sunwoo, M. (2009). State-of-charge estimation of lead-acid batteries using an adaptive extended Kalman filter. *Journal of Power Sources* 188 (2): 606–612.

37 Paul, M. and Josh, F. (2019, August). Switched reluctance motor, the future of modern electric vehicle – a technical review. *Proceedings of International Conference on Recent Trends in Computing,* Communication & Networking Technologies (ICRTCCNT), Chennai, Tamil Nadu, India (18–19 October 2019).

38 Weicker, P. (2013). *A Systems Approach to Lithium-Ion Battery Management.* Artech House.

39 Khaligh, A. and Li, Z. (2010). Battery, ultracapacitor, fuel cell, and hybrid energy storage systems for electric, hybrid electric, fuel cell, and plug-in hybrid electric vehicles: state of the art. *IEEE Transactions on Vehicular Technology* 59 (6): 2806–2814.

40 Chen, L., Xu, Z., Liu, M. et al. (2012). Lead exposure assessment from study near a lead-acid battery factory in China. *Science of the Total Environment* 429: 191–198. https://doi.org/10.1016/j.scitotenv.2012.04.015.

41 Jafari, H. and Rahimpour, M.R. (2020). Pb acid batteries. In: *Rechargeable Batteries: History, Progress, and Applications.* https://doi.org/10.1002/9781119714774.ch2.

42 Pourabdollah, K. (2017). Development of electrolyte inhibitors in nickel cadmium batteries. *Chemical Engineering Science* 160: 304–312. https://doi.org/10.1016/ j.ces.2016.11.038.

43 Jeyaseelan, C., Jain, A., Khurana, P. et al. (2020). Ni-Cd batteries. In: *Rechargeable Batteries: History, Progress, and Applications.* https://doi.org/10.1002/ 9781119714774.ch9.

44 Kurhe, N.T., Nagare, P., Wakchaure, V.D., and Gurnani, A.H.U. Human powered hybrid vehicle: a review of history, design and development of electric bicycles. *Gradiva Review Journal*, ISSN NO: 0363-8057.

45 Szałek, A., Pielecha, I., and Cieslik, W. (2021). Fuel cell electric vehicle (FCEV) energy flow analysis in real driving conditions (RDC). *Energies* 14 (16): 5018.

46 Chayambuka, K., Mulder, G., Danilov, D.L., and Notten, P.H. (2020). From li-ion batteries toward Na-ion chemistries: challenges and opportunities. *Advanced Energy Materials* 10 (38): 2001310.

47 Javadi, M., Liang, X., Gong, Y., and Chung, C.Y. (2022, September). Battery energy storage technology in renewable energy integration: a review. *2022 IEEE Canadian Conference on Electrical and Computer Engineering (CCECE)*, Halifax, NS, Canada (18–20 September 2022), pp. 435–440, IEEE.

48 Rajarathnam, G.P. and Vassallo, A.M. (2016). *The Zinc/Bromine Flow Battery*. In: *Materials Challenges and Practical Solutions for Technology Advancement*. Springer.

49 Tian, T., Zhang, T.W., Yin, Y.C. et al. (2019). Blow-spinning enabled precise doping and coating for improving high-voltage lithium cobalt oxide cathode performance. *Nano Letters* 20 (1): 677–685.

50 Nasara, R.N., Tu, C.H., and Lin, S.K. Perspective on battery research. In: *Green Energy Materials Handbook* (ed. M.-F. Lin and W.-D. Hsu), 341. CRC Press.

51 Mu, Y., Zhang, C., Zhang, W., and Wang, Y. (2021). Electrochemical lithium recovery from brine with high Mg2+/Li+ ratio using mesoporous λ-MnO2/LiMn2O4 modified 3D graphite felt electrodes. *Desalination* 511: 115112.

52 Manane, Y. and Yazami, R. (2017). Accurate state of charge assessment of lithium-manganese dioxide primary batteries. *Journal of Power Sources* 359: 422–426.

53 Yamauchi, H., Park, G., Nagakane, T. et al. (2013). Performance of lithium-ion battery with tin-phosphate glass anode and its characteristics. *Journal of the Electrochemical Society* 160 (10): A1725.

54 Bi, H., Zhu, H., Zu, L. et al. (2020). Environment-friendly technology for recovering cathode materials from spent lithium iron phosphate batteries. *Waste Management & Research* 38 (8): 911–920.

55 Varghese, A.J., Roy, R., and Thirunavukkarasu, S. (2014). Optimized speed control for BLDC motor. *International Journal of Innovative Research in Science, Engineering and Technology* 3 (S1): 1019–1030.

56 Krishnan, R. (2017). *Permanent Magnet Synchronous and Brushless DC Motor Drives*. CRC Press.

57 Kushwah, M. and Patra, A. (2014). Tuning PID controller for speed control of DC motor using soft computing techniques – a review. *Advance in Electronic and Electric Engineering* 4 (2): 141–148.

58 Hawkins, T.R., Singh, B., Majeau-Bettez, G., and Strømman, A.H. (2013). Comparative environmental life cycle assessment of conventional and electric vehicles. *Journal of Industrial Ecology* 17 (1): 53–64.

59 Un-Noor, F., Padmanaban, S., Mihet-Popa, L. et al. (2017). A comprehensive study of key electric vehicle (EV) components, technologies, challenges, impacts, and future direction of development. *Energies* 10 (8): 1217.

60 Yang, L. and Griffin, S.J. (2010). Interventions to promote cycling: systematic review. *BMJ* 2010: 341.

61 Shamami, M.S., Alam, M.S., Ahmad, F. et al. (2020). Artificial intelligence-based performance optimization of electric vehicle-to-home (V2H) energy management system. *SAE International Journal of Sustainable Transportation, Energy, Environment, & Policy* 1 (13-01-02-0007): 115–125.

62 Muthuramalingam, S., Bharathi, A., Rakesh Kumar, S. et al. (2019). IoT based intelligent transportation system (IoT-ITS) for global perspective: a case study. *Internet of Things and Big Data Analytics for Smart Generation* 279–300.

63 Sanguesa, J.A., Torres-Sanz, V., Garrido, P. et al. (2021). A review on electric vehicles: technologies and challenges. *Smart Cities* 4 (1): 372–404.

64 Thomas, C.S. (2012). How green are electric vehicles? *International Journal of Hydrogen Energy* 37 (7): 6053–6062.

10

Electric Vehicle Path Towards Sustainable Transportation: A Comprehensive Structure

Vikas Khare[1], Ankita Jain[2], and Miraj Ahmed Bhuiyan[3]

[1] School of Technology Management and Engineering, NMIMS, Indore, Madhya Pradesh, India
[2] Prestige Institute of Global Management, Indore, Madhya Pradesh, India
[3] Faculty Member School of Economics, Guangdong University of Finance and Economics, Guangzhou, China

Nomenclature

ϕ	Road angle
a	Frontal area
av	vehicle frontal area
d	Coefficient of drag
g	Gravitational constant
m	Mass of the vehicle
P_{max}	Maximum power from the engine
r_f	Rolling resistance
T_f	Tractive force
V	Maximum voltage
v_s	Vehicle speed
γ	Air density

10.1 Introduction

In the twenty-first century, few technological innovations worldwide have resulted in a massive change in the perspective of old or conventional methodology. EVs play a vital role in changing the automobile industry's paradigm and

Interconnected Modern Multi-Energy Networks and Intelligent Transportation Systems: Towards a Green Economy and Sustainable Development, First Edition. Edited by Mohammadreza Daneshvar, Behnam Mohammadi-Ivatloo, Amjad Anvari-Moghaddam, and Reza Razzaghi.
© 2024 The Institute of Electrical and Electronics Engineers, Inc.
Published 2024 by John Wiley & Sons, Inc.

creating pollution-free transportation in the transportation sector. Compared to conventional transport, plug-in hybrid electric vehicle reduces greenhouse gas emissions by approximately 45% and provides a better alternative to gasoline- and diesel-operated vehicles [1, 2]. China's EV market is growing splendidly, followed by Germany, United States, and India. The worldwide EV policy has created awareness among the people for green transportation. Many researchers are paying attention to enhancing the technology of electric motors and batteries in terms of lightweight, increasing the charging capability and efficiency of the device [1, 2]. With the increased use of EVs in developed and developing countries, EV charging station facilities must be developed to provide cheaper charging facilities for all vehicles with more negligible environmental effects. Based on the data of Global EV Outlook in 2019, the global electric car navy reached 5.12 million in 2018, an increase of 2.1 million compared to 2017, almost doubling EV revenue in just one year. However, in 2018, only 2.2% of the world's passenger cars were electric cars, which means there is still a long way to go before electrification becomes a reality. Figure 10.1 shows the top countries' battery EV market by volume till 2019 [2].

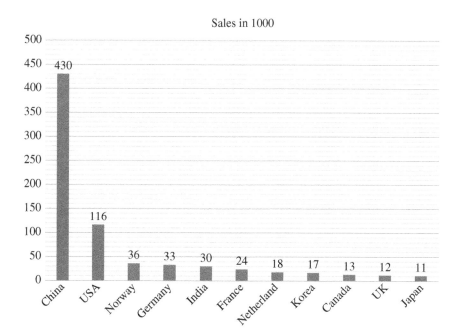

Figure 10.1 Top country's battery EV market by volume till 2019. *Source:* Adapted from [2].

According to the figure, China is the leading country in electric vehicles, followed by the United States, Norway, and Germany. In the recent scenario, India has played a crucial role in the field of electric vehicles, and over the last two to three years, the number of registered electric vehicles has increased splendidly in India. This chapter comprehensively reviews EVs, including design and control systems, battery thermal management and design, reliability analysis, and worldwide policies.

The motivation to write on this topic is that EVs are widely regarded as a path towards sustainable transportation. They run on electricity stored in batteries, which can be charged from the electric grid, and produce zero tailpipe emissions. This makes them an attractive alternative to traditional gasoline- or diesel-powered vehicles, which contribute significantly to air pollution and climate change.

However, there are also some research gaps or challenges to the widespread adoption of electric vehicles. One of the biggest challenges is the high cost of batteries, which makes EVs more expensive than traditional vehicles. Range anxiety is another issue, as many consumers are concerned about the limited range of EVs and the availability of charging infrastructure. Interconnected modern multi-energy networks refer to the integration of various energy sources and infrastructures to meet the energy demands of different sectors, including transportation. In the context of EVs, this concept involves the integration of renewable energy sources, such as solar and wind power, with the electricity grid to support the charging infrastructure for electric vehicles.

To enable the widespread adoption of EVs, it is crucial to establish an efficient and reliable charging infrastructure that can accommodate the increased electricity demand. Interconnected modern multi-energy networks can help optimize the charging process by leveraging renewable energy sources, energy storage systems, and smart grid technologies. These networks can balance the supply and demand of electricity, manage peak loads, and ensure the integration of EV charging with the overall energy system.

Furthermore, interconnected multi-energy networks can facilitate the development of vehicle-to-grid (V2G) technology. This technology allows electric vehicles not only to consume electricity from the grid but also feed excess energy back into the grid when needed. This bidirectional energy flow can enhance the grid's flexibility, support renewable energy integration, and contribute to a more resilient and sustainable energy system.

To address these challenges, governments and private companies are investing in the development of more affordable and efficient batteries, as well as in the expansion of charging infrastructure. Many countries are also offering incentives, such as tax credits and rebates, to encourage consumers to purchase electric vehicles.

The following are the objectives and novelty of this chapter:

1) Assessment of the optimum and design prospectus of the EVs.
2) Analyze the thermal management of the battery system.
3) Reliability assessment of the EV system.
4) Assessment of the EV charging station.

10.2 Optimum Design of EVs

Analyzing the different technical parameters and their relationship to other parameters is the first step in database design to create a conceptual model of various components connected. The electric motor provides propulsion to the wheel as per torque-speed characteristics. As it provides high starting torque, it is necessary to identify appropriate motor with proper specification. EVs' performance and reliability depend on the battery bank's battery system capacity and charging capability. The converter is a key component of EVs, which converts AC to DC, DC to AC, and stepping up/down the DC voltage through buck and boost converters [1–3]. Figure 10.2 shows the framework of the different components of EVs.

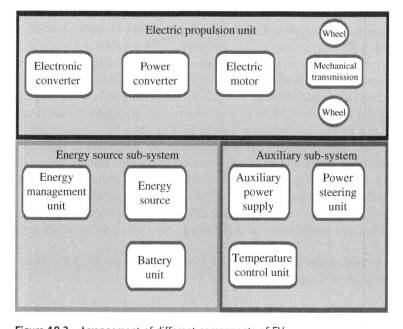

Figure 10.2 Arrangement of different components of EV.

The foundation of ammonia-based plug-in EVs was defined by Chu and Wu [4]. To overcome some of the limitations of conventional EVs, a hydrogen fuel cell hybrid vehicle is proposed, in which ammonia serves as a source of hydrogen, resulting in greater vehicle range capacity. The basic components of this dynamic model were an electric motor, battery, capacitor, and fuel cell. Zhang and Liu [5] suggested an assessment of hybrid green cars using a unique driving system. The outcome indicates a roughly 16.32% reduction in fuel usage when compared to a traditional EV drive system. This study also discusses the vehicle framework evaluation procedure, static and dynamic conditions, and various drive system modes. Tanozzi and Sharma [6] developed a method for determining the thermal characteristics of hybrid electric vehicles.

This paper's three main points of conversation are the layout of heat exchanger networks, optimum heat exchanger sizing, and thermal parameter optimization. Chatterjee and Iyer [7] analyzed the optimum design of a wireless energy transfer system for EVs, which is assessed by calculating passive elements, quality factor, and resonance condition or bandwidth of the coil used in the EVs. Borthakur and Subramanian [8] presented the optimum design and assessment of series-parallel integrated green vehicles and power train mechanisms for heavy four-wheeler EVs. Figure 10.3 shows the block diagram of the series-parallel integrated mechanism. Figure 10.4 shows the key parameters of the design of EVs.

Czogalla and Jumar [9] described the design assessment of electric bus vehicles to determine long-range energy consumption. The author discussed and worked on the optimum route for EV buses to reduce energy consumption and increase system efficiency. The main technical parameters are the static and dynamic traffic conditions, terrain grades, and charging and discharging capacity of the electric bus. Hofman and Janssen [10] described the optimum hybrid design of the transmission system of EVs. The key ingredients of this paper are the splendid transition between the power train, optimum energy utilization and reduction of loss of an electric motor, and overall speed ratio. Vora and Jin [11] explained EV modeling for medium-duty truck applications for large highway distances. Several EV limitations, such as economic viability, risk reluctance, and lower efficiency on the highway, are mitigated in this paper, along with a technical-financial assessment of heavy-duty trucks.

During the modeling of an electric vehicle, the maximum power from the engine for a constant speed is given by

$$P_{\max} = \frac{1}{T_{\text{eff}}} \left(mgr_{\text{f}} + \frac{1}{2}\gamma da V^2 \right) V \tag{10.1}$$

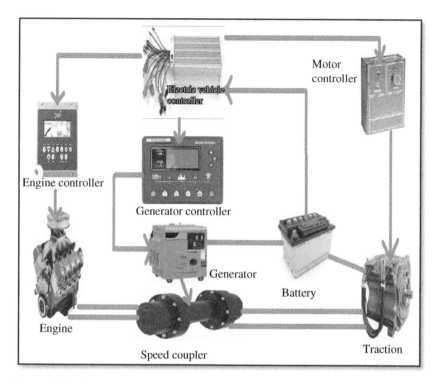

Figure 10.3 Series-parallel integrated mechanism. *Source:* Courtesy of rsvindustries.

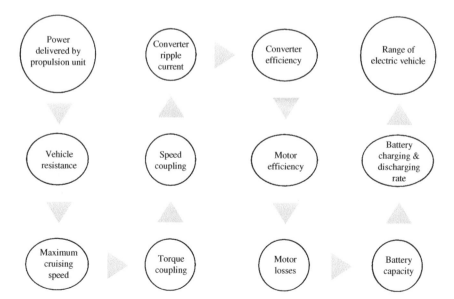

Figure 10.4 Design Parameters of EVs. *Source:* Adapted from [4–7].

Valle and Viera [12] assessed EVs' design and validation and found EVs' energy consumption and performance assessment with real-time static and dynamic data. A software tool is developed to analyze and forecast the power consumption of green vehicles. Delogu and Zanchi [13] investigated composites and materials to further the development of lightweight EVs; it also gave cradle grave assessment of EVs from an environmental point of view, and the result shows that some hurdles persist. Hang and Han [14] proposed a model of a collision-avoidance system for stand-alone four-wheel steering and drive EVs. Zhou and Qin [15] presented an EV power train's multi-optimum design and technical parameter assessment. Parameters such as fuel cost, vehicle energy consumption, vehicle velocity and acceleration, and power train performance assessment are the key point of this paper. The result shows that EVs must be developed with higher efficiency and low fuel cost to perform well.

During the optimum design of the electric vehicle, another essential factor is tractive force, which is given by

$$T_F = r_f mg + da_v \frac{\gamma}{2} v_s^2 + mg\sin(\emptyset) + ma \tag{10.2}$$

Xin and Chengning [16] assessed the perfect range capacity and life cycle and battery of an EV by developing a model using the principle of minimum curb mass. Table 10.1 summarizes the technical aspects of EVs.

Roozegar et al. [24] described the modeling parameters of a novel modular multispeed transmission system for EVs. K-filter, neural network, and Leuenberger observer were used to assess EVs' technical parameters. Giorgio and Trolio [25] described IC engines and fuel cell hybrid-based vehicle modeling. The technical parameters of the hybrid EVs are the rolling inertia, frontal area, drag coefficient, rolling resistance coefficient, battery pack energy density, and fuel cell and IC engine power weight. Guo and Sun [26] explained the design and optimization of energy management of the EV by assessing the technical parameter of hybrid EV using the optimal Latin hypercube algorithm, and error is mitigated by Pontryagin's minimum principle-based controller, which is also used for energy management of hybrid EVs. Li et al. [27] described proton exchange membrane fuel cell-based EVs used for range extension of hybrid vehicles. The paper discussed the interrelationship between the membrane and nanostructured system-based EV characterization. Dimitrova and Maréchal [28] presented the techno-economic design of a hybrid EV using a multi-objective algorithm. A genetic algorithm assesses the technical parameters and estimates power-train equipment cost.

Outcomes: Extensive work has already been reported on EVs; still, there is a need to design an EV that provides 0% greenhouse gas emissions and creates a pollution-free environment. Finding a proper battery system replacement is also necessary due to their invariable charging and discharging nature. Research shows

Table 10.1 Different technical aspects of EVs.

Author	Objective	Technique used	Outcomes
Hodgson and Mecrow [17]	EV speed	Kalman filter Recursive least-square method Genetic algorithm	Improved the speed and range of the EV
Cocron et al. [18]	Assessment through consumer perspective	Psychological founded method	Identified human-machine interaction and traffic safety implications
Bukhari and Alalibo [19]	Switched reluctance motor design	Finite element method	Optimized excitation voltage and switching sequence
Jenal et al. [20]	Lightweight EV	MATLAB/ Simulink	Cogging and output torque identification
Thanapalan and Liu [21]	Modeling and control of fuel cell EV	Simulation toolbox	Decrease the pollution compared to conventional battery
Villa and Montoya [22]	Taxonomy of energy consumption	Comprehensive reviews	Classification of existing models according to their input parameters
Grewal and Darnell [23]	Range prediction	Model-based technique	Description of flow sequence of EV

that fuel cells, electrolyzers, and ultra-capacitors will provide better alternatives to battery systems shortly. Electric vehicles (EVs) have attracted the attention of the automobile industry and researchers to attain zero emissions and low energy usage. The primary requirements for consumers in the EV industry are mileage, speed, performance, efficiency, high storage battery for increased mileage and protection, and, most crucially, EV cost. It is necessary to identify the optimum design of curb weight, gross weight, wheel power and energy, wheel force, and aerodynamic force for the ideal electric vehicle performance. The Internet of things, artificial intelligence, and machine learning may be used to design EVs with higher efficiency, less pollution, and better driving comfort.

10.3 Characterization of EV Battery System

A battery is a device that stores chemical energy and transforms it into electrical energy. In a battery's chemical reactions, electrons go from one substance to another through an external circuit. In rechargeable batteries, the chemical reaction happening inside is reversed, bringing the battery to a charged condition [29, 30]. Following are the different characterizations of the electric vehicle battery system.

10.3.1 Thermal Management of Battery

Lithium-air batteries are a relatively new development in the lithium-ion battery area. The lithium-oxygen ($Li-O_2$) battery (or lithium-air battery) obtains its energy from the interaction of oxygen in the air with lithium, which comprises Li-metal and a porous conductive framework as electrodes. Although the technology is still in its early stages, it has the potential to deliver substantially greater energy storage and better thermal control than a normal lithium-ion battery. Many academics have already worked on battery system heat management. Wiriyasart and Hommalee [31] provided a computational analysis method representing temperature distribution. The EV battery modules include 444 cylindrical lithium-ion cell batteries. Lyu and Siddique [32] discussed the thermal maintenance of EV batteries using the thermal cooling method.

Furthermore, when 40 V is provided to the heater, the battery surface temperature drops from 54°C to 11°C utilizing a cooling water system for a single cell with a copper holder. Thermal control of Li-ion EVs using air cooling and a heat pipe was detailed by Behi and Karimi [33]. Furthermore, the temperature consistency of the battery module has improved by 38.2%, 67.5%, and 74.4%, respectively, for restricted air cooling, heat funnels, and HPCS. Akinlabi and Solyali [34] described a complete review of the configuration, modeling, and assessment of thermal management of the EV battery system. This research examines and compares passive and active air-cooled battery thermal management system (BTMS) methodologies and design parameter optimization strategies (through iteration or algorithms) for achieving various BTMS design goals. In the thermal management system, the battery cooling system is a significant factor and is classified as shown in Figure 10.5.

Chung and Kim [35] used a liquid cooling approach to develop a thermal model for EV batteries. The inadequate warmth transmission from the foot of the cell stack to the cooling plate is one of the major impediments to productive warmth dispersion and deviating plan of the balance cell game plan and has a negative influence on the battery pack's temperature consistency to numerical results. Shen and Gao [36] demonstrated a refrigerant-based EV battery bank heat management system. The difference in temperature between cells and the mean temperature of the battery module is effectively managed, resulting in increased system efficiency. A cabin priority control strategy and a series-connected system configuration are used to develop the thermal management system. Heat pipe-based thermal management methods for EV battery systems were described by Smith and Singh [37]. The heat pipe framework comprises two parts: heat pipe cooling plates for

Figure 10.5 Types of battery cooling system.

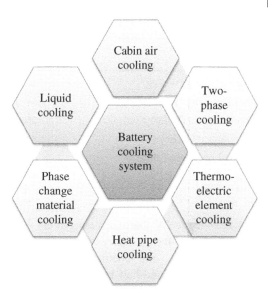

extracting heat from the battery module's electrochemical cells and remote heat transfer funnel to transport heat from the module to fluid-cooled cold plates 300 mm away. Compared to a traditional fluid-cooled system, a two-stage heat pipe-based warm control provides superior cell/module temperature constancy, a cleaner construction, and a more stable framework (no spillage issues in high voltage regions). A two-stage investigation of these warmth pipe segments and a single-stage examination for the virus plate partition were carried out. Cen and Li [38] assessed the EV battery system's thermal management system's air-conditioning system. Experimental findings show that the EV battery thermal management-based device can easily regulate the battery pack's temperature at an acceptable preset value at an intense ambient temperature of up to 41°C. Optimizing the coolant circuit can minimize temperature nonuniformity inside the battery pack. Kim and Oh [39] described a complete overhaul of the temperature management system for EV systems with a battery bank. This work studied the thermal dynamics and important thermal issues of Li-ion batteries. Various battery thermal management system experiments are carefully examined and classified according to selected temperature cycles. Cabin air cooling, second-circuit liquid, and two-stage direct liquid cooling are part of the battery thermal management system with a vapor compression cycle. Tian and Gan [40] described battery cooling and waste heat recovery-based thermal management systems for EV-used battery systems. The outcomes show that the level of cooling limit decrease was in the range of 26.30–32.10%, and the level of waste thermal recuperation was in the range of 18.73–45.17%.

Table 10.2 Properties of lead-acid and lithium-ion battery.

Parameters	Lead acid	Lithium ion
Specific energy (Wh/kg)	20–35	90
Energy density (Wh/L)	54–95	153
Specific power (W/kg)	250	300
Amp hour efficiency (%)	80	87
Internal resistance (ohm)	0.022	0.005
Recharge time (hour)	8	2–3

Moreover, the EV thermal management system (EVTMS) model was created and approved with exploratory information, which anticipated vitality utilization, framework limit, and coefficient of performance (COP) in the range of 0.68–21.05%. Half-and-half reproduction strategies were completed to contemplate the impact of EVTMS on the vehicle's driving extent. Ianniciello and Biwolé [41] described a phase transition material-based temperature management system for EV batteries. This article reviews prior advances and introduces PCMs for battery cooling in electric vehicles. Different systems are examined, and ways to improve PCM efficiency in those networks are given. Table 10.2 shows the properties of lead-acid and lithium-ion batteries.

10.3.2 Assessment of Battery System

Numerous charging methods are the most ordinarily utilized strategy for corrosive lead batteries. This technique charges the battery until the cell voltage reaches a foreordained level. The current is then turned off, the cell voltage is allowed to decline to another foreordained level, and the current is turned on again. Since the late 1980s, battery-powered lithium cells have entered the market. Many researchers are already working in the field of EV battery, and Hong and Wang [42] described the assessment parameters of the EV battery system through a short-term prediction model. In all, 8760 hours of charging station data were retrieved at the provision and organization center for EVs in Beijing to assist the short-term model and verify the model's validity, maintainability, and reliability.

Further, a synchronous multiple fault prognosis is feasible based on reliable battery device parameter prediction. Wang [43] described the performance assessment of the EV battery system. Yang and Zhu [44] detailed the execution, design, and control system of the battery system of an electric vehicle. Nonlinear robust fractional order control costs 78.00%, 85.21%, and 87.76% of proportional integral derivative control, linear control, and sliding-mode control, respectively, in profound load situations. Finally, a hardware-in-loop experiment utilizing the DSpace platform is used to verify the viability of its implementation. Wegmann and Döge [45] conducted a comparative analysis of solid-state lithium metal, high polymer energy, and lithium-ion high-power batteries.

$$\text{Open-circuit voltage} = n \times (2.15DD) \times (2.15 - 2.00) \qquad (10.3)$$

where n is the number of cells in the battery and DD is the depth of discharge.

From the MATLAB simulation in the NiCad battery, open-circuit voltage is given by

$$\text{Open-circuit voltage} = n \times \big(-8.28DD^7 + 23.58DD^6 - 30DD^5$$
$$+ 23.70DD^4 - 12.58DD^3 + 4.13DD^2 - 0.86DD + 1.37 \big)$$
$$(10.4)$$

The device's effectiveness is estimated under ordinary and outrageous driving conditions based on the general vehicle reproduction model, including the thermal model of the lithium metal polymer battery pack. The discoveries show that the hybridization of the battery allows the lithium metal polymer battery to be utilized appropriately in an EV. Saw and Poon [46] explained the mathematical modeling of EV battery systems. The outcomes acquired reveal that the dynamic pressure, top force request, and thermal execution of the battery have been essentially improved by consolidating supercapacitors into the battery pack in HESS. In correlation with the ordinary battery energy stockpiling framework, the peak current requests of the battery in HESS for UDDS and US06 cycles have been decreased by 62%, 71.9%, and 70.7%, individually. This methodology has been demonstrated to be viable in expanding the battery's life expectancy and can improve the well-being and unwavering quality of the traditional battery energy storage system in EVs. The electromechanical and chemical evaluation of lithium-ion batteries, which are utilized in EVs, was presented by Niu and Garg [47]. Five study concerns, including the plan criteria for the exposition assessment of batteries, are attempted in this one-of-a-kind circumstance. The main concern is battery performance under pressure situations, which was addressed by varying the initial pressure, continued loads, and a range of battery pack limits. Wegmann and Döge [48] described the optimum working of EV-used battery through deterministic and stochastic dynamic programming. This research focuses on the optimum feasible functioning of the mixed battery system utilized in EVs. Two algorithms were investigated and tested concerning the provided system and control issue. First, the worldwide ideal control process is inferred using dynamic programming and a causal controller, which ensures adequate continuing control of the battery framework. Figure 10.6 shows different assessment parameters of the battery system.

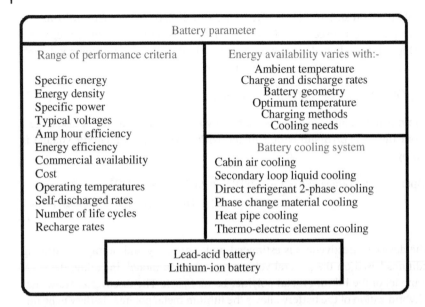

Figure 10.6 Different Battery Parameters.

Outcomes: The above research work on the battery is classified into two categories: the battery's thermal system and the battery bank's properties. Research shows that different cooling processes for the battery bank's thermal system are air cooling, water cooling, cold water cooling, and heat pipe cooling and mentioned that cooling of the battery is equally important for the proper functioning of EV. Attention must be paid to charging and discharging capability, including various properties of batteries, to increase the range of EVs.

10.4 Control System of EVs

The inconsequential functions of the control system are component monitoring and protection, such as battery state of charge, battery temperature, electric motor overheating, and internal combustion engine overheating. Lots of researchers are working towards the proper control mechanism for EV systems. Wu and Xu [49] presented a thermoelectric surge management system for fuel cell-powered automobiles, which lowers heat loss and generates saturated features in the vehicle system. The notion of a neural network brake control system for EV operation was presented by He and Wang [50]. This control method increases energy efficiency, and an adaptive fuzzy controller regulates EV and single-pedal regenerative braking stability; multi-objective neural networks control the brake pedal further. Figure 10.7 shows the control mechanism of EVs.

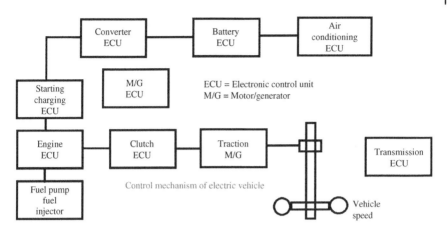

Figure 10.7 Control mechanism of the EV system.

Tan and Shen [51] described applying a non-fragile H infinity control system in the field of EVs. In this paper, Lyapunov theory is also used to find the desired controller gains and feasible solutions to the matrix inequality of an EV. Shi and Cheng [52] explained the concept of adaptive control systems to control the steering mechanism of EVs under different road conditions. The overall system is also simulated in the CORSIM-MATLAB environment, and the results show that the adaptive control system plays a vital role in mitigating the power steering problem in different road circumstances. Balaska and Ladaci [53] adopted an adaptive cruise control system for an EV. An adaptive controller is used for regulating the current in the inner loop and maintaining desired speed tracking for the outer loop of EVs. Qiu and Wang [54] described a control system of regenerative braking for green vehicles under different critical driving conditions. The paper also proposed a serial control strategy for control systems for EVs. EV is tested in four different road conditions and at four different drag coefficients. Yang and Wang [55] used an observer-based control system on supercapacitor-operated EVs. Based on the conventional fifth-order averaged model described, sliding mode control for battery and supercapacitor-operated EVs in which a DC/DC converter is used to isolate the battery and supercapacitor packs from the DC bus. Finally, the hardware in the loop test is utilized to determine whether the suggested control system is successful. Luo and Yu [56] analyzed the control system of the AC motor of the EV drive-train by finite time dynamic surface control system. This paper proposes a neural system-based limited-time dynamic surface position-based control strategy for acceptance engines with input immersion in EV drive frameworks. The neural systems are used to estimate the obscure nonlinear capacities immediately. Dynamic surface control is utilized in conventional backstopping innovation to control the "blast of unpredictability." He and Quan

[57] conducted an error analysis of fuel cell-operated EVs by a hydrogen circulation system-based model predictive control system. This study creates a model prescient control approach for the hydrogen course framework to manage the flowing hydrogen stream. A model of the flexible hydrogen framework that contains a stream control valve, a gracefully complex, an arrival complex, and a hydrogen coursing siphon is right off the bat created to depict the conduct of the hydrogen mass stream elements in the Polymer electrolyte membrane fuel cell. Xu and Chen [58] described the torque optimization control of an EV with a regenerative braking system. When subject to the actuator limitations, this proposed torque dissemination controller can boost the recovery effectiveness by determining the pressure-driven deceleration torque and engine slowing down torque. Absolute slowing down torque of the four wheels tracks the slowing down prerequisite.

Moreover, to guarantee brake well-being, the slowing down torque of the front and back wheels is advanced to improve the chance of vitality recuperation. Alamdari and Voos [59] described the concept of nonlinear predictive control in the field of EVs. The paper's main contribution is the structure of an ongoing nonlinear model prescient controlled with improved disparity requirements taking care of and monetary punishment capacity to design the online-savvy cruising speed. An all-inclusive voyage control driver helps framework controls the longitudinal velocity of the BEV in a safe and vitality proficient way by exploiting street slants, powerful drive around bends, and adhering to the guidelines. Figure 10.8 shows the control variable of electric vehicles.

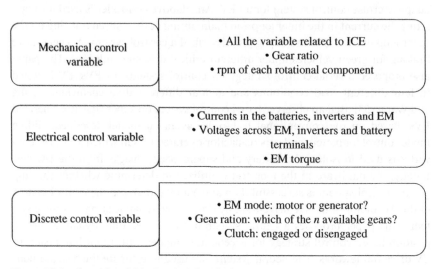

Figure 10.8 Control variable for electric vehicle.

Zhang and Zhao [60] described decoupling control of the steering and driving system of an EV system. A control procedure dependent on the nonlinear decoupling control technique is proposed in this paper to improve the mobility and steadiness of in-wheel-engine drive EVs. The reverse framework decoupling hypothesis is applied to disintegrate the vehicle's nonlinear arrangement, making it conceivable to understand the planned control of every subsystem.

Outcomes: The above literature survey shows that applying the perfect control system on the EV is essential to reduce error and increase efficiency. It is also essential to develop first stable control mechanisms for individual components of the EV and then for the overall vehicle. Some researchers have already worked on the thermo-electric surge control, braking stability, steering control system, torque optimization of electric motor, and speed control mechanism of an electric motor using several methods such as neural network, fuzzy controller, and model predictive control system. In the near future, it is necessary to find the control mechanism of the power flow of a generator or motor, the stable functioning of a battery system, the converter system, and the proper control mechanism of the steering system. Different optimization techniques may also be applied, such as the cuckoo algorithm, chaotic particle swarm optimization technique, teacher learning-based algorithm to develop good dynamic characteristics, and state-space model of the control system of EVs.

10.5 Reliability Assessment of EV

One of the essential characteristics for evaluating any equipment is reliability, which determines the gadget's defect rate. In the case of EVs, the maintainability and validity of various mathematical functions play a critical role in evaluating EV performance [61, 62]. The dependability assessment of EV hosting capacity was provided by Kamruzzaman and Benidris [63]. This study aims to maximize the permissible penetration of electric vehicles on electric power network buses using a reliability-restricted demand response-based strategy. If no remedies are adopted, the further growth in electric vehicles will impact power system efficiency. Implementing demand response systems can be an excellent way to minimize or even eliminate this burden. Anand and Bagen [64] provided a probabilistic evaluation of the EV distribution system's dependability. In this article, the trustworthiness of several technical characteristics of an electric vehicle is analyzed using a Markov chain method, and the modeling of travel patterns and recharging capability is assessed using a Monte-Carlo simulation approach. Gandoman and Ahmadi [65] presented a reliability assessment of the electric motor, battery, traction system, and electronics converter system. The problems and potential prospects of EVs relating to reliability and health, which need to be addressed, have also been studied. Gandoman and Jaguemont [66] explained the safety and reliability assessment of lithium-ion batteries used in EVs. Maintainability and validity assessment of all the technical parameters of battery related to further charging and discharging capability of lithium-ion systems is also assessed.

Additionally, the input of Li-ion battery squalor in five main failure modes and ability and power fade for providing validity evaluation models as solutions to existing challenges have been investigated. Shu and Yang [67] described that the reliability of the entire battery system is much lower than that of any other individual parts, like battery modules. Davidov and Pantoš [68] analyzed the charging station's reliability analysis and presented an improvement model for charging station arrangement to limit the general expense by fulfilling the charging dependability and nature of administration expected by EV proprietors/drivers.

Outcomes: The analysis of reliability of EVs is necessary to assess the validity, availability, maintainability, and fault rates of electric motors, batteries, converters, and EV charging stations. A sampling technique for location assessment for charging stations and fault tree analysis determine EVs' reliability in different circumstances.

10.6 Assessment of EV Charging Station

Electric recharging points, charging points, electronic charging stations (ECS), and EV supply equipment (EVSE) provide electric energy to recharge plug-in EVs such as electric vehicles, neighborhood EVs, and plug-in hybrids, for use at home or business. Some electric vehicles can connect onboard converters to a conventional electrical socket or a high-capacity appliance outlet. Others require or make use of a charging station with electrical conversion, monitoring, or safety features.

10.6.1 Location Assessment for EV Charging Station

Prefeasibility analysis is the first step to identifying suitable locations for installing an EV charging station. Napoli and Polimeni [69] studied the best sites for green vehicle charging stations along highways. The filtering algorithm does a prefeasibility study of an EV and technological input parameters, and the outcome reveals that charging station distribution is based on the capacity of EVs in those places. Multi-criteria fuzzy logic-based optimum placement selection for EV charging stations was developed by Liu and Zhao [70]. Kong and Luo [71] described location identification for charging stations according to the comfort level of operators, consumers, traffic conditions, and power grid supply. Liu and Liu [72] described data-driven, intelligent system-based location identification of an EV charging station. This will reduce the regular CO_2 emissions generated by approximately 0.14–0.37 ha of forest per year and a reduction of 0.85–2.64% of the additional charging criteria for forests each year. Csiszár and Csonka [73] presented a location assessment for charging stations through the land-use approach. The uniqueness of the approach relative to previous studies is that it assesses the feasibility of EV usage at a macro level and the future location of a micro-level charging station with a focus on land use. He and Kockelman [74] described the optimal location identification in the United States to develop a charging station for EVs. The issue is detailed as a blended number program, and an altered stream refueling area model (FRLM) is unravelled by employing a branch-and-bound calculation. Results reveal that the 60-mile-AER rate fluctuates between 31% and 65% as one expands the station tally from 50 to 250 stations. Lee and Hur [75] created a wind-powered EV charging station that adapts to various consumer charging habits. The Autoregressive Integrated Moving Average with Exogenous Variables model forecasts hourly wind power yields. The model is created by adding exogenous variables like wind speed and lattice mix investigation recreation forms to the Autoregressive Integrated Moving Average model. The recommended technique for EV charging dispersion is approved using trial data from breeze cultivates on Jeju Island in South Korea. Sun and Gao [76] proposed a location framework for charging stations considering the consumer's travel discrepancy properties. The model is structured into two parts, one for short-distance traveling and the other for long-distance traveling, keeping in mind that long-distance travelers should take less time to charge. A typical Chinese city is used as an example to illustrate the implementation and results of the model. Kadri and Perrouault [77] studied the charging station's location and financial assessment and discussed the expansion of the charging station's capacity. Hosseini and Sarder [78] identified the optimal location of charging stations through the Bayesian network (BN) model. This article provides a new study idea by including uncertainty and qualitative and quantitative parameters in the location assessment and presents the typical penetration of BN as an influential decision-making tool in electrical power management. Wang and Fe [79] described the location and capacity assessment of EV charging stations by expanding network approaches. They achieved this by planning the areas and limits of charging offices under a fixed spending limitation and handling the streamlining issue with a tweaked neighborhood search system. A lower threshold for the framework cost is developed to assess the characteristics of arrangements gained utilizing our proposed heuristic. Bai and Chin [80] described location assessment for charging stations through a bi-objective model using GPS trajectory data. The issue is defined as a bi-objective blended number, numerical model, with one goal identified with limited expense and the other identified with boosting administration quality. To fathom it, we propose a half-and-half developmental calculation consolidating the non-commanded hereditary calculation with direct programming and neighborhood search.

10.6.2 Characterization of Charging Station

Many researchers are already working in this field, and Mehrjerdi [81] described the concept of the micro-grid integrated EV charging station and gave a plan to increase the capacity of this charging station. Solar, wind, gas turbine, and energy storage systems are the resource components of micro-grid, which can increase its capacity. Results show that if the capacity of the wind energy system is increased by 200%, then the cost of the overall micro-grid system increases by 55%. Cost analysis of micro-grid-based EV charging stations is also done in this paper. Mehrjerdi and Hemmati [82] created a solar-diesel battery-powered charging station that can charge at three different levels: low, medium, and high. Low-, medium-, and high-level charging are favored for two-, three-, and four-wheeled EVs, respectively, mostly at the station. Because the solar energy resource influences the total charging system's performance, uncertainty owing to the intermittency of solar radiation is also evaluated in this work. The solar energy system is thought to help with the pollution problem in the atmosphere.

Ding and Lu [83] provided the best charging station scheduling to meet the demand for two-, three-, and four-wheeled electric vehicles. The linear integer optimization model highlights the link between customer demand and charging station capacity. The results reveal that if the charging procedure is correctly scheduled, the entire capital and operating cost is reduced by 75%. Mehrjerdi and Hemmati [84] described wind-operated EV charging station modeling. The charging station is integrated with a wind energy system as a primary source, a battery as a storage medium, and a multivariable EV charging facility.

The uncertainty due to the wind velocity is also analyzed in this paper, and the capacity of the low-, medium-, and fast-level charging is 116 kW, 84 kW, and 52 kW, respectively. The overall cost of the system is $945/year, in which network development and charging facility consume 15% and 12%, respectively, of the overall cost of the system. Huang and Kockelman [85] analyzed elastic demand, station congestion, and network equilibrium of the location of an EV charging station. The genetic algorithm assesses technical parameters and finds that consumers are willing to pay $6 for 30-minute charging of EVs. Wang and Wang [86] explained the energy trading of EV charging stations using the game theory concept. A multilevel charging facility is considered a play of the game, and the payoff matrix shows that the overall cost is reduced by 4%, and profit is increased in variable mode from 7.5% to 7.9%. Luo and Wu [87] explained the coordination and allocation aspects of an EV charging station in the distributed energy system. This paper proposed an optimization model between charging stations and distributed energy sources. A real urban area fed by a 31-bus distribution network in China is chosen as a test method to check the feasibility of the proposed optimization model, and their numerical results are analyzed. He and Fathabadi [88] proposed a solar-fuel cell-operated EV charging station. To cater to a high charging and discharging frequency and the intermittent nature of the solar energy system, the effective charging station battery is replaced by fuel cells, hydrogen tanks, and electrolyte capacitors. The result shows that the fuel cell is a better alternative to the battery in terms of operation and maintenance cost and overall performance. Schmidt et al. described the utilization of battery-swapping approach to improve electric vehicles' environmental balance and price-performance ratio. The basic design of hybrid vehicles, significant components, and various power train combinations are addressed in applications and constraints. Table 10.3 shows the assessment of different charging stations.

Table 10.3 Assessment of different charging stations.

Author	Objective	Supply source	Technique used
Biya and Sindhu [89]	Design and energy management	Solar energy system	MATLAB-Simulink
Li and Zhang [90]	Optimal design and analysis	Power grid + solar energy system	Monte-Carlo simulation
Yazdi et al. [91]	Design and location assessment	Power grid + solar energy system	Discrete cooperation covering technology
Nizam and Wicaksono [92]	Design and economic assessment	Solar + wind energy system	The derivative-free optimization technique
Divyapriya et al. [93]	Smart charging station	Solar energy system	Raspberry PI, Internet of Things, MATLAB-Simulink
Turan [94]	Dynamic consumption characteristics	Power grid + solar energy system	MATLAB-Simulink
Khalkhali et al. [95]	Location planning and management	Power grid	Data envelopment analysis theorem
Li et al. [96]	Optimal power dispatch	Power grid + solar energy system	Two-stage stochastic optimization
Thakur et al. [97]	Energy trading	Microgrids	Blockchain
Piao et al. [98]	Pricing strategy	Power grid	Game theory

Gampa and Jasthi [99] analyzed all the technical and financial parameters of EV charging stations using the Grasshopper optimization technique. Fuzzy logic also allocates distribution sources, shunt capacitors, and EV charging stations. The result shows that the 51- and 62-bus distribution network offer many advantages over conventional energy system-operated EV charging stations. Mehrjerdi [100] described an off-grid solar-powered charging station that charges hydrogen-fuelled vehicles and normal EVs. This paper's significant contribution is that electric and hydrogen vehicles can be charged simultaneously as one part of the solar system handles the demand of EVs and the other part run a water electrolyzer to produce hydrogen. The result shows that 95% of the overall cost of the charging station is consumed in solar energy systems, and 97% of solar energy is converted into hydrogen and storage media. Kabli and Quddus [101] explained the expansion planning of green vehicle charging stations by stochastic programming techniques. A multistage stochastic programming language is used to expand the capacity of the charging station to cater to the increasing number of EVs day by day. The technical and financial parameter of the charging station is assessed using a Progressive hedging algorithm. The result of this algorithm is also given considering various managerial viewpoints for decision-makers. Quddus and Kabli [102] described grid-green energy-operated green vehicle charging stations' long-term scheduling and short-term functional decision-making. A case study of Washington City is taken to analyze technical and financial parameters by Sample Average Approximation with an enhanced Progressive Hedging Algorithm. The result shows that the grid supply can meet load demand at a lower cost during the peak charging time when a solar energy system is unavailable. Zhang and Liu [103] described the optimum design of a solar power-operated EV charging station. The key points of the case study in Beijing city are "rationalization, modularization and an intelligentization of the fast charging station." Zhang and Tang [104] described the optimal green car charging station capacity expansion using whale optimization techniques. An enhanced whale optimization approach (IWOA) is presented in this research. This study extends the whale optimization algorithm (WOA) by adding the Gaussian operator and simultaneously merging the differential evolution algorithm with the notion of a crowding factor in the artificial fish swarm method's back-end operations. Luo and Wu [105] proposed a collaborative planning model based on the spatial scheduling problem of EV charging requirements and focused on the social cost equilibrium. A suitable second-order cone programming relaxation is utilized to put the suggested model into a type to reduce the complexity of the optimization model. Fathabadi [106] explained the feasibility analysis, modeling, and controlling of totally autonomous nonconventional energy-operated EV charging stations. Due to the intermittent nature of wind velocity and solar radiation, the author proposed a hybrid solar-wind energy system and a fuel cell storage system to avoid the intermittent nature of solar/wind and the high battery system cost. Fuel cell is utilized to have an increased lifetime and an overall expense of about $21,000 in place of a 6.5 kWh Li-particle battery bank, giving a short lifetime of around 500 days and a complete expense of about $70,000 for 30 years. It is stand-alone, fully autonomous, and 100 % green energy due to the lack of any adverse environmental effects and the prospect of building charging points in remote areas with no grid. The overall construction costs of the charging station are roughly $99,500, with a payback period of just 16 months. Huang and Ma [107] explained the concept of geographical information system-based charging station in a highly dense area of Hongkong city using nonconventional energy resources. Choosing the ideal areas and an ideal number of sustainable ones fuelled accusing stations of contemplating the current charging stations and inexhaustible possibilities. Hafez and Bhattacharya [108] assessed the optimal design of EV charging stations using various energy sources. Various energy sources, such as nonconventional energy systems and diesel generation, have practical inputs on their corporeal, operational, and financial characteristics. The configuration of the EVCS along the highways, as an independent microgrid, is being examined to resolve the "range anxiety" issues of EV owners regarding the distance the vehicle will travel. Luo and Gu [109] planned to enhance the capacity of EV charging stations through a multilevel charging facility. To illustrate the viability and usefulness of the proposed solution, a natural urban area fed by a 31-bus delivery network in China has been used as a testing ground, and the mathematical results are accessed and analyzed.

Outcomes: Much work has already been reported on EV charging stations. The grid-solar hybrid system is a feasible solution for a charging station. The literature reported is categorized into two categories: location identification for charging stations and capacity calculation and capacity expansion of charging stations. However, the concept of big data and decision science+ can be applied to location assessment for charging stations, and a drone-based system may also be developed to find a suitable location where sufficient amounts of solar radiation/wind velocity exist for solar/wind energy systems. Specific optimization techniques may also be used for the optimum design of an EV charging station.

10.7 Worldwide Policy Framework for EV

Several assessment parameters such as fuel prices, incentives for zero- and low-emission green vehicles, and financial instruments help bridge the cost gap between electric and ordinary vehicles and offer incentives for electric mobility in developed and developing countries. Policy support is gradually being applied to tackle the strategic significance of the value chain of battery technology. Hu and

Wang [110] described EV diffusion due to the different policies in China. The findings show that the government's purchasing subsidy and restricted travel policies will encourage a 60% diffusion rate for electric vehicles, and the short-term impact is remarkable. Output subsidies and infrastructure-building policies will support EVs' 70% diffusion rate. Offering incentives to producers have a more significant effect than offering buying incentives to customers. At the same time, low energy and high oil rates would push the delivery rate of EVs to 60% and 70%, respectively. Kong and Xia [111] explained the effect of the subsidy policy on the EV market in China. This study examines the impact of several approaches, including commodity subsidies, carbon emissions, trading and license plate restrictions, and policies on EV sales as soon as the purchasing subsidy scheme is phased out in China. Therefore, a system dynamics model is proposed, explicitly considering the relationship among government, businesses, and consumers.

Fang and Wei [112] analyzed China's EV charging station-related policy framework and incentives. The findings demonstrate the benefits of a balanced, integrated subsidy and tax strategy for developing electricity charging infrastructures. It is noteworthy that investment is not the biggest obstacle to installing charging stations. The main driving forces are EV penetration rates and the charging price. Melton and Axsen [113] studied the policy framework for EVs in Canada. The study provides a more detailed assessment method, taking into account five criteria: (i) efficiency in long-term PEV adoption, (ii) government expenditure, (iii) public holds up, (iv) policy flexibility, and (v) a transformational indication, which is a measure of the policy's ability to increase interest and investment in the PEV transition. Table 10.4 shows the policy assessment of different countries.

Wang and Tang [114] compared global incentive policies on the EV market. In 2015, the worldwide milestone of 1 million electric cars on the road surpassed 1.26 million, thanks to the political dividend. The subsidy system was deemed the most

Table 10.4 Policy assessment of different countries.

Parameters	Canada	China	European Union	India	Japan	United States
Fuel economy standard	Yes	Yes	Yes	Yes	Yes	Yes
Fiscal incentive	Yes	Yes	Yes	Yes	No	Yes
Subsidy	Yes	Yes	Yes	No	Yes	No
Hardware standard	Yes	Yes	Yes	Yes	Yes	No
Charging station standard	Yes	Yes	Yes	Yes	Yes	Yes

relevant and successful incentive strategy. Numerous countries, like China, the United States, and Germany, plan to eliminate electric car incentives. The findings reveal that the number of charging stations, charging rate, and road conditions are vital positive variables in the country's EV market share.

On the other hand, fiscal incentives are no longer the primary cause of significant disparities in promoting electric vehicles between nations. Capuder and Sprčić [115] comprehensively reviewed the limitations and advantages of worldwide EV policy. This paper offers a detailed overview of the political, cultural, social, technological, law-making, and ecological aspects and rigorously assesses the achievement of the objectives of EV integration. In addition to a comprehensive literature review of all relevant factors, this paper introduces a hybrid approach focused on finances and incorporates hazard supervision techniques. Wee and Coffman [116] explained the pros and cons of the EV policy implemented in the United States. This report outlines the length and importance of US State and Local EV (EV) policies in force from 2010 to 2015. While the emphasis is on state-level policies, municipal governments and energy company policies are reported as they collectively represent most of the state's population or energy customers. Carley and Zirogiannis [117] highlighted the shortcomings of EV policy in the United States and how to overcome them. Our research demonstrates a trade-off between growing PEV market penetration and reducing greenhouse gas emissions from the fleet of light-duty cars. These two goals are in conflict, at least in the near term, because of how federal and state policy interact. Egner and Trosvik [118] outlined Sweden's EV policy framework. The author discovers that providing public charging sites enhances adoption rates, particularly in metropolitan areas, and indicates that public procurement of battery electric vehicles might be a useful policy tool.

A qualitative evaluation of the policy process on EVs in the Nordic Region was presented by Kester and Noel [119]. To address the thinking and justifications underlying EV incentives and policy frameworks, this article presents a comprehensive comparative analysis based on 221 semi-structured conversations with 251 transport and energy professionals from 203 organizations across 18 cities in the Nordic area. Finally, this study advocates for clear, consistent, coherent goals and pricing incentives, a more substantial commitment to EV awareness efforts, and local freedom to give extra advantages. The multilevel policy mechanism of EV in Germany was presented by Taefi and Kreutzfeldt [120]. By comparing and evaluating the groups' rating findings, it is clear that their differences might be significant, providing valuable information and room for future research and practice. Figenbaum [121] elaborated EV policy framework in Norway.

Norway has achieved an incredible milestone for electric car batteries. The market share was 16.9% in 2014, and the total fleet was 2.6%, with around

71,505 vehicles. Norwegian purchase incentives are significant enough to make EVs a competitively priced option for car purchasers. Improved model range, new technology, reduced vehicle prices, and robust marketing have driven further sales. In an experimental form, Kwon and Son [122] studied the EV policy on Jeju Island. Experiments have shown that, relative to prospective customers, the actual owners of EVs demonstrated a greater inconvenience to charging time. For incentives, EV owners have shown different preferences for different incentive measures.

Outcomes: The above data show that untill 2019, only 2–5% countries worked in the field of EVs and their charging station, but in those countries, the perception of their people is not very favorable regarding purchasing an EV at a higher cost compared to the conventional vehicles. All over the world, it should be mandatory for the individual government to create awareness about the EV and develop a policy for EVs and charging stations in a form that is helpful to the people and creates lots of employment opportunities for the young generation in the field of green mobility.

10.8 Electric Vehicles on the Sustainability and Reliability of Transportation Network

Electric vehicles can significantly improve the sustainability of the transportation network. Electric vehicles emit fewer greenhouse gases than traditional gas-powered vehicles. According to the Union of Concerned Scientists, on average, electric vehicles emit 60% less carbon dioxide than gas-powered vehicles. This reduction in emissions can help to mitigate climate change and improve air quality in urban areas. Electric vehicles are typically more energy-efficient than traditional gas-powered vehicles. Electric motors convert more energy stored in the battery into motion, whereas gas-powered engines waste energy through heat and friction. Increased efficiency translates into reduced energy consumption and lower costs. Electric vehicles can be used in conjunction with renewable energy sources, such as solar panels, to create a more resilient and decentralized transportation network. This can help to reduce the impact of power outages and other disruptions on the transportation network. Adopting electric vehicles can play an important role in making the transportation network more sustainable. However, there are still challenges to overcome, such as the availability of charging infrastructure, the cost of batteries, and the need for more sustainable sources of electricity.

Electric vehicles (EVs) can positively and negatively impact the reliability of the transportation network. Here are some potential ways in which EVs could impact transportation reliability.

Positive impacts

- **Reduced dependence on fossil fuels:** Since EVs run on electricity, they don't rely on fossil fuels for energy. This means that they can reduce the vulnerability of the transportation network to disruptions in oil supply, such as oil price spikes or geopolitical conflicts.
- **Lower maintenance costs:** EVs have fewer moving parts than internal combustion engine vehicles, requiring less maintenance. This could reduce the breakdowns and delays on the transportation network caused by mechanical problems.
- **More efficient use of road space:** EVs are generally smaller and more compact than internal combustion engine vehicles. This could allow for more efficient use of road space, as more EVs can fit on the same stretch of road than would be possible with larger vehicles.

Negative impacts

- **Dependence on the electricity grid:** Since EVs require electricity to operate, they depend on the grid. This means that disruptions to the grid, such as power outages or equipment failures, could impact the reliability of the transportation network.
- **Range anxiety:** EVs typically have shorter ranges than internal combustion engine vehicles, which means that drivers may need to plan their trips more carefully to avoid running out of power. This could increase the risk of delays or disruptions if drivers cannot find a charging station or if charging stations are not functioning properly.
- **Congestion at charging stations:** If there are not enough charging stations to meet demand, EV drivers may experience delays or disruptions while waiting for a charging spot to become available. This could be particularly problematic in urban areas where there is already limited space for parking and infrastructure.

10.9   Recent Trends and Future Challenges

In the recent scenario, EV technology is ice-breaking in the transportation sector, and research shows that lots of work is needed to change people's perceptions. Further advanced research is required for the charging and discharging capability of the battery system to increase the range of EVs. The government must take initiatives to develop a renewable energy-based EV charging station that do not create an extra load on the power grid. The following points are proposed to create a splendid environment for EVs.

- Solar energy-operated EVs with backup must be developed to mitigate environmental problems such as pollution through greenhouse gases.
- Use of switched reluctance motors in place of conventional electric motors.
- A market basket model and fault tree analysis may be used for the reliability analysis of EVs.

References

1 Khare, V., Nema, S., and Baredar, P. (2013). Status of solar-wind renewable energy in India. *Renewable & Sustainable Energy Reviews* 27: 1–10.

2 https://www.iea.org/reports/global-ev-outlook-2019

3 Khare, V. and Bunglowala, A. (2019). Design and assessment of solar-powered EV by different techniques. *International Transaction of Electrical Energy System* https://doi.org/10.1002/2050-7038.12271.

4 Chu, Y. and Wu, Y. (2019). Design of energy and materials for ammonia-based extended-range electric vehicles. *Energy Procedia* 158: 3064–3069.

5 Zhang, L.P. and Liu, W. (2019). Innovation design and optimization management of a new drive system for plug-in hybrid electric vehicles. *Energy* 186: 115823.

6 Tanozzi, F. and Sharma, S. (2019). 3D design and optimization of heat exchanger network for solid oxide fuel cell-gas turbine in hybrid electric vehicles. *Applied Thermal Engineering* 163: 114310.

7 Chatterjee, S. and Iyer, A. (2017). Design optimisation for an efficient wireless power transfer system for electricEVs. *Energy Procedia* 117: 1015–1023.

8 Borthakur, S. and Subramanian, S.C. (2018). Optimized design and analysis of a series-parallel hybrid electric vehicle powertrain for a heavy duty truck. *IFAC-Papers On Line* 51 (1): 184–189.

9 Czogalla, O. and Jumar, U. (2019). Design and control of electric bus vehicle model for estimation of energy consumption. *IFAC-Papers On Line* 52 (24): 59–64.

10 Hofman, T. and Janssen, N.H.J. (2017). Integrated design optimization of the transmission system and vehicle control for electric vehicles. *IFAC-Papers On Line* 50 (1): 10072–10077.

11 Vora, A.P. and Jin, X. (2017). Design-space exploration of series plug-in hybrid electric vehicles for medium-duty truck applications in a total cost-of-ownership framework. *Applied Energy* 202: 662–672.

12 del Valle, J.A. and Viera, J.C. (2018). Design and validation of a tool for prognosis of the energy consumption and performance in electric vehicles. *Transportation Research Procedia* 33: 35–42.

13 Delogu, M. and Zanchi, L. (2017). Innovative composites and hybrid materials for electric vehicles lightweight design in a sustainability perspective. *Materials Today Communications* 13: 192–209.

14 Hang, P. and Han, Y. (2018). Design of an active collision avoidance system for a 4WIS-4WID electric vehicle. *IFAC-Papers On Line* 51 (31): 771–777.

15 Zhou, X. and Qin, D. (2017). Multiobjective optimization design and performance evaluation for plug-in hybrid electric vehicle power-trains. *Applied Energy* 208: 1608–1625.

16 Xin, X. and Chengning, Z. (2017). Optimal design of electric vehicle power system with the principle of minimum curb mass. *Energy Procedia* 105: 2629–2634.

17 Hodgson, D. and Mecrow, B.C. (2013). Effect of vehicle mass changes on the accuracy of Kalman filter estimation of electric vehicle speed. *IET Electrical Systems in Transportation* 3 (3): 67–78.

18 Cocron, P., Bühler, F., and Neumann, I. (2011). Methods of evaluating electric vehicles from a user's perspective – the MINI E field trial in Berlin. *IET Intelligent Transport Systems* 5 (2): 127–133.

19 Bukhari, A.A.S. and Alalibo, B.P. (2019). Switched reluctance motor design for electric vehicles based on harmonics and back EMF analysis. *The Journal of Engineering* 2019 (17): 4220–4225.

20 Jenal, M., Sulaiman, E., and Kumar, R. (2016). Effects of rotor pole number in outer rotor permanent magnet flux switching machine for light weight electric vehicle. *4th IET Clean Energy and Technology Conference (CEAT 2016)*, 79 (6).

21 Thanapalan, K. and Liu, G.-P. (2010). Modelling and control of fuel cell hybrid electric vehicle systems. *UKACC International Conference on CONTROL 2010*: 1100–1105.

22 Villa, D. and Montoya, A. (2018). A taxonomy of energy consumption models for electric vehicles. *MOVICI-MOYCOT 2018: Joint Conference for Urban Mobility in the Smart City*, 16 (7 pp.).

23 Grewal, K.S. and Darnell, P.M. (2013). Model-based EV range prediction for electric hybrid vehicles. *IET Hybrid and Electric Vehicles Conference 2013 (HEVC 2013)*, 3.3.

24 Roozegar, M., Setiawan, Y.D., and Angeles, J. (2017). Design, modelling and estimation of a novel modular multi-speed transmission system for electric vehicles. *Mechatronics* 45: 119–129.

25 Di Giorgio, P. and Di Trolio, P. (2018). Model based preliminary design and optimization of Internal Combustion Engine and Fuel Cell hybrid electric vehicle. *Energy Procedia* 148: 1191–1198.

26 Guo, H. and Sun, Q. (2018). A systematic design and optimization method of transmission system and power management for a plug-in hybrid electric vehicle. *Energy* 148: 1006–1017.

27 Li, Y., Yang, J., and Song, J. (2017). Structure models and nano energy system design for proton exchange membrane fuel cells in electric energy vehicles. *Renewable and Sustainable Energy Reviews* 67: 160–172.

28 Dimitrova, Z. and Maréchal, F. (2015). Techno-economic design of hybrid electric vehicles using multi objective optimization techniques. *Energy* 91: 630–644.

29 Khare, V., Nema, S., and Baredar, P. (2016). Optimization of Hydrogen based hybrid renewable energy system using HOMER, BB BC AND GAMBIT. *International Journal of Hydrogen Energy* 41 (38): 16743–16751.

30 Khare, V., Nema, S., and Baredar, P. (2015). Optimization of hybrid renewable energy system by HOMER, PSO and CPSO for the study area. *International Journal of Sustainable Energy* 36 (4): 326–343.

31 Wiriyasart, S. and Hommalee, C. (2020). Thermal management system with nanofluids for electric vehicle battery cooling modules. *Case Studies in Thermal Engineering* 18: 100583.

32 Lyu, Y. and Siddique, A.R.M. (2019). Electric vehicle battery thermal management system with thermoelectric cooling. *Energy Reports* 5: 822–827.

33 Behi, H. and Karimi, D. (2020). A new concept of thermal management system in Li-ion battery using air cooling and heat pipe for electric vehicles. *Applied Thermal Engineering* 174.

34 Hakeem Akinlabi, A. and Solyali, D. (2020). Configuration, design, and optimization of air-cooled battery thermal management system for electric vehicles: a review. *Renewable and Sustainable Energy Reviews* 125: 109815.

35 Chung, Y. and Kim, M.S. (2019). Thermal analysis and pack level design of battery thermal management system with liquid cooling for electric vehicles. *Energy Conversion and Management* 196: 105–116.

36 Shen, M. and Gao, Q. (2020). System simulation on refrigerant-based battery thermal management technology for electric vehicles. *Energy Conversion and Management* 203: 112176.

37 Smith, J. and Singh, R. (2018). Battery thermal management system for electric vehicle using heat pipes. *International Journal of Thermal Sciences* 134: 517–529.

38 Cen, J. and Li, Z. (2018). Experimental investigation on using the electric vehicle air conditioning system for lithium-ion battery thermal management. *Energy for Sustainable Development* 45: 88–95.

39 Kim, J. and Jinwoo, O. (2019). Review on battery thermal management system for electric vehicles. *Applied Thermal Engineering* 149: 192–212.

40 Tian, Z. and Gan, W. (2018). Investigation on an integrated thermal management system with battery cooling and motor waste heat recovery for electric vehicle. *Applied Thermal Engineering* 136: 16–27.

41 Ianniciello, L. and Biwolé, P.H. (2018). Electric vehicles batteries thermal management systems employing phase change materials. *Journal of Power Sources* 378: 383–403.

42 Hong, J. and Wang, Z. (2019). Synchronous multi parameter prediction of battery systems on electric vehicles using long short-term memory networks. *Applied Energy* 254: 113648.

43 Wang, Y.N. (2019). Power battery performance detection system for electric vehicles. *Procedia Computer Science* 154: 759–763.

44 Yang, B. and Zhu, T. (2020). Design and implementation of Battery/SMES hybrid energy storage systems used in electric vehicles: a non-linear robust fractional-order control approach. *Energy* 191: 116510.

45 Wegmann, R. and Döge, V. (2018). Assessing the potential of an electric vehicle hybrid battery system comprising solid-state lithium metal polymer high energy and lithium-ion high power batteries. *Journal of Energy Storage* 18: 175–184.

46 Saw, L.H. and Poon, H.M. (2019). Numerical modeling of hybrid supercapacitor battery energy storage system for electric vehicles. *Energy Procedia* 158: 2750–2755.

47 Niu, X. and Garg, A. (2019). A coupled electrochemical-mechanical performance evaluation for safety design of lithium-ion batteries in electric vehicles: an integrated cell and system level approach. *Journal of Cleaner Production* 222: 633–645.

48 Wegmann, R. and Döge, V. (2017). Optimized operation of hybrid battery systems for EVs using deterministic and stochastic dynamic programming. *Journal of Energy Storage* 14 (Part 1): 22–38.

49 Wu, X.-l. and Xu, Y.-w. (2020). Extended-range EV-oriented thermoelectric surge control of a solid oxide fuel cell system. *Applied Energy* 263: 114628.

50 He, H. and Wang, C. (2020). An intelligent braking system composed single-pedal and multiobjective optimization neural network braking control strategies for electric vehicle. *Applied Energy* 259: 114172.

51 Tan, H. and Shen, B. (2020). Non-fragile H∞ control for body slip angle of electric vehicles with onboard vision systems: the dynamic event-triggering approach. *Journal of the Franklin Institute* 357 (4): 2008–2027.

52 Shi, K. and Cheng, D. (In Press). Interacting multiple model-based adaptive control system for stable steering of distributed driver electric vehicle under various road excitations. *ISA Transactions*.

53 Balaska, H. and Ladaci, S. (2019). Adaptive cruise control system for an electric vehicle using a fractional order model reference adaptive strategy. *IFAC-Papers On Line* 52 (13): 194–199.

54 Qiu, C. and Wang, G. (2018). A novel control strategy of regenerative braking system for electric vehicles under safety critical driving situations. *Energy* 149: 329–340.

55 Yang, B. and Wang, J. (2020). Applications of battery/supercapacitor hybrid energy storage systems for electric vehicles using perturbation observer based robust control. *Journal of Power Sources* 448: 227444.

56 Luo, H. and Yu, J. (2019). Finite-time dynamic surface control for induction motors with input saturation in electric vehicle drive systems. *Neurocomputing* 369: 166–175.

57 He, H. and Quan, S. (In Press). Hydrogen circulation system model predictive control for polymer electrolyte membrane fuel cell-based electric vehicle application. *International Journal of Hydrogen Energy*.

58 Wei, X. and Chen, H. (2019). Torque optimization control for electric vehicles with four in-wheel motors equipped with regenerative braking system. *Mechatronics* 57: 95–108.

59 Sajadi-Alamdari, S.A. and Voos, H. (2019). Non-linear model predictive control for ecological driver assistance systems in electric vehicles. *Robotics and Autonomous Systems* 112: 291–303.

60 Zhang, H. and Zhao, W. (2018). Decoupling control of steering and driving system for in-wheel-motor-drive electric vehicle. *Mechanical Systems and Signal Processing* 101: 389–404.

61 Khare, V., Nema, S., and Baredar, P. (2018). Reliability analysis of hybrid renewable energy system by fault tree analysis. *Energy & Environment* 30 (3): 542–555.

62 Dubarry, M. and Devie, A. (2017). Durability and reliability of electric vehicle batteries under electric utility grid operations: bidirectional charging impact analysis. *Journal of Power Sources* 358: 39–49.

63 Kamruzzaman, M.D. and Benidris, M. (2020). A reliability-constrained demand response-based method to increase the hosting capacity of power systems to electric vehicles. *International Journal of Electrical Power & Energy Systems* 121: 106046.

64 Anand, M.P. and Bagen, B. (2020). Probabilistic reliability evaluation of distribution systems considering the spatial and temporal distribution of electric vehicles. *International Journal of Electrical Power & Energy Systems* 117: 105609.

65 Gandoman, F.H. and Ahmadi, A. (2019). Status and future perspectives of reliability assessment for electric vehicles. *Reliability Engineering & System Safety* 18: 1–16.

66 Gandoman, F.H. and Jaguemont, J. (2019). Concept of reliability and safety assessment of lithium-ion batteries in electric vehicles: basics, progress, and challenges. *Applied Energy* 25: 113343.

67 Shu, X. and Yang, W. (2020). A reliability study of electric vehicle battery from the perspective of power supply system. *Journal of Power Sources* 45: 106046.

68 Davidov, S. and Pantoš, M. (2017). Planning of electric vehicle infrastructure based on charging reliability and quality of service. *Energy* 118: 1156–1167.

69 Napoli, G. and Polimeni, A. (2019). Optimal allocation of electric vehicle charging stations in a highway network: part 2: the Italian case study. *Journal of Energy Storage* 26: 101015.

70 Liu, A. and Zhao, Y. (2020). A three-phase fuzzy multi-criteria decision model for charging station location of the sharing electric vehicle. *International Journal of Production Economics* 22: 107572.

71 Kong, W. and Luo, Y. (2019). Optimal location planning method of fast charging station for electric vehicles considering operators, drivers, vehicles, traffic flow and power grid. *Energy* 186: 115826.

72 Liu, Q. and Liu, J. (2019). Data-driven intelligent location of public charging stations for electric vehicles. *Journal of Cleaner Production* 232: 531–541.

73 Csiszár, C. and Csonka, B. (2019). Urban public charging station locating method for electric vehicles based on land use approach. *Journal of Transport Geography* 74: 173–180.

74 He, Y. and Kockelman, K.M. (2019). Optimal locations of US fast charging stations for long-distance trip completion by battery electric vehicles. *Journal of Cleaner Production* 214: 452–461.

75 Lee, Y. and Hur, J. (2019). A simultaneous approach implementing wind powered electric vehicle charging stations for charging demand dispersion. *Renewable Energy* 144: 172–179.

76 Sun, Z. and Gao, W. (In Press). Locating charging stations for electric vehicles. *Transport Policy*.

77 Kadri, A.A. and Perrouault, R. (2020). A multi-stage stochastic integer programming approach for locating electric vehicle charging stations. *Computers & Operations Research* 117: 104888.

78 Hosseini, S. and Sarder, M.D. (2019). Development of a Bayesian network model for optimal site selection of electric vehicle charging station. *International Journal of Electrical Power & Energy Systems* 105: 110–122.

79 Wang, C. and He, F. (2019). Designing locations and capacities for charging stations to support intercity travel of electric vehicles: an expanded network approach. *Transportation Research Part C: Emerging Technologies* 10: 210–232.

80 Bai, X. and Chin, K.-S. (2019). A bi-objective model for location planning of electric vehicle charging stations with GPS trajectory data. *Computers & Industrial Engineering* 128: 591–604.

81 Mehrjerdi, H. (2020). Dynamic and multi-stage capacity expansion planning in microgrid integrated with electric vehicle charging station. *Journal of Energy Storage* 29.

82 Mehrjerdi, H. and Hemmati, R. (2019). Electric vehicle charging station with multi-level charging infrastructure and hybrid solar-battery-diesel generation incorporating comfort of drivers. *Journal of Energy Storage* 26: 100924.

83 Ding, Z. and Lu, Y. (2020). Optimal coordinated operation scheduling for electric vehicle aggregator and charging stations in an integrated electricity-transportation system. *International Journal of Electrical Power & Energy Systems* 121: 106040.

84 Mehrjerdi, H. and Hemmati, R. (2020). Stochastic model for electric vehicle charging station integrated with wind energy. *Sustainable Energy Technologies and Assessments* 37: 100577.

85 Huang, Y. and Kockelman, K.M. (2020). Electric vehicle charging station locations: elastic demand, station congestion, and network equilibrium. *Transportation Research Part D Transport and Environment* 78: 102179.

86 Wang, Y. and Wang, X. (2020). Distributed energy trading for an integrated energy system and electric vehicle charging stations: a Nash bargaining game approach. *Renewable Energy* 155: 513–530.

87 Luo, L. and Wu, Z. (2020). Coordinated allocation of distributed generation resources and electric vehicle charging stations in distribution systems with vehicle-to-grid interaction. *Energy* 192: 116631.

88 He, F. and Fathabadi, H. (2020). Novel stand-alone plug-in hybrid electric vehicle charging station fed by solar energy in presence of a fuel cell system used as supporting power source. *Renewable Energy* 156: 964–974.

89 Biya, T.S. and Sindhu, M.R. (2019). Design and power management of solar powered electric vehicle charging station with energy storage system. *2019 3rd International Conference on Electronics, Communication and Aerospace Technology (ICECA)* (12–14 June 2019), Coimbatore, India.

90 Li, T. and Zhang, J. (2018). An optimal design and analysis of a hybrid power charging station for electric vehicles considering uncertainties. *IECON 2018 - 44th Annual Conference of the IEEE Industrial Electronics Society* (21–23 October 2018), Washington, DC, USA.

91 Yazdi, L., Ahadi, R., and Rezaee, B. (2019). Optimal electric vehicle charging station placing with integration of renewable energy. *2019 15th Iran International Industrial Engineering Conference (IIIEC)* (23–24 January 2019), Yazd, Iran.

92 Nizam, M. and Rian Wicaksono, F.X. (2018). Design and optimization of solar, wind, and distributed energy resource (DER) hybrid power plant for electric vehicle (EV) charging station in rural area. *2018 5th International Conference on Electric Vehicular Technology (ICEVT)* (30–31 October 2018).

93 Divyapriya, S., Amutha, A., and Vijayakumar, R. (2018). Design of residential plug-in electric vehicle charging station with time of use tariff and IoT technology. *2018 International Conference on Soft-computing and Network Security (ICSNS)*, 14–16 February 2018, Coimbatore, India.

94 Turan, M.T. (2019). Effect of distributed generation based campus model combined with electricvehicle charging stations on the distribution network. *2019 International Conference on Smart Energy Systems and Technologies (SEST)* (9–11 September 2019), Porto, Portugal.

95 Khalkhali, K., Abapour, S., Moghaddas-Tafreshi, S.M., and Abapour, M. (2015). Application of data envelopment analysis theorem in plug-in hybrid electric vehicle charging station planning. *IET Generation, Transmission and Distribution* 9 (7): 666–676.

96 Li, W.(.J.)., Tan, X., Sun, B., and Tsang, D.H.K. (2019). Optimal power dispatch of a centralised electric vehicle battery charging station with renewables. *IET Communications* 12: 579–585.

97 Thakur, S., Hayes, B.P., and Breslin, G. (2018). A unified model of peer to peer energy trade and electric vehicle charging using blockchains. *Mediterranean Conference on Power Generation, Transmission, Distribution and Energy Conversion (MEDPOWER 2018)*.

98 Piao, L., Ai, Q., and Fan, S. (2015). Game theoretic based pricing strategy for electric vehicle charging stations. *International Conference on Renewable Power Generation (RPG 2015)*.

99 Gampa, S.R. and Jasthi, K. (2020). Grasshopper optimization algorithm based two stage fuzzy multiobjective approach for optimum sizing and placement of distributed generations, shunt capacitors and electric vehicle charging stations. *Journal of Energy Storage* 27: 101117.

100 Mehrjerdi, H. (2019). Off-grid solar powered charging station for electric and hydrogen vehicles including fuel cell and hydrogen storage. *International Journal of Hydrogen Energy* 44 (23): 11574–11583.

101 Kabli, M. and Quddus, M.A. (2020). A stochastic programming approach for electric vehicle charging station expansion plans. *International Journal of Production Economics* 220: 107461.

102 Quddus, M.A. and Kabli, M. (2019). Modeling electric vehicle charging station expansion with an integration of renewable energy and Vehicle-to-Grid sources. *Transportation Research Part E: Logistics and Transportation Review* 128: 251–279.

103 Zhang, J. and Liu, C. (2019). Design scheme for fast charging station for electric vehicles with distributed photovoltaic power generation. *Global Energy Interconnection* 2 (2): 150–159.

104 Zhang, H. and Tang, L. (2019). Locating electric vehicle charging stations with service capacity using the improved whale optimization algorithm. *Advanced Engineering Informatics* 41: 100901.

105 Luo, L. and Wei, G. (2019). Joint planning of distributed generation and electric vehicle charging stations considering real-time charging navigation. *Applied Energy* 242: 1274–1284.

106 Fathabadi, H. (2020). Novel stand-alone, completely autonomous and renewable energy based charging station for charging plug-in hybrid electric vehicles (PHEVs). *Applied Energy* 260: 114194.

107 Huang, P. and Ma, Z. (2019). Geographic Information System-assisted optimal design of renewable powered electricvehicle charging stations in high-density cities. *Applied Energy* 255: 113855.

108 Hafez, O. and Bhattacharya, K. (2017). Optimal design of electric vehicle charging stations considering various energy resources. *Renewable Energy* 107: 576–589.

109 Luo, L. and Gu, W. (2018). Optimal planning of electric vehicle charging stations comprising multi-types of charging facilities. *Applied Energy* 226: 1087–1099.

110 Yi, H. and Wang, Z. (2020). Impact of policies on electric vehicle diffusion: an evolutionary game of small world network analysis. *Journal of Cleaner Production* 265: 121703.

111 Kong, D. and Xia, Q. (2020). Effects of multi policies on electric vehicle diffusion under subsidy policy abolishment in China: a multi-actor perspective. *Applied Energy* 266: 114887.

112 Fang, Y. and Wei, W. (2020). Promoting electric vehicle charging infrastructure considering policy incentives and user preferences: an evolutionary game model in a small-world network. *Journal of Cleaner Production* 258: 120753.

113 Melton, N. and Axsen, J. (2020). Which plug-in electric vehicle policies are best? A multi-criteria evaluation framework applied to Canada. *Energy Research & Social Science* 64: 101411.

114 Wang, N. and Tang, L. (2019). A global comparison and assessment of incentive policy on electric vehicle promotion. *Sustainable Cities and Society* 44: 597–603.

115 Capuder, T. and Sprčić, D.M. (2020). Review of challenges and assessment of electric vehicles integration policy goals: integrated risk analysis approach. *International Journal of Electrical Power & Energy Systems* 119: 105894.

116 Wee, S. and Coffman, M. (2019). Data on US state-level electric vehicle policies, 2010–2015. *Data in Brief* 23: 103658.

117 Carley, S. and Zirogiannis, N. (2019). Overcoming the shortcomings of US plug-in electric vehicle policies. *Renewable and Sustainable Energy Reviews* 113: 109291.

118 Egnér, F. and Trosvik, L. (2018). Electric vehicle adoption in Sweden and the impact of local policy instruments. *Energy Policy* 121: 584–596.

119 Kester, J. and Noel, L. (2018). Policy mechanisms to accelerate electric vehicle adoption: a qualitative review from the Nordic region. *Renewable and Sustainable Energy Reviews* 94: 719–731.

120 Taefi, T.T. and Kreutzfeldt, J. (2016). Supporting the adoption of electric vehicles in urban road freight transport – a multi-criteria analysis of policy measures in Germany. *Transportation Research Part A: Policy and Practice* 91: 61–79.

121 Figenbaum, E. (2017). Perspectives on Norway's supercharged electric vehicle policy. *Environmental Innovation and Societal Transitions* 25: 14–34.

122 Kwon, Y. and Son, S. (2018). Evaluation of incentive policies for electric vehicles: an experimental study on Jeju Island. *Transportation Research Part A: Policy and Practice* 116: 404–412.

11

Electric Vehicle Charging Management in Parking Structures

Tania Panayiotou, Michalis Mavrovouniotis, and Georgios Ellinas

KIOS Research and Innovation Center of Excellence, Department of Electrical and Computer Engineering, University of Cyprus, Nicosia, Cyprus

11.1 Introduction

In recent years, electric vehicles (EVs) have become very popular, thus, leading to an increased demand for EV charging stations available to the drivers [1]. Apart from charging the EVs at home, drivers can now charge their electric cars either while parking on street or while parking within large parking structures. This creates a new business opportunity for the operators of the parking structures; nevertheless, it also creates challenging problems in terms of scheduling the charging of EVs within the parking structure. For example, when several EVs are simultaneously charging, this may add a significant strain on the power grid. Thus, intelligent scheduling techniques are needed in order to manage both the charging load as well as the expectations of the customers and the parking structure operators. These expectations can be expressed as specific quality-of-service (QoS) requirements and can include metrics such as delay in departure time, satisfaction of charging demand, operator monetary gain, etc., subject to the constraints imposed by the system under consideration. This is the main motivation of this work, which tries to develop intelligent charging scheduling algorithms for EVs that use large parking structures.

In the state of the art, several models for scheduling the charging of EVs have been presented. These models, for the most part, fall under two control architectures, namely decentralized and centralized [2]. In general, decentralized control architectures are more flexible, as they allow the drivers themselves to make decisions on when to commence and end their charging. However, such flexibility comes at a price, as such a decentralized approach does not account for the total

Interconnected Modern Multi-Energy Networks and Intelligent Transportation Systems: Towards a Green Economy and Sustainable Development, First Edition. Edited by Mohammadreza Daneshvar, Behnam Mohammadi-Ivatloo, Amjad Anvari-Moghaddam, and Reza Razzaghi.

charging load, potentially causing unacceptable outcomes, ranging from overload of the power system to the waste of available resources. This, in turn, will have an impact on all actors, i.e., parking operators will experience monetary losses and EV drivers will be dissatisfied with their charging capacity or their departure time. Due to the aforementioned drawbacks of the decentralized control architectures, nowadays, centralized control architectures are regarded as the most favorable techniques for scheduling the charging of EVs within large parking structures.

When centralized control architectures are utilized, a central decision management (CDM) unit (for example, a computing unit hosted in the cloud) makes all scheduling decisions using the input given by the drivers of the EVs [3,4]. Centralized solutions, since they have all necessary information regarding charging needs, arrival and departure times, etc., can allow for the efficient charging of all EVs via the implementation of smart scheduling techniques. Further, they can be utilized for the development of novel costing approaches that also increase the revenue of the parking operators.

Even though there are several approaches in the literature for addressing the charging scheduling problem, fairness is not a common metric that is examined (as described in the later section). Note that fairness can be investigated in terms of several attributes such as departure delay, charged capacity, charging cost, etc. Thus, this book chapter focuses on developing fair and efficient EV scheduling schemes. We first briefly discuss existing scheduling schemes, and then we analytically describe scheduling schemes specifically targeting the fair allocation of the available system resources to all the EVs that contend for these resources.

The main contributions of this work are the development of the "delay-fair" and "QoS-fair" integer linear program (ILP) formulations and the design of delay-fair and QoS-fair ant colony optimization (ACO)-based heuristic algorithms for addressing the EV static (i.e., all arrivals/departures are known a priori) charging scheduling problem in parking structures, with both ILP formulations and heuristics leveraging the α-fairness scheme. Extensive performance evaluation results demonstrate a trade-off between fairness and efficiency, that gives flexibility to the parking operator (i.e., through the value of α) to achieve a balance between its needs and the needs of the car owners, subject to the system's constraints.

11.2 EV Charging Management Schemes

Monitoring and controlling the EV charging process within a charging station is important to achieve EV user satisfaction, reduce operational costs, and guarantee power system stability [5]. EV charging management schemes are characterized by the following three attributes: (i) mobility, (ii) coordination, and (iii) control [6].

First, the mobility of an EV charging scheme is classified into *static* and *dynamic*. Static charging schemes assume that all EVs are present in a station while charging takes place [7,8], whereas dynamic charging schemes allow different system "disruptions" (e.g., new EV arrival or parked EV departure) while charging [9,10]. Although dynamic charging is more challenging and requires an advanced charging infrastructure, it is closer to the real-world requirements [11].

Second, an EV charging scheme can be either *coordinated* or *uncoordinated*. The coordinated method refers to a charging scheme that aims to optimize different aspects of the system (e.g., reduce charging times, satisfy power demand, reduce power losses, minimize daily electricity costs, etc.) [12,13] by coordinating the decisions taken by the various system entities (in a centralized or a distributed manner). In contrast, the uncoordinated method refers to a charging scheme in which different tasks are independently decided by various system entities (e.g., EVs can start charging as soon as they are plugged) [14]. Clearly, the uncoordinated method may satisfy some EV users; however, it is prone to power loss and affects the reliability and efficiency of the system, as the various actions are taken in an uncoordinated manner.

Third, there are two control strategies to manage the power flow of an EV charging station, namely *centralized* and *decentralized* [15]. The former strategy processes the information provided by the EVs and provides an optimal solution for the overall system considering the user and system constraints [8, 9, 10, 16]. On the contrary, in the latter strategy, the EV users decide their charging schedule depending on either their convenience, electricity cost, or both [17]. It should be noted that the difference between coordinated/uncoordinated and centralized/distributed lies in the fact that distributed approaches may also be (partially) coordinated (assuming that information is exchanged between [neighboring] system entities), whereas uncoordinated methods are always distributed, as they do not consider the use of a centralized control unit that would coordinate the actions taken.

Furthermore, EV charging management schemes vary mainly according to their optimization objectives. For example, authors in Refs. [8, 9, 10, 18] aim to minimize the total tardiness of the EVs, authors in Refs. [19,20] focus on minimizing the total electricity cost for charging the EVs, while the flattening of the overall load profile (i.e., grid congestion) is examined in Ref. [21]. Multiple objectives have also been considered in the state of the art; for example, the authors in Ref. [22] aim to both minimize parking cost and concurrently maximize the charging station's efficiency for the case of plug-in EVs' daily scheduling. In addition, other works in the literature consider variable EV charging rates [21], variable station charging power [23], or variable energy prices [24].

Recently, the problem of scheduling EV charging has been examined in [25] in terms of fairness, where the aim is to obtain a resource allocation that is fair concerning the charging levels that can be achieved by the EVs that contend for energy resources over a specified time interval (i.e., our planning horizon, defined in hours, days, etc.). Even though fairness in EV charging is viewed as a significant policy by operators, in general currently the EV owners unilaterally make decisions on their charging levels that mainly depend on human behavior (i.e., to forego charging if their capacity level is satisfactory, or to only charge up to a specific level, thus allowing the charging of other vehicles that have lower capacity) [26]. Nevertheless, explicitly addressing the fair resource allocation problem in parking structures is very important for the charging infrastructure operators, as it directly relates to the satisfaction of the EV users and their decisions concerning re-using the specific infrastructure. Thus, in this work we focus on addressing the gap in the literature concerning the fair scheduling problem for charging EVs within parking structures through the development of fair charging techniques that consider specifically two attributes, namely delay and QoS.

11.3 Fair Charging Management

A centralized EV charging management scheme is considered in this work, with the assumption of a coordinated power flow and static charging, where in-advanced reservations are possible. It is worth mentioning, however, that the static scheduling scheme proposed, can also be used in dynamic systems by sequentially solving a number of static problem instances that are shifted in time (i.e., as a multi-period planning problem that can be solved, for example, hourly) [18]. The reader should also note that EV uncertainties (in terms of arrival times, charging times, etc.) have not been considered in this work and are left as an interesting avenue for future research work.

In particular, we consider a parking structure environment where a central power source is connected through a power line with multiple parking/charging spots (Figure 11.1). Specifically, the figure illustrates a simple charging station comprised of two charging lines, and for each charging line there are several spots where the owners can park their cars. Each parking spot doubles also as a charging spot and is centrally controlled by the CDM. Thus, at each spot, the CDM can provide charging services to an EV by activating/deactivating the spot (active spots are shown in orange color, while inactive spots are shown in black color in the figure) to charge the EV up to a specified level. Thus, the role of the CDM is to coordinate the activation/deactivation decisions for all EVs requesting charging services.

In this environment, due to limited power resources, and to avoid power link overload (i.e., congestion) and potential power outages, it may be impossible to fully charge all EVs within their finite stay time. While a solution to this problem

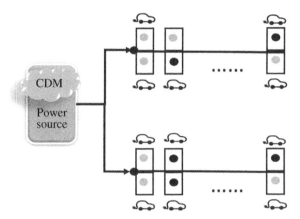

Figure 11.1 Illustration of a simple charging station comprised of multiple charging lines and parking spots.

is to maximize system efficiency, subject to the constraints imposed by the system, this may leave some EV users unsatisfied (i.e., their EVs are not charged within their stay time or they are experiencing extensive departure delays while waiting to be charged). This clearly results in an unfair solution from the user's point of view, leading to the need for fair charging scheduling algorithms. Specifically, fair charging scheduling algorithms are necessary to effectively control both power link congestion and the achievable service quality of end-users.

As the fair scheduling problem and the rising efficiency-fairness trade-off [27] have received considerable attention for various types of networks and applications [28], including smart grids [29,30], communication networks [31–35], transportation networks [36–38], cloud and edge computing networks [39,40] amongst others, several fairness measures have been proposed to capture the diverse fairness-efficiency criteria in different environments, e.g., proportional fairness, max-min fairness, Jain's index, and Gini coefficient [28], with the most widely considered measures being proportional and max-min fairness [28,41]. In general, proportional fairness is considered when the target is to balance the efficiency-fairness trade-off, while the max-min fairness is considered when the target is the fairest allocation (i.e., achieving an equitable distribution of resources across the contending end-users). Limitations, however, exist when committing to either one of these measures. Max-min fairness is likely to lead to inefficient system utilization (e.g., waste of energy resources) and to an end-user's maximum achievable service level that is heavily degraded for the sake of equity (e.g., all EVs are experiencing extensive delays for the sake of equity). Proportional fairness, on the other hand, is likely to lead to conservative fairness levels, while fairer resource allocation solutions may be possible that negligibly affect system efficiency and achievable service level.

Such limitations are alleviated by the *α-fairness scheme* [27]. In general, the α-fairness scheme constitutes a unified framework where fairness is controlled by changing a single parameter; that is, the *inequality aversion parameter α*. Due to its flexibility, this scheme constitutes the basis for many resource allocation problems encountered in various types of networks and applications [28].

11.3.1 Preliminaries on α-Fairness

The α-fairness scheme aims to maximize the constant elasticity welfare function F_α, parameterized by $\alpha \geq 0$ [27], given by

$$F_\alpha(u) = \begin{cases} \sum_{i=1}^{n} \dfrac{u_i^{1-\alpha}}{1-\alpha} & \alpha \geq 0, \alpha \neq 1 \\ \sum_{i=1}^{n} \log(u_i) & \alpha = 1 \end{cases} \tag{11.1}$$

In Eq. (11.1), $u \in \mathbb{R}_+^n$ denotes a *utility allocation* and u_i equals the utility derived by end-user i. An *α-fair* allocation (named $\pi(\alpha)$), is an allocation where $\pi(\alpha) \in \text{argmax}_{u \in \mathcal{U}} F_\alpha(u)$, with $\mathcal{U}$ being the set that contains all utility allocations that are feasible.

In general, the α parameter is used to control the rate at which marginal increases diminish. In other words, with an increasing α fairer allocations can be obtained; however, these allocations are likely to have a negative effect (i.e., reduce) the efficiency of the system. There are certain α values that are noteworthy, namely $\alpha = 0$, $\alpha = 1$, and $\alpha \to \infty$. Specifically, $\alpha = 0$ corresponds to maximizing system efficiency, which, may, however, lead to unequal treatment among the end-users contending for resources. For $\alpha = 1$, the scheme corresponds to proportional fairness, and for $\alpha \to \infty$ the scheme converges to max-min fairness; that is, it converges to an allocation ensuring that the minimum allocated utility (i.e., satisfaction of an end-user) is as high as possible.

11.3.2 Generic-Fair Energy Allocation Algorithm

Initially, we assume a simplified system, where n EVs contend for the available resources (i.e., charging time slots) over a single power line, with this power line having a capacity of m resource units (i.e., charging time slots or equivalently possible resource allocations). Each EV requests a number of charging time slots over this power line. Nevertheless, due to power line constraints, up to N charging spots can be activated simultaneously (i.e., to avoid congestion), thus, leading to the possibility that all EVs cannot be fully charged or that some EVs may experience delays in order for all EVs to be fully charged. Given that, the utility function $U = [u_{ij}] \in \mathbb{R}^{n \times m}$, over which the F_α function operates, is defined in such a way so as to declare the EVs' preferences over the m possible resource allocation decisions. Specifically, u_{ij} captures the satisfaction of EV i when resource allocation decision j is taken over the fairness attribute of interest. As an example, when we are interested in deriving a fair allocation concerning the delay that each EV will experience for the sake of fully charging all EVs (i.e., *delay-fair*), then decision j may represent the first time slot when charging is activated (i.e., to ultimately capture delay). Alternatively, when we are interested in deriving a fair allocation as it concerns the charging level achieved, or equivalently according to the achievable quality-of-service, of each EV (i.e., *QoS-fair*), then decision j may represent the number of charging slots allocated to EV i.

Clearly, utility function U depends on the services provided by the system and on how each EV "perceives" each possible resource allocation decision (i.e., how the resource decision affects the satisfaction of the users). Hence, it must be carefully designed. In this work, we examine both the delay-fair and the QoS-fair resource allocation problems, for which U is appropriately designed. Specifically, we design the following optimization problem:

$$\max W_\alpha(u) = \sum_{i=1}^{n} \sum_{j=1}^{m} x_{ij} F_\alpha(u_{ij}) \tag{11.2}$$

subject to:

$$\text{End} - \text{User and System Constraints}$$

where $x = [x_{ij}] \in \mathbb{R}^{n \times m}$ are the Boolean variables to be optimized to find an energy allocation that is either delay-fair or QoS-fair. In general, x_{ij} takes the value 1 if the j-th resource allocation is selected for the i-th EV; otherwise, it takes the value 0. Hence, a feasible allocation corresponds to a specific $x \in \mathbb{R}^n$ that satisfies all the end-user and system constraints.

11.4 Delay-Fair Charging Management

For the delay-fair charging management problem, the assumption is that EV requests are initially dynamically accepted by the CDM (e.g., utilizing a 5G network). Subsequently, reservations of parking spots and scheduling of charging slots are performed for a specified future planning horizon. Thus, a request for service, denoted as S_i, is defined over the tuple $S_i = \{t_i, t'_i, c_i\}$, where t_i and t'_i are the corresponding times of arrival and departure, respectively, of EV i, and c_i is the EV's charging level demand requirement.

The allocation of a parking spot to an EV by the CDM, without loss of generality, is performed using a first come first serve (FCFS) approach, provided that the specific charging spot assigned is free and can be used for the entire time period specified by the EV via the information provided through its reservation request. For the delay-fair charging slot scheduling, the generic-fair energy allocation algorithm (Section 11.3.2) is specifically designed taking into consideration two (main) constraints for the EV and the system: (i) the number of EVs being charged simultaneously cannot surpass a pre-defined value to avoid congestion and (ii) the charging level that each EV can reach must be up to the demand that it has requested.

Given this, it is clear that to charge all EVs up to their requested demand may be impossible (i.e., due to congestion). As such, the delay-fair energy allocation algorithm is designed to find a fair allocation concerning the delay experienced by EVs contending for resources.

11.4.1 Optimal Algorithm: Delay-Fair

We first formulate the problem as an ILP utilizing the constants/variables as described below:

Constants

- n: number of EVs to be serviced within the specified planning horizon.
- m: number of charging time slots within the specified planning horizon.
- N: maximum number of simultaneously active charging spots.
- $A = [a_{ij}] \in \mathbb{R}^{n \times m}$: the arrival matrix; $a_{ij} = 1$ when the jth time slot is equal or greater to the arrival time slot of the i-th EV, otherwise $a_{ij} = 0$. Hence, A is formed according to the arrival time t_i of each EV i (i.e., converted to the corresponding time slot).
- $D = [d_i] \in \mathbb{R}^n$: the departure vector, where d_i is a constant indicating the last time slot that EV i prefers to be present in the parking structure. Accordingly, D is created based on the preferred departure time t'_i of each EV i.
- $R = [r_i] \in \mathbb{R}^n$: the charging demand vector, where r_i is a constant indicating the number of requested charging times slots by EV i. Hence, R is formed according to the requested charging time c_i of each EV i.
- $H = [h_i] \in \mathbb{R}^n$: the tolerance delay vector, where h_i is a constant indicating the number of time slots that EV i can tolerate after the departure time has elapsed (i.e., after time slot d_i).
- $U = [u_{ij}] \in \mathbb{R}^{n \times m}$: the utility matrix indicating the preferences of each EV over the possible time slots that charging may be activated. Specifically, u_{ij} indicates the level of satisfaction of EV i when charging starts at time slot indicated by decision j. As the level of satisfaction of each EV i depends on the delay that the EV user will experience, u_{ij} is computed according to Algorithm 11.1.

Algorithm 11.1

Computing $U = [u_{ij}]$: the preferences of EVs over the possible time slots when charging may be activated.

```
for i = 1 : n do
    k = 0
    for j = 1 : m do
        if j ≤ dᵢ - rᵢ then
            uᵢⱼ = 1
        else
            k = k + 1
            uᵢⱼ = hᵢ - k
                  ─────
                   hᵢ
        end if
        if uᵢⱼ ≤ 0 then
            uᵢⱼ = ε
        end if
    end for
end for
```

According to Algorithm 11.1, utility values range between 1 and ε, where ε is a small value used to ensure that $u_{ij} \in \mathbb{R}_+$ for all i and j. Specifically, u_{ij} equals to 1 if the charging procedure for EV i is activated upon a time slot indicated by decision j that does not violate the EV's preferred departure time d_i (i.e., the EV user is fully satisfied as it is fully charged without experiencing any delay). Otherwise, utility value u_{ij} starts decreasing up to ε to express that, as delay increases the EV user is less satisfied. The worst value, ε, is used to express the fact that the delay of an EV user i exceeds its tolerance delay h_i.

Variables

- x_{ij}: Boolean variable; takes the value 1 if charging time slot indicated by decision j is the first time slot used for charging the i-th EV amongst a set of contiguous time slots allocated to that EV. Otherwise, it takes the value 0.
- y_{ij}: Boolean variable; takes the value 1 if charging time slot indicated by decision j is used by the i-th EV. Otherwise, it takes the value 0.

Objective

$$\max_{\mathbf{x}} W_\alpha(u) \tag{11.3}$$

subject to:

$$\sum_j x_{ij} = 1, \quad \forall \quad i = 1, ..., n \tag{11.4}$$

$$\sum_j y_{ij} = r_i, \quad \forall \quad i = 1, ..., n \tag{11.5}$$

$$\sum_i y_{ij} \leq N, \quad \forall \quad j = 1, ..., m, \tag{11.6}$$

$$a_{ij} \geq y_{ij}, \quad \forall \quad i = 1, ..., n, \quad \forall \quad j = 1, ..., m \tag{11.7}$$

$$y_{ij} - y_{i(j-1)} \leq x_{ij}, \quad \forall \quad i = 1, ..., n, \quad j = 2, ..., m \tag{11.8}$$

$$y_{ij} = 0 \quad \text{for} \quad j = 1, \quad \forall \quad i = 1, ..., n \tag{11.9}$$

The ILP aims to maximize $W_\alpha(u)$ [Eq. (11.3)], thus obtaining a utility allocation that is delay-fair. Constraint (11.4) guarantees the selection of one utility allocation per EV (i.e., charging is activated within one time slot), while constraint (11.5) makes certain that the number of time slots allocated for charging by each EV is the same as the ones requested. Constraint (11.6) is utilized to make sure that no more than N charging spots are simultaneously activated, while constraint (11.7) ensures the consideration of an EV's arrival time during the allocation of its corresponding charging time slots. Finally, constraints (11.8) and (11.9) ensure that contiguous charging time slots are assigned for each EV.

11.4.2 Heuristic Algorithm: Delay-Fair

ACO is a popular metaheuristic that is known to provide high-quality outputs efficiently when applied to challenging ($\mathcal{NP}$-hard) combinatorial optimization problems [42]. The sections that follow describe the different ACO components that are specifically designed to coordinate the charging process of EVs while maximizing the welfare function of Eq. (11.1).

Initialization
Firstly, each artificial ant chooses in a random manner an EV (e.g., the i-th EV) that in turn selects a utility allocation, i.e., based on decision j. Each choice of utility allocation, u_{ij}, is related to a value for the pheromone trail, which is initialized following a uniform distribution at the beginning of the execution as

$$\tau_{ij} \leftarrow \tau_0, \quad \forall i = 1, ..., n, \quad \forall j = 1, ...m, \tag{11.10}$$

with τ_0 denoting the initial pheromone trail value and τ_{ij} denoting the pheromone trail value associated with EV i for utility allocation j.

Constructing Solutions

A complete solution of allocated utilities for all EVs is represented by each ant, which is initially empty. Then, each ant selects a utility for each EV and incrementally appends the selection to the solution. The constructed solution is completed until all EVs are activated. Specifically, ant z selects EV i with a utility not yet allocated in a random manner and an allocation value j based on the following distribution:

$$p_{ij}^z = \frac{\tau_{ij}\left[\eta_{ij}\right]^\beta}{\sum\limits_{l \in \mathcal{N}_i^z} \tau_{il}[\eta_{il}]^\beta}, \text{if } j \in \mathcal{N}_i^z, \tag{11.11}$$

where $\mathcal{N}_i^z$ denotes the set of feasible utility choices for the i-th EV, satisfying the aforementioned constraints (11.6) and (11.7), η_{ij} denotes the heuristic information values that are associated with the i-th EV for feasible utility allocation value j (i.e., the "attractiveness" of assigning a utility value j for the i-th EV), and β (constant) is used for controlling the influence of the heuristic information. Specifically, η_{ij} is computed by utilizing the welfare function as:

$$\eta_{ij} = \begin{cases} 1 - \left(\dfrac{1}{u_{ij}^{1-\alpha}/(1-\alpha)}\right), & \text{if } \alpha \geq 0, \alpha \neq 1 \\[3mm] 1 - \left(\dfrac{1}{\log u_{ij}}\right), & \text{if } \alpha = 1 \end{cases} \tag{11.12}$$

Pheromone Update

In every iteration, the pheromone trails are updated based on the quality of the constructed solutions. Specifically, because of pheromone evaporation, all pheromone trail values are initially decreased by a small value:

$$\tau_{ij} \leftarrow (1-\rho)\tau_{ij}, \quad \forall i = 1, ..., n, \quad \forall j = 1, ...m, \tag{11.13}$$

with $\rho \in (0, 1]$ is a constant denoting the evaporation rate. Subsequently, the pheromone trail values for the utility allocation selections of T^{best} (i.e., the solution that is constructed by the best ant [BA]), are reinforced as

$$\tau_{ij} \leftarrow \tau_{ij} + \Delta\tau_{ij}^{\text{best}}, \tag{11.14}$$

where $\Delta\tau_{ij}^{\text{best}} = 1 - \left(1/C^{\text{best}}\right)$ is the pheromone amount deposited by the BA and C^{best} is the quality of T^{best}, which is computed by maximizing the objective function (i.e., welfare function $W_\alpha(u)$ – Eq. (11.1)). Since it is possible that $W_\alpha(u)$ may take negative values, windowing is utilized (i.e., adding the lowest quality value solution to all remaining solution quality values and adding a very small constant so as not to have zero values). The "best" ant that can deposit pheromone is either the best-so-far ant (BSFA) (i.e., the ant representing the best-so-far solution amongst all iterations and, thus, may not necessarily be part of the constructing colony at the current iteration), hence $C^{best} = C^{bs} = C^{best\ so\ far}$, or the iteration-best ant (IBA), hence $C^{best} = C^{ib} = C^{iteration\ best}$. By default, the IBA deposits pheromone at each iteration, while the BSFA deposits pheromone from time to time (in this work, this is done every 25 iterations) [43].

MAX-MIN Ant System (MMAS) [44], one of the best-performing ACO algorithms, is the basis for the pheromone update policy. For this policy, lower and upper limits are imposed on the pheromone trail values as follows:

$$\tau_{ij} \leftarrow \begin{cases} \tau_{max}, & \text{if } \tau_{ij} > \tau_{max}, \\ \tau_{min}, & \text{if } \tau_{ij} < \tau_{min}, \quad \forall i = 1, ..., n, \quad j = 1, ..., m, \\ \tau_{ij}, & \text{otherwise}, \end{cases} \quad (11.15)$$

where τ_{max} and τ_{min} are the maximum and minimum pheromone trail values, respectively. Thus, excessive accumulation of pheromone trail values will not occur, preventing stagnation behavior. In addition, pheromone trail re-initialization is implemented whenever there is no improvement in the solution quality.

11.4.3 Results and Main Outcomes

The delay-fair energy allocation algorithm was evaluated according to the EV profiles (i.e., requests S_i) of a real-world benchmark [18] that includes arrival times (t_i), departure times (t'_i), and demands c_i for a 24-hour planning horizon. The planning horizon is in this work discretized according to a 10-min interval. Note that the exact instances considered for the results are available at [45]. Briefly, both algorithms are evaluated for $n = 15, 20$ EVs, $m = 196$ charging time slots, $N = 3, 5$, respectively, and $\varepsilon = 10^{-6}$. Charging demand vector R is formed according to the discretized real c_i demands, arrival matrix A is formed according to t_i, and departure vector D is formed according to t'_i. Finally, the tolerance delay vector H is formed by randomly sampling an integer number between the intervals [5, 29] for each EV. For the ACO approach, the colony size (ω) is set to n, while the rate of evaporation (ρ) and the decision rule parameter (β) are explored for a number of values, with $\rho = 0.2$ and $\beta = 2$ having a performance that is close to the optimal delay-fair allocations.

Specifically, Table 11.1 shows that ACO and ILP only slightly deviate in the optimization (W_α) results for most of the values of α explored. In the table, "best" denotes the minimum value, "worst" denotes the maximum value, "mean" denotes the average value, and "stdev" denotes the standard deviation value, with these values obtained from 10 independent runs for the ACO approach. Note that

Table 11.1 $W_\alpha(u)$ results for the optimal (ILP) and algorithmic (ACO) approaches (for the case of ACO 10 independent runs are executed).

			ACO		
α	ILP	Best	Mean ± stddev	Worst	
$n = 15$	0.0	122.69	122.69	119.82±0.66	116.01
	0.5	81.90	81.90	78.72±0.84	74.52
	1.0	18.10	15.34	12.82±0.51	9.84
	1.5	−642.36	−678.63	−756.44±9.21	−865.42
	2.0	−1.0000e5	−1.0349e5	−1.1055e5±5169.42	−1.2060e5
	2.5	−2.1082e7	−2.2360e7	−2.4578e7±5.88e5	−2.6353e7
$n = 20$	0.0	166.08	166.08	160.24±0.71	154.91
	0.5	109.08	109.08	105.74±0.63	103.54
	1.0	27.62	25.33	21.54±0.93	18.51
	1.5	−646.51	−683.54	773.31±7.8	854.81
	2.0	−1.0001e5	−1.0395e5	−1.1067e5±5261.65	−1.2108e5
	2.5	−2.1082e7	−2.2381e7	−2.4593e7±6.12e5	−2.6375e7

the maximum deviation encountered is approximately 15% for $n = 15$ and $\alpha = 1$, while the overall deviation is negligible given the significant savings in processing time that ACO achieved. Specifically, regarding the processing time, ILP requires 21 minutes (2.5 hours) for $n = 15$ ($n = 20$), while ACO requires 0.5 minutes (0.9 minutes) for $n = 15$ ($n = 20$), with ACO resulting up to 99% savings in processing time.

Importantly, both ILP and ACO are evaluated for various other measures, including fairness with respect to the delay that each EV encounters (Figure 11.2) and total delay (Figure 11.3). Specifically, delay-fairness is evaluated according to the coefficient of variations metric [46] over α (CV_α), with CV measuring the variability of delays the EVs experience around the mean delay value. In brief, Figure 11.2 illustrates that delay-fairness improves with an increasing α for both (ILP and ACO) approaches (i.e., lower CV values imply fairer allocations). Hence, higher α values are shown to schedule the time that each EV starts to charge in a fairer manner.

However, fairness comes at the expense of high total delay times that may be encountered as α increases (Figure 11.3). Note that the total delay is computed as the sum of delays for all EVs, given the EVs' starting charging time (i.e., the allocation) and the EVs' demand (i.e., r_i). Specifically, the impact of α on the total

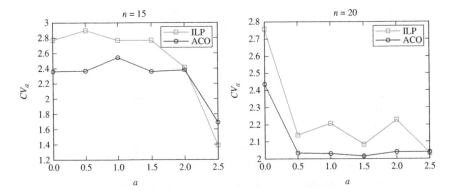

Figure 11.2 CV_α vs. α. Results obtained for both the optimal (ILP) and algorithmic (ACO) techniques (for the ACO case, results are averaged over 10 runs).

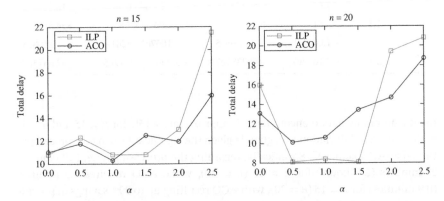

Figure 11.3 Total delay vs. α. Results obtained for both the optimal (ILP) and algorithmic (ACO) techniques (for the ACO case results are averaged over 10 runs).

delay is shown in Figure 11.3, illustrating that fairer allocations result in extensive delays, consequently also affecting the individual end-user delays (i.e., a fairer allocation may increase the individual delays for the sake of equity).

Thus, it is important that the operator examines both fairness and delay over several α values to effectively balance the delay-fairness trade-off. As an example, the operator may choose the scheduling solution for $\alpha = 2$ when $n = 15$ EVs are considered, since, for this case, delay is negligibly affected (Figure 11.3), while fairness is shown to improve (Figure 11.2). In general, the presence of delay may not be considered acceptable by end-users; thus, in the next section, we also examine the QoS-fair charging management in the presence of hard constraints on the departure time of the EVs.

11.5 QoS-Fair Charging Management

For the QoS-fair charging management scheme, similar to the delay-fair case, we assume that the EV requests are dynamically accepted by the CDM that subsequently carries out both reservations of parking spots and scheduling of charging slots for a given future planning horizon. Thus, in this case a service request S_i is defined over the tuple $S_i = \left\{ t_i, t'_i, c_i^{mn}, c_i^{mx} \right\}$, where t_i and t'_i are as previously defined in Section 11.4, and c_i^{mn}, c_i^{mx} are the minimum and maximum charging level requirements of EV i, respectively.

Without loss of generality, the CDM again makes an allocation of parking spot to an EV on an FCFS basis, provided it is not used by another vehicle for the entire time period stated within the EV's reservation information, and that it is possible to charge the EV, at least up to its minimum capacity requirement. The CDM, before accepting an EV request, has first to examine whether the minimum charging requirement can be achieved, given the previously admitted EVs into the system. For the admission step, the algorithm analytically described in Ref. [25] can be used, which was shown to be capable of deciding in real time whether an EV request should be admitted to the system or not.

Regarding the QoS-fair energy allocation algorithm, this is executed periodically, to consider for the EVs that are admitted into the system for a predefined future planning horizon. Specifically, the CDM runs the QoS-fair algorithm in order to (in-advance) schedule when the charging spots for each accepted EV are activated/deactivated, subject to the following (main) EV/system constraints: (i) the number of charging spots that are activated simultaneously cannot be greater than a predefined number (to avoid congestion) and (ii) each EV's capacity must at least reach its minimum charging demand requirement. In order to avoid delays, the time period designated for an EV's charging must be within the EV's arrival and departure times.

Given the aforementioned constraints, it is clear that to charge all EVs up to their requested demand, without any delays, may be impossible (i.e., due to congestion). Therefore, the QoS-fair energy allocation algorithm is designed to find a fair allocation as it concerns the achievable QoS that each EV enjoys (i.e., according to the satisfaction of EVs over their state-of-charge after service).

11.5.1 Optimal Algorithm: QoS-Fair

In the case of the ILP-based QoS-fair energy allocation algorithm, constants n and N are as previously defined in Section 11.4, while additional *constants* are defined as follows:

- m: the maximum number of charging time slots that an EV may request.
- k: the number of time slots in the planning horizon considered.

- v_i^{mn}: the minimum charging demand (in terms of the number of time slots used for charging) of EV i, computed according to c_i^{mn}.
- v_i^{mx}: the maximum charging demand (in terms of the number of time slots used for charging) of the i -th EV, computed according to c_i^{mx}.
- $L = [l_{is}] \in \mathbb{R}^{n \times k}$: the contention matrix, where $l_{is} = 1$ if $t_i \leq s \leq t_i'$, and 0 otherwise, $\forall i = 1, ..., n$, $\forall s = 1,., k$. Hence, L indicates which EVs contend for the same charging slots (i.e., are simultaneously present in the parking structure).
- $V = [v_{ij}] \in \mathbb{R}^{n \times m}$: the resource matrix, where $v_{ij} = j$ if $j \leq v_i^{mx}$, otherwise $v_{ij} = 0$ (i.e., utilized to make certain that an EV is not assigned a number of charging time slots that is greater than the ones requested).
- $U = [u_{ij}] \in \mathbb{R}_+^{n \times m}$: the normalized resource matrix, with

$$u_{ij} = \frac{v_{ij}}{v_i^{mx}} \quad \text{if } v_{ij} > 0, \quad u_{ij} = \varepsilon \quad \text{otherwise,} \tag{11.16}$$

where ε is a small value, so that any $u_{ij} \in \mathbb{R}_+$. This utility function represents the percentage of satisfaction of an EV i, when allocated v_{ij} charging time slots with respect to the maximum resources requested. Specifically, $0 < u_{ij} \leq 1$, with 1 signifying that EV i is fully satisfied when allocated resources that are equal to or exceed its maximum charging demand v_i^{mx}.

Regarding the *variables*, these are defined as follows:

- x_{ij}: Boolean variable; takes the value 1 if the j-th allocation is selected for the i-th EV. Otherwise, it takes the value 0.
- y_{is}: Boolean variable; takes the value 1 if charging time slot s is used by the i-th EV. Otherwise, it takes the value 0.
- z_{is}: Boolean variable; takes the value 1 if the s-th charging time slot is the first time slot used for charging the i-th EV amongst a set of contiguous time slots allocated to that EV. Otherwise, it takes the value 0.

To this end, the optimization algorithm is formulated as follows.

Objective

$$\max_x W_\alpha(u) \tag{11.17}$$

subject to:

$$\sum_j x_{ij} = 1, \quad \forall i = 1, ..., n \tag{11.18}$$

$$\sum_s y_{is} = \sum_j v_{ij} x_{ij}, \quad \forall i = 1, ..., n \tag{11.19}$$

$$\sum_j v_{ij}x_{ij} \geq v_i^{mn}, \quad \forall i = 1, ..., n \tag{11.20}$$

$$\sum_j v_{ij}x_{ij} \leq v_i^{mx}, \quad \forall i = 1, ..., n \tag{11.21}$$

$$\sum_i y_{is} \leq N, \quad \forall s = 1, ..., k, \tag{11.22}$$

$$l_{is} \geq y_{is}, \quad \forall i = 1, ..., n, \quad \forall s = 1, ..., k \tag{11.23}$$

$$\sum_s z_{is} \leq 1, \quad \forall i = 1, ..., n \tag{11.24}$$

$$y_{is} - y_{i(s-1)} \leq z_{is}, \quad \forall i = 1, ..., n, \quad \forall s = 2, ..., l \tag{11.25}$$

$$y_{is} = 0 \quad \text{for } s = 1, \quad \forall i = 1, ..., n \tag{11.26}$$

The ILP aims to maximize $W_\alpha(u)$ (11.17), thus obtaining an energy allocation that is QoS fair. In this formulation, with constraint (11.18) we ensure that we choose one energy allocation decision for each EV, with constraint (11.19) we ensure that the number of charging time slots allocated is the same as the one specified by the resource allocation decision implemented, and with constraints (11.20) and (11.21) we make certain that the allocations of charging time slots for the EVs at least meet (i.e., they are greater than or equal to) their required minimum charging level without, however, exceeding their maximum charging level requirement. Further, constraint (11.22) is used to ensure that no more than N charging spots can be in use at the same time, constraint (11.23) is used to ensure that the charging time for an EV is within a time window that is specified by the arrival- and departure-time information for that EV, and constraints (11.24), (11.25), and (11.26) make certain that contiguous charging time slots are assigned for each EV.

11.5.2 Heuristic Algorithm: QoS-Fair

The same ACO algorithm described in Section 11.4.2 is applied in this scenario as well, with the only difference being that the utility allocation now represents service time rather than activation time. Furthermore, the $\mathcal{N}_i^z$ set of feasible utility choices for EV i in (11.11) must now satisfy constraints (11.20), (11.21), and (11.22). Finally, the pheromone update policy of the ACO is the same as described in Eqs. (11.13)–(11.15).

11.5.3 Results and Main Outcomes

The QoS-fair energy allocation algorithm is again evaluated according to the EV profiles (i.e., requests S_i) of the same real-world benchmark [18] that includes arrival times (t_i), departure times (t'_i), and demands c_i^{mx} for a 24-hour period. Information on the parameters considered for solving both the ILP and ACO are analytically given in Ref. [25], while the exact instances are available in Ref. [45]. Briefly, both algorithms are evaluated for $n = 20(25)$ EVs, with $k = 144$, $m = 44$ (53), $N = 4(5)$, and $\varepsilon = 10^{-4}$. The maximum charge demands v_i^{mx} correspond to the discretized real c_i^{mx} demands, while the minimum charge demands v_i^{mn} are computed as the 10% of v_i^{mx} for all EVs. Finally, matrix L is created in accordance with the discretized planning horizon, given t_i and t'_i. For the ACO, the colony size (ω) takes the value n, while the evaporation rate (ρ) and decision rule parameter (β) are examined for various values, with $\beta = 5$ and $\rho = 0.02$ providing a near-optimal solution [25].

Specifically, the ACO allocations are compared with the optimal ILP allocations, for various measures of interest, including the processing time, fairness (Figure 11.4), and efficiency (Figure 11.5) of the QoS-fair allocations over several values of α. Fairness is evaluated according to the coefficient of variations (CV_α) metric [25, 46], measuring the fairness of an α-fair allocation with respect to the unserved charging demand that EVs experience. Efficiency (E_α), is measured as the sum of utilized charging time slots (i.e., $E_\alpha(u) = \sum_i u_i$, where $u_i \in \pi(\alpha)$).

Regarding the processing time, ILP requires 18.3 hours (54 hours) for $n = 20$ ($n = 25$), while ACO requires 1.6 minutes (3.1 minutes) for $n = 20$ ($n = 25$), with ACO resulting in up to 99% savings in processing time. Importantly, for the case of the ILP, the processing time does not scale with the number of the EVs, while ACO is shown to be executed within minutes, for all n values. Thus, ACO constitutes a good alternative approach for problem sizes where ILP is rendered impractical,

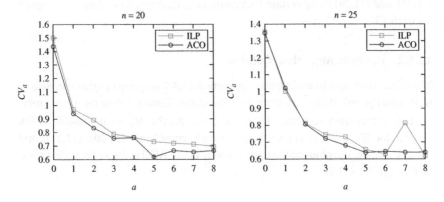

Figure 11.4 CV_α vs. α. Results obtained for both the optimal (ILP) and algorithmic (ACO) techniques (for the ACO case results are averaged over 10 runs).

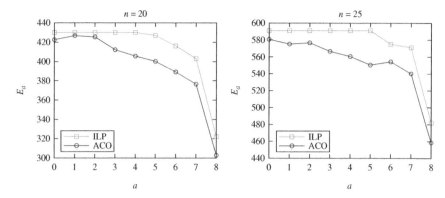

Figure 11.5 $E_\alpha(u)$ vs. α. Results obtained for both the optimal (ILP) and algorithmic (ACO) techniques (for the ACO case results are averaged over 10 runs).

especially as it obtains near-optimal results (deviating from the ILP by only up to 7%) in both QoS-fairness (Figure 11.4) and efficiency (Figure 11.5) measures. Note that the processing time signifies the time it takes to derive allocations for α up to 8 (i.e., the value at which efficiency (Figure 11.5) starts degrading), and thus higher α values would not be desirable by the parking operator and are thus not examined.

Importantly, Figure 11.4 illustrates that QoS-fairness improves with increasing α for both the ILP and ACO approaches. Larger α values are shown to divide the available charging time slots among the EVs that contend for resources in a fairer manner; this is contrary to the greedy approach for allocating resources (i.e., for $\alpha = 0$), which only aims at maximizing system efficiency. The impact of α on the system efficiency is shown in Figure 11.5, illustrating that fairer allocations (in this case, $\alpha > 5$ for both instances) result in efficiency degradation, consequently affecting the operator's revenue (i.e., this is the price to pay for fairness).

In general, by investigating both the fairness and efficiency metrics over various values of α, a parking structure operator can effectively balance the trade-off between efficiency and fairness through an appropriate selection of a value for α that is the most suitable for meeting the requirements set by both the operator as well as the owners of the EVs.

Finally, the reader should note that, concerning a direct comparison between the delay- and QoS-fair algorithms, this cannot provide any additional information, as different metrics are targeted. However, as demonstrated by the discussion in both Sections 11.4.3 and 11.5.3, CV_α follows the same trend for both cases as α increases, the savings in processing time when ACO is used is similar, and so is the deviation of ACO from the ILP. Further, the trade-off between fairness and efficiency is shown in both, albeit at different α values.

11.6 Closing Remarks

While several EV charging management schemes for parking structures have been previously examined in the literature, under diverse technologies and optimization objectives (e.g., optimizing total delay, energy cost, balancing congestion), in this work we explicitly examine the fair charging management problem. Specifically, we examine the delay-fair and QoS-fair charging management problems, ultimately aiming to fairly divide the available energy resources to the EVs that contend for these resources, that is, targeting to improve the end-user satisfaction over the services they enjoy.

Overall, the main achievements of this work are the development of the delay-fair and QoS-fair ILP formulations and ACO-based heuristic algorithms leveraging the α-fairness scheme, and the performance evaluations of both the optimal and heuristic-based solutions in terms of fairness and also system efficiency. In general, for both targets, results reveal the trade-off between fairness and efficiency. In essence, a parking operator can choose the solution (i.e., by an appropriate selection of the α value) that best meets its needs (i.e., end-user satisfaction, operator's revenue, etc.).

As a fairly simplistic scenario for the charging station is utilized in the performance results that are presented in this work, a possible future direction is to include charging stations that are connected via several lines and with different charging rates. Further, novel pricing schemes can be developed that reflect end-user satisfaction, and the multi-period planning problem can be examined over planning horizons that are relatively short (e.g., hours). Finally, there are a number of metrics that can be used when examining fairness in charging. Thus, an interesting avenue for future research is to capture fairness and end-user satisfaction over other attributes such as charging cost for each EV owner.

Acknowledgment

This work was supported by the European Union's Horizon 2020 research and innovation programme under grant agreement No 739551 (KIOS CoE – TEAMING) and from the Republic of Cyprus through the Deputy Ministry of Research, Innovation, and Digital Policy.

References

1 Wrede, M. (2020). Going electric: the future of electric driving, charging stations and parking. https://www.intertraffic.com/news/parking/future-electric-driving-charging-stations-and-parking (accessed 15 January 2023).

2 García-Villalobos, J., Zamora, I., San Martín, J. et al. (2014). Plug-in electric vehicles in electric distribution networks: a review of smart charging approaches. *Renewable and Sustainable Energy Reviews* 38: 717–731.

3 Intercomp. Smart parking systems. https://smartparkingsystems.com/en/management-of-charging-stations-for-electric-cars (accessed 15 January 2023)

4 Vitra for parking operators. https://www.virta.global/commercial-parking (accessed 10 December 2022).

5 Uddin, M., Fakhizan, M.R., Abdullah, M.F. et al. (2018). A review on peak load shaving strategies. *Renewable and Sustainable Energy Reviews* 82: 3323–3332.

6 Das, H., Rahman, M., Li, S., and Tan, C. (2020). Electric vehicles standards, charging infrastructure, and impact on grid integration: a technological review. *Renewable and Sustainable Energy Reviews* 120: 109618.

7 García Álvarez, J., González, M.A., Rodríguez Vela, C., and Varela, R. (2018). Electric vehicle charging scheduling by an enhanced artificial bee colony algorithm. *Energies* 11 (10).

8 Mavrovouniotis, M., Ellinas, G., and Polycarpou, M. (2019). Electric vehicle charging scheduling using ant colony system. *Proceedings of the IEEE Congress on Evolutionary Computation (CEC)*, Wellington, New Zealand (10–13 June 2019), pp. 2581–2588.

9 García-Álvarez, J., González, M.A., Vela, C.R., and Varela, R. (2017). Electric vehicle charging scheduling using an artificial bee colony algorithm. *Proceedings of the International Work-Conference on the Interplay Between Natural and Artificial Computation*, Corunna, Spain (19–23 June 2017).

10 García-Álvarez, J., González, M.A., and Vela, C.R. (2015). A genetic algorithm for scheduling electric vehicle charging. *Proceedings of the Annual Conference on Genetic and Evolutionary Computation (GECCO)*, Madrid, Spain (11–15 July 2015).

11 Mukherjee, J.C. and Gupta, A. (2015). A review of charge scheduling of electric vehicles in smart grid. *IEEE Systems Journal* 9 (4): 1541–1553.

12 Qian, K., Zhou, C., Allan, M., and Yuan, Y. (2011). Modeling of load demand due to EV battery charging in distribution systems. *IEEE Transactions on Power Apparatus and Systems* 26 (2): 802–810.

13 Masoum, M.A., Moses, P.S., and Hajforoosh, S. (2012). Distribution transformer stress in smart grid with coordinated charging of plug-in electric vehicles. *Proceedings of the IEEE PES Innovative Smart Grid Technologies (ISGT)*, Washington, DC, USA (16–20 January 2012), pp. 1–8.

14 Vliet, O., Sjoerd, A.B., Kuramochi, T. et al. (2011). Energy use, cost and CO2 emissions of electric cars. *Journal of Power Sources* 196 (4): 2298–2310.

15 Wang, R., Xiao, G., and Wang, P. (2017). Hybrid centralized-decentralized (HCD) charging control of electric vehicles. *IEEE Transactions on Vehicular Technology* 66 (8): 6728–6741.

16 Bañol, N.A., Franco, J.F., Lavorato, M., and Romero, R. (2017). Metaheuristic optimization algorithms for the optimal coordination of plug-in electric vehicle charging in distribution systems with distributed generation. *Electric Power Systems Research* 142: 351–361.

17 Ma, Z., Callaway, D.S., and Hiskens, I.A. (2013). Decentralized charging control of large populations of plug-in electric vehicles. *IEEE Transactions on Control Systems Technology* 21 (1): 67–78.

18 Hernández-Arauzo, A., Puente, J., Varela, R., and Sedano, J. (2015). Electric vehicle charging under power and balance constraints as dynamic scheduling. *Computers and Industrial Engineering* 85: 306–315.

19 He, Y., Venkatesh, B., and Guan, L. (2012). Optimal scheduling for charging and discharging of electric vehicles. *IEEE Transactions on Smart Grid* 3 (3): 1095–1105.

20 Wu, H., Kwok-Hung Pang, G., Lun Choy, K., and Yan Lam, H. (2018). Dynamic resource allocation for parking lot electric vehicle recharging using heuristic fuzzy particle swarm optimization algorithm. *Applied Soft Computing* 71: 538–552.

21 Gan, L., Topcu, U., and Low, S. (2011). Optimal decentralized protocol for electric vehicle charging. *Proceedings of the IEEE Conference on Decision and Control and European Control Conference (CDC & ECC)*, Orlando, FL, USA (12–15 December 2011).

22 Janjic, A., Velimirovic, L., Stankovic, M., and Petrusic, A. (2017). Commercial electric vehicle fleet scheduling for secondary frequency control. *Electric Power Systems Research* 147: 31–41.

23 Han, J., Park, J., and Lee, K. (2017). Optimal scheduling for electric vehicle charging under variable maximum charging power. *Energies* 10 (7).

24 Su, W. and Chow, M.-Y. (2012). Performance evaluation of an EDA-based large-scale plug-in hybrid electric vehicle charging algorithm. *IEEE Transactions on Smart Grid* 3 (1): 308–315.

25 Panayiotou, T., Mavrovouniotis, M., and Ellinas, G. (2021). On the fair-efficient charging scheduling of electric vehicles in parking structures. *Proceedings of the IEEE International Intelligent Transportation Systems Conference (ITSC)*, Indianapolis, IN, USA (19–22 September 2021).

26 D. o. S. S. Northwestern University (2023). Electric vehicle charging. https://www.northwestern.edu/transportation-parking/evanston-parking/policies/electric-vehicle-charging.html (accessed 15 December 2022).

27 Bertsimas, D., Farias, V.F., and Trichakis, N. (2012). On the efficiency-fairness trade-off. *Management Science* 58 (12): 2234–2250.

28 Karsu, O. and Morton, A. (2015). Inequity averse optimization in operational research. *European Journal of Operational Research* 245 (2): 343–359.

29 Zou, H., Mao, S., Wang, Y. et al. (2019). A survey of energy management in interconnected multi-microgrids. *IEEE Access* 7: 72 158–72 169.

30 Ardakanian, O., Rosenberg, C., and Keshav, S. (2013). Distributed control of electric vehicle charging. *Proceedings of the ACM e-Energy*, Berkeley, California, USA (21–24 May 2013), pp. 101–112.

31 Zhu, Z., Peng, J., Gu, X. et al. (2018). Fair resource allocation for system throughput maximization in mobile edge computing. *IEEE Access* 6: 5332–5340.

32 Allybokus, Z., Avrachenkov, K., Leguay, J., and Maggi, L. (2018). Multi-path alpha-fair resource allocation at scale in distributed software-defined networks. *IEEE Journal on Selected Areas in Communications* 36 (12): 2655–2666.

33 Altman, E., Avrachenkov, K., and Garnaev, A. (2008). Generalized α-fair resource allocation in wireless networks. *Proceedings of the IEEE Conference on Decision and Control (CDC)*, Cancun, Mexico (9–11 December).

34 Lan, T., Kao, D., Chiang, M., and Sabharwal, A. (2010). An axiomatic theory of fairness in network resource allocation. *Proceedings of the IEEE International Conference on Computer Communications (INFOCOM)*, San Diego, CA, USA (14–19 March 2010).

35 Panayiotou, T. and Ellinas, G. (2021). Optimal and near-optimal alpha-fair resource allocation algorithms based on traffic demand predictions for optical network planning. *IEEE/OSA Journal of Optical Communications and Networking* 13 (3): 53–68.

36 Samá, M., D'Ariano, A., D'Ariano, P., and Pacciarelli, D. (2017). Scheduling models for optimal aircraft traffic control at busy airports: tardiness, priorities, equity and violations considerations. *Omega* 67: 81–98.

37 Huang, M., Smilowitz, K., and Balcik, B. (2012). Models for relief routing: equity, efficiency and efficacy. *Transportation Research Part E Logistics and Transportation Review* 48 (1): 2–18.

38 Tzeng, G.-H., Cheng, H.-J., and Huang, T. (2007). Multi-objective optimal planning for designing relief delivery systems. *Transportation Research Part E Logistics and Transportation Review* 43 (6): 673–686.

39 Lin, F., Zhou, Y., An, X. et al. (2018). Fair resource allocation in an intrusion-detection system for edge computing: ensuring the security of Internet of things devices. *IEEE Consumer Electronics Magazine* 7 (6): 45–50.

40 Jiang, Y., Huang, Z., and Tsang, D.K. (2018). Towards max-min fair resource allocation for stream big data analytics in shared clouds. *IEEE Transactions on Big Data* 4 (1): 130–137.

41 Bertsimas, D., Farias, V.F., and Trichakis, N. (2011). The price of fairness. *Operations Research* 59 (1): 17–31.

42 Mavrovouniotis, M., Yang, S., Van, M. et al. (2020). Ant colony optimization algorithms for dynamic optimization: a case study of the dynamic travelling salesperson problem. *IEEE Computational Intelligence Magazine* 15 (1): 52–63.

43 Stützle, T. and Hoos, H.H. (2000). MAX–MIN ant system. *Future Generation Computer Systems* 16 (8): 889–914.

44 Stutzle, T. and Hoos, H. (1997). MAX-MIN ant system and local search for the traveling salesman problem. *Proceedings of the IEEE International Conference on Evolutionary Computation (ICEC)*, Indianapolis, IN, USA (13–16 April 1997).

45 Mavrovouniotis, M. (2023). EV station instances database. https://github.com/Mavrovouniotis/ev_station_instances (accessed 5 February 2023).

46 Jain, R., Chiu, D., and Hawe, W. (1984). *A Quantitative Measure of Fairness and Discrimination for Resource Allocation in Shared Computer System*. Eastern Research Laboratory, Digital Equipment Corporation.

12

Multi-Energy Management Schemes for the Sustainability of Intelligent Interconnected Transportation Systems

M. Edwin[1], M. C. Eniyan[1], M. Saranya Nair[2], and G. Antony Miraculas[3]

[1] Department of Mechanical Engineering, University College of Engineering, Nagercoil, Anna University Constituent College, Nagercoil, Tamilnadu, India
[2] School of Electronics Engineering, Vellore Institute of Technology, Chennai, Tamilnadu, India
[3] Department of Mechanical Engineering, St. Xavier's Catholic College of Engineering, Chunkankadai, Nagercoil, Tamilnadu, India

Nomenclature

AI	Artificial Intelligence
AVI	Automated Vehicle Identification
BDA	Big Data Analysis
EMAS	Expressway Monitoring and Advisory System
GLIDE	Green Link Determination System
GT	Green Transportation
ICT	Information and Communication Technology
IITS	Intelligent Interconnected Transportation System
IoT	Internet of Things
TRA	Telecommunication Regulatory Authority

12.1 Introduction

It is vital to look for new potential sources of energy due to the destruction of the environment and the scarcity of petroleum-based fuels. Additionally, although traditional fuels are finding it challenging to achieve these criteria, cars that utilize petroleum-based fuels must comply with increasingly strict emission rules. The amount of CO_2 in the atmosphere has grown from the pre-industrial era's 280 ppm to the current level of 350 ppm as a result of the use of different fossil

Interconnected Modern Multi-Energy Networks and Intelligent Transportation Systems: Towards a Green Economy and Sustainable Development, First Edition. Edited by Mohammadreza Daneshvar, Behnam Mohammadi-Ivatloo, Amjad Anvari-Moghaddam, and Reza Razzaghi.
© 2024 The Institute of Electrical and Electronics Engineers, Inc.
Published 2024 by John Wiley & Sons, Inc.

fuels like coal and petroleum products [1]. The use of conventional fuels continues to increase CO_2 levels, which causes the greenhouse effect, acid rain, pollution, and global climate change [2].

To prevent the catastrophic effects of climate change, according to the IPCC's Fifth Assessment Report, the world must phase out the use of fossil fuels in energy generation by the end of thiscentury and cut their usage to 20% by 2050 [3]. Additionally, it was said that if nations do not reduce greenhouse gas emissions to almost zero by the year 2100, the effects of climate change will be severe, widespread, and permanent. To keep the increase in the world temperature below 2°C from the level of 1800, a decrease in emissions of 40–70% is also anticipated globally between 2010 and 2050, with a drop to zero or below zero by 2100 [3]. This inherent problem has motivated academics to look for an alternative, environment-friendly fuel that would cut down on the import of crude oil from oil-importing countries like India, thereby improving the economics of the nation and greatly reducing emissions.

The main goal of the "multi-energy system" is to develop an open, interactive, and smart energy supply system by resolving challenges with numerous energies interaction, coordination, and sharing. The "multi-energy system" actually organizes and optimizes linkages on energy production, transportation, distribution, transfer, storage, and consumption via integrating the various energy and consumer sectors is primarily intended to completely decarbonize the environment [4]. Different energy sources will cohabit in this way to achieve the objectives of comprehensive combining and high-efficiency utilization.

To reduce CO_2 emissions and protect the environment, multi-energy systems are the only option. It is a more advanced technique of energy supply. It is a pattern that offers a variety of energy products, including power, heating, and cooling. The multi-energy system does not only combine or add different resource types. It is a development based on conventional energy uses that have lessened its drawbacks.

12.1.1 Objectives

Most of the IITS prevailing nowadays use electricity from the grid to operate the system. To provide a sustainable IITS incorporation of the renewable energy system is necessary and the main issue is the energy management of such systems. Energy management solutions for heat-power systems, power management in hydro-power systems, and gas-power systems were examined by researchers. It was also reported that connection between transportation networks and electric vehicles [5]. The investigation on multi-energy system management in transportation aspects was quite minimal and further research needs to be carried out on transportation energy management, particularly in terms of decreasing economic and environmental issues confronting the energy sectors. To provide some solutions to the above issues, objectives are defined as follows.

- To assess the energy needs for IITS and the potential of renewable energy resources based on the data collection and travelerinformation in the particular field.
- To identify the influencing factors for IITS in terms of both financial and nonfinancial.
- To analyze the various perspectives of multi-energy management schemes for the sustainability of IITS.

12.1.2 Distributed Energy Generation

It primarily consists of advanced heat pump technology, low-cost commercial fuel cells, micro-gas turbine power generation technology, variable-speed constant-frequency wind power generation technology, and large-scale photovoltaic and solar thermal power generation technology [6]. More focus is being placed on the development of energy conversion stations, energy hubs, smart energy, metering equipment at the customer end, and transportation networks based on smart electric cars.

12.1.3 Energy Storage Technology

In multi-energy systems, energy storage technology addresses the discrepancy between producing power and load power as well as the mismatch between reaction times of various power suppliers. Battery storage, compressed air storage, and other forms of power storage are used as forms of energy storage in multi-energy systems [7]. Energy storage technologies also cover the conversion and storage of electrical energy as well as other energy types such as electrochemical energy storage, heat storage, hydrogen storage, and electric vehicles. Multiple energy networks, including electricity grids, natural gas pipeline networks, cooling and heating energy networks, and transportation networks, can be connected by using different energy storage techniques.

12.1.4 Integrated Energy Management System

The three primary management components of integrated energy management are transaction center, distributed power supply, and user load. To guarantee the efficient and reliable operation of multi-energy systems, the energy flow is controlled by the information flow. The core modules for multi-energy systems include forecasting, analysis, and decision-making links, grid, renewable energy, non-renewable energy, energy storage systems, and various energy loads.

This is done through the establishment of a system platform for analytical processing and global optimization management. An acceptable conversion plan for energy output and various energy sources is created at the energy supply end to carry out organic integration, and multi-dimensional comprehensive decision-making [2, 8].

To increase the capacity of renewable energy consumption, meet energy conservation and emission reduction goals, and encourage a change in energy structure, a multi-energy complementary integrated optimization system may be constructed.

12.2 History of Transportation System – Overview

The history of transportation traces back to the prehistoric era and has evolved. The human foot served as the initial mode of transportation. To get anywhere, people used to travel long miles on foot. Travel times dropped as innovative solutions to transportation issues were found while carrying greater and heavier cargo became possible. Researchers in the field of transportation are constantly coming up with new ideas to save costs and improve efficiency [9]. The development of international trade in the pre-modern era was the primary driver of advances in global transportation [9].

In the twenty-first century, smart cities have gained global attention and many cities have started initiatives to transform into smart cities. For example, the Indian government started the Smart Cities Mission, in which they chose 100 cities for the Smart City Mission Project; as of now, 5002 of the 7742 projects have been completed [9]. China has already more than 500 smart cities [10]. IITS plays a vital role in the transportation sector of smart cities. Therefore, eventually, IITS or smart transport, or smart mobility improve the quality of life. Consequently, energy and greenhouse gas reduction are key benefits, and it is taken in cost analysis to justify IITS projects.

12.3 Concept of IITS

IITS integrates people, road, and vehicle design to improve road safety, and comfort as well as environmental conservation with the help of information and communication technologies [11]. IITS uses information and communication technology to enhance transportation services and lowering traffic, accidents, and air pollution through the effective utilization of existing transportation facilities [11]. The IITS network is focused on meeting the transportation demands of

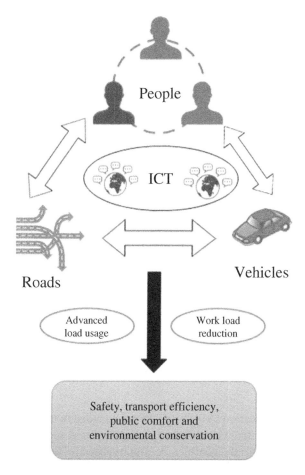

Figure 12.1   Overview of IITS.

current and future generations to support economic growth and minimize its negative environmental impacts. Based on information control and electronics technology, IITS includes a wide range of wireless and wireline communication networks [12]. Figure 12.1 presents a schematic representation of the overview of IITS.

12.3.1   Components of IITS

The components of IITS are as follows [13]; Figure 12.2 presents the components of IITS.

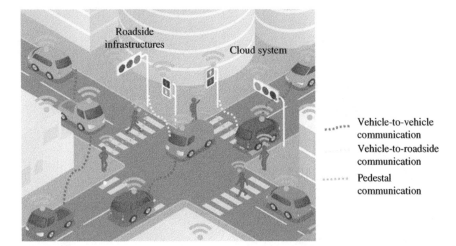

Figure 12.2 Components of IITS.

- Data Acquisition System
- Data Communication System
- Data Management System
- Display System

12.3.1.1 Data Acquisition System
Sensors, automated vehicle identification (AVI), and GPS are included in the data acquisition system. Traffic metrics like vehicle count, occupancy, and speed are obtained via sensors [12]. To specifically identify a car and its speed on the road, AVI systems are employed. The real-time location and speed of the vehicle are determined by GPS technology. With the use of GPS systems, travel time, speed, distance, and delay are estimated.

12.3.1.2 Data Communication System
Information gathered by the data acquisition system must be properly communicated to the control centers and display devices. The majority of the time, data communication systems use wireless technology [14].

12.3.1.3 Data Management System
For producing clean signals, data captured from acquisition systems need to have the error signals removed. Once the data is cleaned of error signals, it can be aggregated and analysis is then carried out to produce effective traffic management plans and predict traffic conditions [13].

12.3.1.4 Display System

Information is provided to travelers utilizing display systems, such as message signs, radio, SMS, etc. Information on journey times, speed, delays, accidents, and other topics is available from IITS [15].

12.4 Barriers to Successful Implementation of IITS

The IITS network will add a new level of monitoring and control to the transport system with adequate planning and coordination. However, for the successful implementation of IITS, one needs to overcome financial constraints and non-financial constraints, namely, institutional, technical, and physical constraints. Figure 12.3 represents the major divisions of barriers related to IITS.

12.4.1 Financial Constraints

The major financial constraint for IITS is the huge investments needed to install the system in society. The infrastructure, such as traffic detectors, roadside information displays, IoT-based communication systems, GPS tracking systems, etc., would require significant investment. Later investment might also be essential for the system's ongoing upgrading, as well as for its proper use in operation and maintenance.

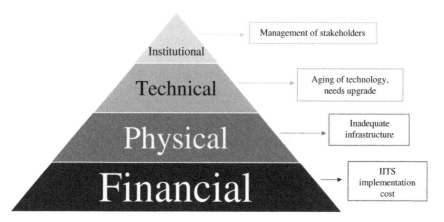

Figure 12.3 Major divisions of barriers related to IITS.

12.4.2 Non-Financial Constraints

Non-financial constraints include Institutional (focuses on organizational, legal, and policy-forming aspects), technical, and physical constraints (including infrastructure, equipment, and devices) [16].

12.4.2.1 Institutional Constraint

Due to the participation of various stakeholder groups, a structured framework that outlines the duties of each stakeholder is necessary. All stakeholders should be notified regarding the standard operating procedure to avoid wrong communication [10].

12.4.2.2 Technical Constraint

To deploy smart mobility solutions, technical capability is required for data collection, data integration, data management, data analysis, and information service provision. Data collected through sensors, cameras, and field operators' survey devices contain errors and, according to officials in Barcelona, error correction devices are quite expensive [8]. For instance, AutoNavi, a navigation software used in China, uses real-time speed data collected by the company personnel using GPS [8]. Hence, a standardized data format and procedure is needed to reduce the data integration process. Otherwise, adopt a common platform as in the case of City SDK in Europe. Furthermore, keeping in touch with current technology development and maintenance of the existing established system is the key technical constraint faced by the IITS [10].

12.4.2.3 Physical Constraint

The basic construction for the operation of IITS solutions is a coherent road architecture. The cost of implementing intelligent mobility solutions would be high if the road network was fragmented, far-flung, and disconnected.

The rest of the physical conditions are rather simple. There must be a specialized physical infrastructure in place for data gathering and communication for a given smart mobility application, say Wi-Fi on the bus. All the physical systems and equipment need regular maintenance [17].

12.5 Intelligent Modern Energy Transport Systems

Modern wireless, electronic, and automated technologies are used in intelligent transportation systems (ITSs). These technologies can combine infrastructure, users, and vehicles all at once. Automated guideways, collision avoidance systems, and precise bus docking are just a few examples of automated and in-vehicle technologies. They can aid in trip optimization, reducing unnecessary miles traveled, increasing the usage of other modes of transportation, decreasing time spent in traffic, reducing reliance on foreign oil, and improving air quality. Transportation authorities must get ready for the future by taking into account how roads are used now and how connected and autonomous vehicles will affect them. Imagine if every vehicle sent a signal when it encountered a pothole, providing transportation authorities with precise information about the state of the roads. It will give knowledge that will assist in solving immediate problems in the short term. This knowledge could eventually help with the construction of better roadways.

Developing technologies are opening up new opportunities in the field of IoT and 5G communication technologies. Any physical machine can be equipped with low-cost sensors and controllers that can be used for remote control and management. They offer the quick connectivity required for running real-time, transportation systems with minimal lag.

Modern ITS systems manage the secure processing of streaming time-series data, prompted by the connected real-time data sources, withstanding the advantages of integrating the many computer models available today to support transportation management. Sensors in intelligent vehicles and transportation systems are two examples of these streaming data sources. The institution directly in charge of managing transportation in a specific area might use its IT infrastructure and resources to pre-process the gathered data locally. However, the data and metadata are frequently transferred to external systems such as cloud computing together with the preliminary analysis findings so that they can perform a more thorough study and return alerts on potential threats and anomalies. All of this highlights the complexity of IITS systems and highlights their potential for localized attacks.

12.5.1 Implementation of Futuristic Transportation Technologies

12.5.1.1 Collection of Data

Real-time precise observation and quick data collecting are required for the planning of an intelligent transport system. The information is gathered using a variety of hardware tools that serve as the foundation for future ITS operations. These include sensors, cameras, automatic vehicle locators with GPS, automatic vehicle identifiers, and more. The hardware primarily keeps track of data like traffic volume, speed, position, vehicle type, and delays among other things. These physical components are linked to the servers, which are often found at the data collection center and store a significant amount of data for later analysis.

12.5.1.2 Data Communication

This aspect of ITS involves transmitting data collected in the field to management centers for analysis, and then management centers providing that information back to travelers. Quick and real-time information transfer is the key to proficiency in ITS implementation. Travelers are informed of announcements about traffic via the Internet, SMS, or onboard units of vehicles.

12.5.1.3 Analysis of Data

The information that was gathered and received by management centers is then processed further in several phases. These processes include data cleansing, data synthesis, error correction, and adaptive logical analysis. Data inconsistencies are found using specialist tools and fixed. The data is then further modified and gathered for analysis. This corrected aggregate data is further examined to forecast traffic scenarios that are available to provide consumers with pertinent information.

12.5.1.4 Travelers' Information

The management system provides real-time data on things like trip time, travel speed, delay, traffic accidents, route changes, detours, and the state of work zones, among other things. A variety of technology devices, including variable message signs, highway advisory radio, the Internet, SMS, and automated cell phones, are used to distribute this information. The number of vehicles on the road is rising because of urbanization. The only answer is to apply an ITS to create a situation where both residents and city officials benefit.

12.6 Role of Multi-Energy Management Schemes for the Sustainability of Transportation Networks

In general, all energy systems are considered multi-energy systems meaning that numerous energy sectors interact at various levels. For instance, considering a conventional power plant, coal is used as fuel to produce electricity and must be transported to every plant, which describes the interconnection between the coal and electricity. Similarly, in the case of air conditioning also one can observe these kinds of connections [18–20]. Nevertheless, the energy connection across multiple systems are typically minimal when compared to the relationship inside a single energy system and that is the main reason past studies on power system focused only on electrical energy. In recent days, coordination between multiple energy systems such as the electric-gas energy system, the transportation-power system, and management focused frequently [5, 21]. Therefore, traditional energy management for a single energy system may become obsolete in the future, which motivates multi-energy management study.

Researchers investigated the energy management strategies for heat-power systems [22], power management in hydro-power systems, and gas-power systems [23, 24]. Furthermore, the connection between transportation systems and electric vehicles was also reported [25, 26]. The preceding investigation has provided a new viewpoint on multi-energy management analysis, especially in the view of reducing economic and environmental problems faced by the transportation sectors. The main benefits of multi-energy management schemes for the sustainability of transportation networks are as follows:

1) Improves utilization of primary energy sources and the overall energy efficiency of the system by the effect of multi-energy management. Let us consider an example; waste heat generated during conventional electricity production can be used for heating services which enhances the system's efficiency.
2) Improved use of diverse energy resources across many system levels.
3) The coordination of multiple energy systems increases system flexibility.

However, the multi-energy system has diverse administrative features and their coordination within a single-energy system is quite complex. Therefore, to assist their functioning, appropriate modeling methodologies and control procedures should be developed.

To frame proper management of a multi-energy system, it is classified into four groups, namely spatial, fuel, service, and network.

The spatial orientation of multi-energy management shown in Figure 12.4 highlights the various degrees of aggregation in terms of physical or by components that connect the sectors to the district, and even countries. In service-based multi-energy management shown in Figure 12.5 points out the provision of different services such as electricity, water supply, gas filling stations, transportation, heating, and cooling services. Fuel-based multi-energy management shown in Figure 12.6 focuses on the integration of various fuels including both renewable and non-renewable sources to provide optimal energy services in view of both economically and environmentally. The network-based multi-energy management shown in Figure 12.7 describes the multiple energy networks like electricity, gas, and transportation networks from one area to another area in view of the sustainability of multi-energy management and its development.

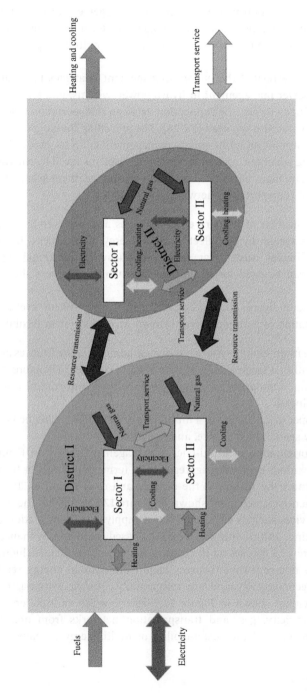

Figure 12.4 Spatial-based multi-energy management system.

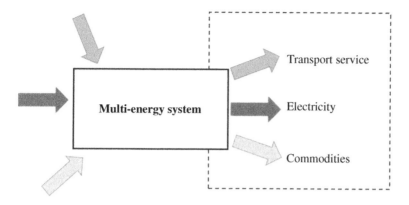

Figure 12.5 Service-based multi-energy management system.

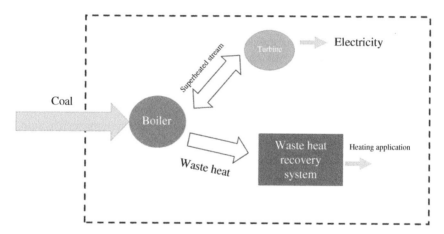

Figure 12.6 Fuel-based multi-energy management system.

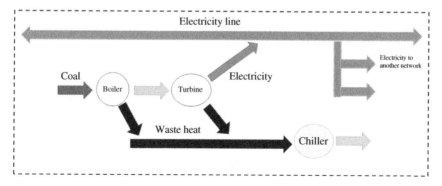

Figure 12.7 Network-based multi-energy management system.

12.7 Result Discussion, Current Challenges, and Future Research Opportunities

12.7.1 Result Discussion

Considering the case study of smart technology in Singapore's transport system, in Singapore smart mobility, sustainable, safe, and smart are the key elements in the smart mobility policy formulation [27]. Before the implementation of IITS, adequate smart transport planning is necessary since it bridges the present and upcoming generations for the growth of sustainable economic activity by reducing ecological problems caused by the current transport system. In this case, the Singapore government took four key areas, namely integration of land use and transport planning, transport supply measures, transport demand management, and incorporation of environment-friendly technologies for vehicles. Based on the identified area they formulate policies and strategies to implement IITS.

On the basics of the function performed, the Smart transport system in Singapore's broadly classified into four parts: control systems, monitoring and enforcement systems, information management systems, and revenue management systems [15].

12.7.1.1 Control System

Singapore's smart transport system uses smart traffic control systems such as automatic signals, namely the Green Link Determination System (GLIDE), B signal, countdown signal, and intelligent road studs [22]. The GLIDE adaptive traffic signal system automatically adjusts signal timing based on the amount of traffic in each direction while also allowing signal coordination along a corridor by simultaneously gathering traffic data by recognizing cars and pedestrians. Furthermore, the transit signal priority scheme was also installed, which is nothing but a B-signal that detects approaching buses and facilitates their movements by extending green time as well as turning on the B-signal to allow an earlier start [28].

12.7.1.2 Monitoring and Enforcement System

Technological improvements allow the transport system to monitor transport facilities, spot interruptions in traffic flow, and detect traffic offenses continuously. IITS operates 24 hours throughout the year and keeps monitoring traffic flow based on the information gathered. To maintain smooth and safe operation, a smart incident management system called Expressway Monitoring and Advisory System (EMAS) was installed [27].

More than 320 major advanced surveillance cameras called J-eyes were installed in Singapore to help spot irregular traffic flow including congestion, illegal parking, and loading/unloading. These cameras automatically detect speeding and red light-running vehicles and take snapshots of the registration plates for identification.

12.7.1.3 Information Management System

Information management system collects, processes, and shares real-time traffic flow with drivers using fixed display platform or mobile platforms, based on the information drivers can modify their paths in advance [27].

12.7.1.4 Revenue Management System

Managing fast and accurate transactions of public transport fares and automobile toll payments is important for a transport system to be efficient [27]. Singapore is utilizing various smart technologies for better management of revenue systems to collect public transport fares, parking charges, and tolls [22].

The EEZ-link card allows users to pay for fares in local transportation modes like buses. It is nothing but contactless tap-and-go fare cards paying fares [22]. The Symphony Fore Payment, a penny improvement to that smart card, now makes it easier to pay for other usages, namely parking fees.

12.7.2 Challenges in the Implementation of IITS

The following list includes both technical and non-technical challenges

1) AI – Enhancing automatic management and control of the system and lowering costs [29].
2) IoT – With more modern networks, it can handle more data at a faster rate. We can use intelligent IoT designs and analytics software in the transportation sector to address issues with security, revenue collecting, traffic control, and user satisfaction.
3) Energy – Experiments are still being conducted on the idea of using alternative energy sources instead of fossil fuels.
4) Big data – Gathered information, IoT, and AI are identified as the major technical challenges to perform.

The main technical challenge for implementing IITS is the bridging of various homogeneous components such as IoT devices, applications, peoples, and processes in different sectors [30]. Moreover, lots of attention was needed in proposing web services, security concerns, and maintenance of the system beyond the uniform management and control of the system. Many of the complicated system tools currently available on the market such as IoT, Cloud storage, and network assets are created and integrated for each unique case, leading to rigid designs that cannot be modified to accommodate updating systems and remain unclear topologies.

Also, the non-technical challenge, a new guiding environment that is compatible with autonomous vehicles, network safety, and security should exist. Furthermore, communities must realize the new challenges to their routine and accept the impacts on skills and employment [30].

12.7.3 Future Research Opportunities

- Market trends – Since the sensor is getting cheaper every day, suppliers are making the most of them, which will increase the reliability of the control system.
- New infrastructure – The advent of ancillary services like bike sharing has been overshadowed by the current revolution in transportation such as the metro. Transportation freedom in the area continues to be a problem. The ability to travel between locations, towns, and sectors in handy ponds is still a desire shared by residents of both urban areas and rural areas [31].
- Behavioral change – Competitors in the market may develop a system that increases customer satisfaction by offering a quicker and more flexible service at automobile facilities. The most recent advancement in big data analytics has added new entrants in understanding the shift in client behavior and requests for more flexible trip planning on demand and better vehicle monitoring services.
- Allowing customers to select the seating and standing capacity on public transit will increase the standardization of transportation as the demand for customized mobility grows daily [30]. The specialized service that places a strong emphasis on client satisfaction will assist in getting past the privacy barrier. Additionally, customers will place more trust in users as a result of tailored services, and agents will have the opportunity to provide the client with superior service [31].
- 5G network – It starts an unrestricted, extensive wireless network. The 5G network has strong support for the World Wide Web. The introduction of the 5G network in Dubai, which has already been announced by the Telecommunications Regulatory Authority (TRA), would transform Internet consumption in the United Arab Emirates [32].
- Industrial revolution–The industrial revolution had a significant impact on IITS. First, Mobility as a Service has focused on providing a customer-centric transportation system. Then, big data has also come a long way, and BDA currently has significant influence on prediction and decision-making. As a result, it is anticipated that smart roads would maximize benefits and improve the driving experience. The networked communication of automobiles, especially with emergency services, has recently been the trend revolution [14].

12.8 Conclusion

More thorough and organized research is required for IITS policy-making. With a host of key policy recommendations, autonomous vehicles will change the transportation industry. To determine the extent to which IITS can reduce bottlenecks and contamination and, as a result, improve the sustainability of the environment, more research is needed. A similar in-depth study is also needed to improve safer and more effective rail systems to improve expedition effectiveness and cross-border enablement. To facilitate easy trade in goods and the growth of tourism in the province, the Economic and Social Commission for Asia and the Pacific (ESCAP) secretariat may also do more research on mixed-mode commuting and integrated transportation networks. Domestic rivers, navigation, air-traffic management, and maritime transportation should all be the subjects of comparable research.

References

1 Kessel, D.G. (2000). Global warming – facts, assessment, countermeasures. *Journal of Petroleum Science and Engineering* 26 (1–4): 157–168. https://doi.org/10.1016/S0920-4105(00)00030-9.

2 Wang, Z., Zhang, X., and Rezazadeh, A. (2021). Hydrogen fuel and electricity generation from a new hybrid energy system based on wind and solar energies and alkaline fuel cell. *Energy Reports* 7: 2594–2604.

3 Jarraud, M. and Steiner, A. (2012). *Summary for Policymakers*, vol. 9781107025. https://doi.org/10.1017/CBO9781139177245.003.

4 Dincer, I. and Ishaq, H. (2022). Integrated systems for hydrogen production. In: *Renewable Hydrogen Production*, 289–335. Elsevier https://doi.org/10.1016/B978-0-323-85176-3.00013-5.

5 Mancarella, P. (2014). MES (multi-energy systems): an overview of concepts and evaluation models. *Energy* 65: 1–17. https://doi.org/10.1016/j.energy.2013.10.041.

6 Alanne, K. and Saari, A. (2006). Distributed energy generation and sustainable development. *Renewable and Sustainable Energy Reviews* 10 (6): 539–558. https://doi.org/10.1016/j.rser.2004.11.004.

7 Mahlia, T.M.I., Saktisahdan, T.J., Jannifar, A. et al. (2014). A review of available methods and development on energy storage; technology update. *Renewable and Sustainable Energy Reviews* 33: 532–545. https://doi.org/10.1016/j.rser.2014.01.068.

8 Di Zhong, Q.L.X.Z. (2018). Research status and development trends for key technologies of multi-energy complementary comprehensive utilization system. *Thermal Power Generation* 47 (2): 56–60.

9 Smart City Mission, Ministry of Housing and Urban Affairs, Government of India, 2021. https://smartcities.gov.in (accessed 6 May 2022).

10 Chen, Y., Ardila-Gomez, A., and Frame, G. (2017). Achieving energy savings by intelligent transportation systems investments in the context of smart cities. *Transportation Research Part D: Transport and Environment* 54: 381–396. https://doi.org/10.1016/j.trd.2017.06.008.

11 Shaheen, S.A. and Finson, R. (2004). Intelligent transportation systems. In: *Encyclopedia of Energy*, 487–496. Elsevier https://doi.org/10.1016/B0-12-176480-X/00191-1.

12 Cornea, T., Gosman, C., Constanda, R. et al. (2017). Cloud services for smart city applications. In: *Adaptive Mobile Computing*, 3–28. Elsevier https://doi.org/10.1016/B978-0-12-804603-6.00001-2.

13 Paul, A., Chilamkurti, N., Daniel, A., and Rho, S. (2017). Intelligent transportation systems. In: *Intelligent Vehicular Networks and Communications*, 21–41. Elsevier https://doi.org/10.1016/B978-0-12-809266-8.00002-8.

14 Pavlova, L. (2017). Wi-fi and IoT in focus. *LastMile* 69 (8): 56–60.

15 Kumar Debnath, A., Haque, M.M., Chin, H.C., and Yuen, B. (2011). Sustainable urban transport. *Transportation Research Record: Journal of the Transportation Research Board* 2243 (1): 38–45. https://doi.org/10.3141/2243-05.

16 Sheren, T. and Elazb, A. (2016). Challenge of intelligent transport system. *International Journal of Modern Engineering* 6 (10): 1–4. https://www.researchgate.net/publication/314156806_Challenge_of_Intelligent_Transport_System (accessed 6 May 2022).

17 Tomaszewska, E.J. (2021). Barriers related to the implementation of intelligent transport systems in cities –the polish local government's perspective. *Engineering Management in Production and Services* 13 (4): 131–147. https://doi.org/10.2478/emj-2021-0036.

18 Banakar, H., Luo, C., and Ooi, B.T. (2008). Impacts of wind power minute-to-minute variations on power system operation. *IEEE Transactions on Power Apparatus and Systems* 23 (1): 150–160. https://doi.org/10.1109/TPWRS.2007.913298.

19 Jiang, Z., Liu, Y., Kang, Z. et al. (2022). Security-constrained unit commitment for hybrid VSC-MTDC/AC power systems with high penetration of wind generation. *IEEE Access* 10: 14029–14037. https://doi.org/10.1109/ACCESS.2022.3148316.

20 Alguacil, N., Motto, A.L., and Conejo, A.J. (2003). Transmission expansion planning: a mixed-integer LP approach. *IEEE Transactions on Power Apparatus and Systems* 18 (3): 1070–1077. https://doi.org/10.1109/TPWRS.2003.814891.

21 Gabrielli, P., Gazzani, M., Martelli, E., and Mazzotti, M. (2018). Optimal design of multi-energy systems with seasonal storage. *Applied Energy* 219: 408–424. https://doi.org/10.1016/j.apenergy.2017.07.142.

22 Dai, Y., Chen, L., Min, Y. et al. (2017). Dispatch model of combined heat and power plant considering heat transfer process. *IEEE Transactions on Sustainable Energy* 8 (3): 1225–1236. https://doi.org/10.1109/TSTE.2017.2671744.

23 Wen, Y., Qu, X., Li, W. et al. (2018). Synergistic operation of electricity and natural gas networks via ADMM. *IEEE Transactions on Smart Grid* 9 (5): 4555–4565. https://doi.org/10.1109/TSG.2017.2663380.

24 Qiao, Z., Guo, Q., Sun, H. et al. (2017). An interval gas flow analysis in natural gas and electricity coupled networks considering the uncertainty of wind power. *Applied Energy* 201: 343–353. https://doi.org/10.1016/j.apenergy.2016.12.020.

25 Yao, S., Wang, P., Liu, X. et al. (2020). Rolling optimization of Mobile energy storage fleets for resilient service restoration. *IEEE Transactions on Smart Grid* 11 (2): 1030–1043. https://doi.org/10.1109/TSG.2019.2930012.

26 Liu, J., Zhang, J., Yang, Z. et al. (2013). Materials science and materials chemistry for large scale electrochemical energy storage: from transportation to electrical grid. *Advanced Functional Materials* 23 (8): 929–946. https://doi.org/10.1002/adfm.201200690.

27 Haque, M.M., Chin, H.C., and Debnath, A.K. (2013). Sustainable, safe, smart – three key elements of Singapore's evolving transport policies. *Transport Policy* 27: 20–31. https://doi.org/10.1016/j.tranpol.2012.11.017.

28 OM (2011). On the Roads. One Motoring, Land Transport Authority, Singapore. www.onemotoring.com.sg/publish/onemotoring/en/on_the_roads (accessed 6 May 2022).

29 Elloumi, M. and Kamoun, S. (2017). Adaptive control scheme for large-scale interconnected systems described by Hammerstein models. *Asian Journal of Control* 19 (3): 1075–1088. https://doi.org/10.1002/asjc.1443.

30 Mathew, E. (2020). Intelligent transport systems and its challenges. *Proceedings of the International Conference on Advanced Intelligent Systems and Informatics 2019*, pp. 663–672. doi: 10.1007/978-3-030-31129-2_61.

31 Uhlemann, E. (2016). Transport ministers around the world support connected vehicles [connected vehicles]. *IEEE Vehicular Technology Magazine* 11 (2): 19–23. https://doi.org/10.1109/MVT.2016.2541582.

32 Lavanya, S.R. (2017). A smart information system for public transportation using IoT. *International Journal of Recent Trends in Engineering & Research* 3 (4): 222–230. https://doi.org/10.23883/IJRTER.2017.3138.YCHJE.

13

Blockchain-Based Financial and Economic Analysis of Green Vehicles: Path Towards Intelligent Transportation

Ankita Jain[1] and Vikas Khare[2]

[1] Prestige Institute of Global Management, Indore, Madhya Pradesh, India
[2] School of Technology Management and Engineering, NMIMS Indore, Madhya Pradesh, India

13.1 Introduction

Electric vehicles (EVs) differ from internal combustion engines (ICEs) in that they use electric motors for propulsion, while combustion engines generate power through burning fuel and gases. As a promising alternative to conventional cars, EVs have gained attention due to concerns such as pollution, global warming, and the depletion of natural resources. Although the concept of EVs has been around for some time, it has recently gained traction in response to the environmental impact of fuel-powered vehicles. EVs are equipped with an onboard energy storage system and can travel up to 400 km before requiring a recharge. Additionally, they produce zero exhaust emissions [1]. The recharge time for EV batteries typically ranges from four to eight hours, depending on their design. However, a super-charger can reduce the time needed to charge half the battery to just 20 minutes. Electric motors have higher efficiency compared to ICEs, which can be further improved by implementing advanced technologies such as active suspension and regenerative braking [2]. Despite the fact that EVs have been around longer than combustion engine cars and are, in many ways, superior to them, their high cost has hindered their widespread adoption. Battery costs, which make up around 35% of the total price of an electric car, have decreased by 85% over the last decade. Currently, EVs are still more expensive than ICE cars, but their Total Cost of Ownership (TCO) is frequently lower and is expected to reach parity in the near future [3].

Interconnected Modern Multi-Energy Networks and Intelligent Transportation Systems: Towards a Green Economy and Sustainable Development, First Edition. Edited by Mohammadreza Daneshvar, Behnam Mohammadi-Ivatloo, Amjad Anvari-Moghaddam, and Reza Razzaghi.
© 2024 The Institute of Electrical and Electronics Engineers, Inc.
Published 2024 by John Wiley & Sons, Inc.

Many potential EV buyers are concerned about range, as they believe that EVs either offer too little or too much driving range compared to their actual driving patterns. For example, an average urban driver in the United States travels only 20 miles per day by car, which increases to 30 miles per day for driving-intensive groups [4]. Based on the current battery efficiency, an ideal energy level for urban drivers is around 25 kWh, which would allow them to travel about 100 miles on a single charge. However, for occasional suburban or rural trips, a battery capacity of around 40 kWh, or about 160 miles, would be optimal. This range would be sufficient for the majority of users, particularly those in urban areas, without disrupting their daily routine. By reducing battery capacity from 50 to 40 kWh, buyers could save $1900–$2100 today [5]. Over the next five to seven years, automakers should seriously consider collaborating with their competitors as the industry transitions to electrification and struggles with profitability challenges. By working together, Original Equipment Manufacturers (OEMs) can reduce the fixed costs of research and development, tooling, and factories, particularly during the process of retooling models and platforms for electrification [6]. Sharing EV platforms and factories would enable a variety of model variants and substantial benefits could be realized. Additionally, purchasing the same battery cells and power electronics in larger volumes would provide scale benefits that are difficult to achieve alone. Many automakers have already announced international agreements aimed at lowering the cost of developing and manufacturing EVs [7].

According to a study by McKinsey and other industry experts, EVs have the potential to achieve cost parity with ICE vehicles by around 2025, and may even become more profitable than ICE vehicles [8]. This is due to advancements in battery cost and efficiency, scale economies in power electronics, and indirect cost reduction based on higher manufacturing volumes. Researchers have conducted financial and economic analyses of EVs, including their environmental impact through CO_2 emission reductions [9]. The financial sustainability of EVs and their environmental benefits vary by area and the time-of-use rate programs used by various electric companies. Financial incentives, charging infrastructure, and local production facilities are important factors for promoting EV adoption, as evidenced by socioeconomic analyses. Blockchain-based financial and economic analyses are also emerging as distinct fields of study that leverage the same underlying technology [10]. Blockchain-based financial analysis involves the analysis of financial data stored on a blockchain. This can include analyzing transactions, identifying patterns, and forecasting trends. The focus is on the financial data of a particular entity or group of entities, such as a company or market segment. On the other hand, blockchain-based economic analysis involves the study of economic activity on a blockchain. This includes analyzing economic transactions, identifying patterns, and forecasting trends. The focus is on the broader economic activity of a blockchain network, including the behavior of participants, the performance of decentralized applications, and the impact on the wider economy. In summary, blockchain-based financial analysis is concerned with analyzing financial data within a specific entity or group of entities, while blockchain-based economic analysis is focused on analyzing the broader economic activity of a blockchain network.

The main objective of this chapter is to provide financial and economic analysis of the EVs. The chapter starts with the introduction in Section 13.1. Section 13.2 describes the country-wise financial analysis of the EVs. Some financial parameters, which are also the part of EVs, are presented in Section 13.3. Section 13.4 describes the data assessment of different parameters of the EVs by the NCSS Tool. Section 13.5 shows the blockchain-based financial analysis of the EVs.

13.2 Country-Wise Financial Analysis of EVs

Manufacturers of EVs and batteries have been handsomely rewarded by financial markets over the past two years. Until 2020, the financial performance of EV equities at the portfolio level was similar to that of other automakers and in line with market performance. The combined market capitalization of the 10 largest car manufacturers accounted for approximately 13% of the total market capitalization of the 14 chosen EV shares. All indicators were negatively impacted by the COVID-19 pandemic, but the extent of recovery varied. In the first half of 2020, both the EV and battery indexes showed strong growth, with increases of 70% and 40%, respectively. Currently, major auto markets are announcing green recovery plans and committing to net zero emissions, further solidifying EVs as the future of transportation in regions like Europe, Japan, and Korea, among others [11]. Table 13.1 shows the major automaker announcements on electrification for the years 2021–2022.

By the end of 2021, the collective market value of the leading 10 automobile manufacturers had decreased by 60% compared to that of EV producers, largely due to Tesla, which accounted for 80% of the market value of the 14 pure-play EV companies. Throughout 2021, EVs and battery indices remained higher than those for automakers and the overall market. The decline in returns observed in the EV index can be attributed to the increased competitiveness of traditional automakers in the market for passenger EVs. As a result, it has become more challenging to differentiate between the top 10 automakers and pure-play EV producers since many of the top 10 automakers have expanded their EV businesses. This trend may have caused environmentally and socially conscious investors to redirect their investments from pure-play EV companies towards a broader portfolio of automakers. To capitalize on the significant market value experienced

Table 13.1 Major Automaker Announcements on Electrification, 2021–2022.

Toyota	2021	Sales of 3.5 million EVs annually and the introduction of 30 BEV types by 2030.
Volkswagen	2021	By 2030, sales of all-EVs will account for more than 70% of sales in Europe, 50% in China, and nearly 100% in the United States.
Ford	2022	One-third of sales to be fully electric by 2026 and 50% by 2030, with all-electric sales in Europe by 2030.
BMW	2021	By 2030 or sooner, 50% of new cars sold will be electric-only vehicles.
Geely	2021	By 2025, 20% of new cars sold will be electric.
General Motor	2022	By 2025, North America will be able to produce 1 million BEVs, 30 EV types, and be carbon neutral.

Source: Adapted from Novo [12].

by pure-play EV manufacturers, OEMs are contemplating the launch of separate initial public offerings for their EV-related activities. According to financial markets, the future is electric [12]. When compared to the number of vehicles manufactured, pure-play EV companies have significantly higher market valuations than traditional OEMs. However, despite these high valuations, many EV producers have failed to meet profitability expectations, with a majority reporting either marginal or negative returns on total assets [13]. Figure 13.1 shows the list of nations where plug-in EV sales made up the largest portion of new passenger cars sold in 2020.

Following are the financial subsidies provided by different nations:

Norway: The Norwegian parliament has made a decision to mandate that all newly sold cars in Norway must be electric or hydrogen-powered by 2026.

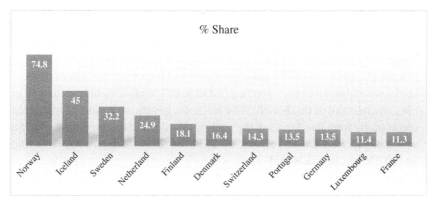

Figure 13.1 List of countries with the highest share of plug-in EVs in new passenger car sales in 2020.

Consequently, by the conclusion of 2025, the number of EVs on the road is projected to exceed 470,000, with 64% of all new cars sold in Norway being battery electric. The government has also resolved to retain current incentives for EVs until the end of 2021, with updates and modifications being made to the incentives in response to market changes [14].

The Norwegian EV incentives:

- No import/purchase taxes apply to EVs (1990).
- A 25% VAT exemption for purchases (2001).
- No yearly road tax (1996–2021).
- There are no fees on toll roads (1997–2017).
- Ferries are free to ride (2009–2017).
- The maximum ferry fee for EVs is 50% of the whole cost (2018).
- On toll highways, no more than 50% of the total is allowed (2018).
- Free public parking (1999–2017).
- A 25% VAT exemption for leasing (2015).
- The Norwegian Parliament set the national objective that by 2025, all new vehicles sold must be zero-emission.
- For residents of apartment complexes, "charging right" was established (2017).
- Publicly purchased vehicles must be ZEVs starting in 2022. From 2025, city buses must follow the same rules.

Iceland: Decades ago, Iceland laid the foundation for low-carbon transportation, and it has now emerged as one of the world's leading markets for EVs. The country's abundance of renewable energy, affordable electricity rates, high fossil fuel costs, and urbanization make it an ideal market for EVs. In 2017, EVs accounted for 8.7% of new car registrations in Iceland, second only to Norway in Europe. Of these registrations, 2.0% were battery EVs (BEVs) and 6.8% were plug-in hybrid EVs (PHEVs). By the end of 2018, the EV share had increased to 13.5%. This is in contrast to the European Union, where the average EV percentage was only 1.4% in 2017. Iceland exempts all EVs from VAT up to IKr 7,000,000, with the standard rate applying to vehicles above that threshold. Additionally, shorter than 5 m electric and hydrogen-powered vehicles are eligible for up to 90 minutes of free parking in the city center. Hybrid vehicles weighing less than 1600 kg and with CO_2 emissions of less than 50 g/km can also avail of this benefit. The Icelandic government aims to transition to electricity from fossil fuels in the coming decades and has proposed a ban on new gasoline and diesel vehicles by 2030. By 2026, Iceland targets having 30,000 EVs on its roads [15].

Sweden: Sweden has been at the forefront of the shift to EVs, offering a plethora of attractive incentives to EV owners. With a remarkable 26% market share and an impressive 253% surge in sales in 2020, Sweden is driving the electric revolution forward. The increasing popularity of EVs can be attributed to a range of incentives, such as federal grants, state and local tax incentives, and local perks [16].

- **Bonus Malus scheme:** A grant of 60,000 SEK (€6000) is available for up to 26% of the cost of the car's initial purchase for low-emission vehicles serviced before 2021 with CO_2 emissions of up to 59.9 g/km and vehicles serviced during 2021 or after with CO_2 emissions of up to 69.9 g/km. The grant is open to both private citizens and companies.
- **Grants for EV buses:** Municipalities, limited corporations, or public transportation authorities may be eligible for a grant worth 20% of the cost of an EV bus. The reward for private transportation businesses is equal to 40% of the price difference between an equivalent diesel bus and an electric bus.
- High occupancy vehicle (HOV) and bus lanes are sometimes available for free.

Netherlands: The infrastructure ministry has confirmed that the government has utilized its €71 million budget for this year's program, which aims to incentivize the purchase of new EVs among the general population. A total of 22,000 individuals were granted €3350 each to assist in the purchase of a more affordable electric car. To qualify, the vehicle must have a maximum price of €45,000 and a minimum range of 120 km. In addition, subsidies of €2000 are available for used EVs, and the €20.4 million fund that supports them remains fully available. In the Netherlands, customers who purchase a new BEVs with a list price between €13,000 and €46,000 and a minimum range of 125 km are eligible for a €5000 subsidy in 2021 and 2022. The subsidy is only available when purchasing or leasing through an authorized dealer. If the leasing agreement is broken before the four-year mark, the subsidies will be discontinued, and if consumers purchase another vehicle within three years, they will be required to repay some of the subsidies. BPM must be paid when purchasing an electric car or motorcycle. The BPM rate is calculated using the World Harmonized Light Vehicle Testing Procedure CO_2 testing method, and additional information on how the rate is determined can be found here [17]. The following BPM tax incentives are available to EV owners:

- Until 2024, all purchases of pure EVs are tax-free. Owners will pay a levy of €360 per vehicle in 2025. The tax levy will rise yearly beginning in 2025.
- Consumers must annually pay the MRB (Motor Vehicle Tax) in order to own a motorcycle, car, or truck.

Additionally, owners of EVs benefit from the following reductions in motor vehicle taxes:

- Up to 2024, completely EVs are exempt from MRB. Owners will receive a 74% tax break on MRB in 2026. The full tax charge will be applied starting in 2025.

- Up to 2024, PHEVs must receive a 50% reduction on MRB. The MRB discount rate will be lowered by the Dutch government to 25% in 2025. The MRB fee will be fully applied starting in 2026.
- Bijtelling is a type of tax that drivers must pay if they use a work vehicle for personal purposes. Your taxable income base is increased by a portion of your vehicle's list price based on its emissions.

Finland: Currently, the Finnish government is providing a purchase subsidy of €2000 ($2260) for consumers who buy new electric cars or enter into long-term leasing agreements for EVs, up to a maximum of €50,000 or $56,000. If a private individual buys a new EV, they may be eligible for a subsidy of up to €2000, provided that the vehicle's list price does not exceed €50,000. Additionally, the Finnish government frequently introduces scrappage programs, which offer private individuals bonuses of up to €2000 for trading in old diesel or gasoline vehicles when purchasing a new EV [18].

- Currently, a fully electric car that costs no more than €50000 can receive an assistance payment of €2000. Private individuals may apply for the subsidy.
- Depending on the size of the vehicle, funding in the amount of €3000, €5000, or €7000 may be provided. Private individuals and businesses can both apply for the subsidy.
- Depending on the size of the vehicle, financial assistance ranging from €7000 to €51,000 may be given for the purchase of an electric truck. Private individuals and businesses can both apply for the subsidy.

Denmark: Denmark is trailing behind its neighboring countries like Sweden and Norway in terms of its EVs and EV charging infrastructure. In 2015, the Danish government started phasing out subsidies for EVs, resulting in a decline in their sales in the country. As a matter of fact, in 2017, EVs only accounted for 0.4% of all cars sold in Denmark. However, Denmark's outlook is positive as they have implemented a new climate law and removed tax increases on EVs in 2019. Currently, their market share for EVs stands at approximately 4% [19].

Tax benefits: As a result of the removal of registration tax incentives for EVs in Denmark, sales of EVs in the country decreased, but only for vehicles valued below DKK 400,000 (approximately €54,000). The tax exemption for these vehicles will gradually decrease from 20% in 2020 to 65% in 2021, 90% in 2022, and 100% in 2023. Additionally, in April 2018, the Danish government introduced a registration tax deduction based on the battery capacity of the EV. Circulation taxes in Denmark are based on the weight and fuel use of the vehicle, and both plug-in hybrids and BEVs pay lower taxes than comparable diesel or gasoline vehicles [20].

Switzerland: While the Swiss Federal Government does not provide official subsidies for EVs, it is worth noting that some regions in the country do offer them.

In the Thurgau region, the government provides a subsidy of CHF3932.73 for the purchase of an EV. Similarly, in St. Gallen, a contribution of CHF4915.91 is provided for EV purchases [21].

Portugal: The Environmental Fund of the Portuguese Government has provided €4,000,000 in financing towards EV ownership incentives in 2021:

- **Private individuals:**
  - €4000 for the acquisition of a brand-new BEV, one vehicle per person only.
  - €6000 for the acquisition of a brand-new, completely electric van; one car per individual only.
  - One electric freight bicycle may be purchased per person for 55% of the buying value price (up to a maximum of €1000).
  - A maximum of one completely electric motorcycle, moped, or bicycle may be purchased for 50% of the purchase price.
- **Companies:**
  - A maximum of two electric vans or light goods vehicles may be purchased for €6000 each.
  - Up to four electric freight bicycles may be purchased per candidate for 50% of the buying value price.
  - Up to four completely electric motorcycles, mopeds, or bicycles may be purchased at 50% of the purchase price (up to a maximum of €350) per candidate.
- **VAT benefits**
  - Both hybrid plug-in vehicles and fully EVs with list prices under €51,000 and €61,400, respectively, are eligible for a full VAT deduction.
  - BEV company vehicles are exempt from corporate income tax.

Germany: Germany is highly dedicated to achieving its target of having 11 million EVs and 1.5 million charging stations by 2032, and as a result, it offers some of the most generous EV subsidies in Europe. Thanks to the significant financing for EV incentives included in the €135 billion post-COVID-19 stimulus plan this summer, an electric Seat Mii can now be purchased for less than €12,000. Germany's "Umweltbonus" program encourages the purchase of EVs and the retirement of diesel and gasoline vehicles, providing additional incentives for customers [22].

- Until 2032, a one-time subsidy of up to 55% of the purchase costs of fully electric cars used for commercial deliveries is available.
- This bonus is available provided the acquired EV has an Acoustic Vehicle Alert System (AVAS), which costs an extra €100.

Tax benefits:

Kfz-Steuer (motor vehicle tax):

- Fully EVs registered between 2012 and 2032 are exempt from this tax for a period of 12 years. This means that EV owners in Germany can save, on average, about €194 in ownership tax for each car and year, though this amount may be significantly greater depending on the type of vehicle. If you decide to drive an ICE vehicle, you can use this calculator to determine how much car tax you would pay [23].

- Due to their lower CO_2 emissions, PHEVs are taxed, although at a lower rate than diesel or gasoline vehicles.

Company car tax:

- A completely electric car with a list price under €59,000 is only taxed at 0.25% of the list price per month for private use. ICE vehicles are taxed at 1%, in contrast.
- A hybrid or completely electric company automobile with a list price over €59,000 is only taxed on private consumption at 0.5% of the list price per month. ICE cars are taxed at 1.2%, in contrast.

VAT:

- Will briefly decrease from 20% to 16.5% between 1 July and 31 December 2020. Both BEVs and PHEVs benefit from this tax relief.

Local incentives for BEVs:

- Depending on where you live, towns and municipalities may provide you up to €1475 in additional purchase grants (you check this here).
- Local energy providers frequently provide additional EV incentives to both new and existing customers.

Luxembourg: Luxembourg, which holds a 3% market share in 2020, maintains its position in the list of the top 15 EV markets in Europe. However, the country's EV market is still in the early stages of growth and falls towards the lower end of the spectrum, in contrast to Norway where the market share is much higher at 48%.

France: France is widely recognized for its dedication to promoting the use of EVs. President Emmanuel Macron has recently made a series of announcements to support this initiative, such as a $9 billion rescue package for the domestic auto-mobile industry and the establishment of targets for the upcoming year. These targets comprise installing 100,000 or more public charging stations, manufac-turing at least 1 million EVs annually by 2027, and encouraging greater adop-tion of EVs. To help achieve these objectives, the government has allocated €1.3 billion in incentives, which enable citizens to save up to €21,000 when purchas-ing an EV. [24]

Subsidies:

- Purchase grant:
  - Bonus of up to €8000 for cars generating no more than 20 g CO_2/km.
  - Up to €2000 in bonuses are available for plug-in hybrid vehicles with CO_2 emissions of between 21 and 50 g/km.

- Scrappage program (conversion bonus): Up to €6000 is available for the purchase of pre-owned or new BEVs and PHEVs if you trade in your diesel or gasoline vehicle that is older than 2001.
- Low emission zone bonus: If you reside or work in a low emission zone, you are eligible for a €1100 subsidy towards the purchase of an EV.

Tax benefits:
- License plate registration fee: Depending on the location, both fully EVs and plug-in hybrids either are eligible for a 55% reduction or are completely exempt from paying the license plate registration fee (carte grise) in Metropolitan France.
- Fully EVs are exempt from the company car tax.

Local incentives:
- In some locations, up to €7000 in additional purchasing allowances are available; for a complete list of regional incentives, see our comprehensive guide to France's EV incentives.
- Green card holders in some towns are entitled to up to two hours of free parking (eligible for EVs).
- The 3264 parking spaces formerly reserved for the Autolib program, a public EV car-sharing program that was offered in Paris, Lyon, and Bordeaux between 2012 and 2019 but is now inactive will be made available for free parking for Paris EV drivers, according to the mayor of Paris.

13.3 Key Financial Ratio for Financial Analysis of EVs

Financial ratio analysis is a technique used to evaluate the correlation (or ratio) between two or more financial data points obtained from a company's financial statements. Its primary objective is to facilitate accurate comparisons over time and among different businesses or sectors. The subsequent financial ratios are significant metrics that analysts and investors consider while assessing the automobile industry. The important financial ratios that analysts and investors take into account while evaluating the auto industry are as follows:

- **Debt to equity (D/E) ratio:** EV firms require significant capital investment, making the D/E ratio a crucial metric for assessing their financial health and ability to meet financial obligations. An increasing D/E ratio indicates greater reliance on debt financing rather than equity financing. Consequently, investors and lenders prefer a lower D/E ratio. The D/E ratio of a company with an equal proportion of assets and liabilities is 1. Comparing the D/E ratios of companies within the same industry is essential since debt requirements vary across industries.

- **Inventory turnover ratio:** The inventory turnover ratio is a significant evaluation metric in the automotive industry, particularly for dealerships that specialize in EVs. When dealerships hold onto inventory for more than 60 days, it is considered a risk to EV sales. The inventory turnover ratio measures how often a company's inventory is sold or "turned over" during a year and is a useful indicator of a company's inventory and order management. For dealerships selling EVs, the inventory turnover ratio is especially important since it indicates the speed at which they are selling their current stock of EVs on the lot.
- **Return on equity (ROE) ratio:** The ROE is a critical financial metric that is useful in assessing the performance of most firms. It is an especially vital statistic for evaluating companies in the electric car industry. The ROE gauges how profitable a company is for its investors by measuring the net profit in relation to shareholder equity. Ideally, analysts and investors prefer higher returns on equity. In the first quarter of 2022, the average ROE for the sector was 15.86%, underscoring its significance to investors.
- **EBITDA margin (%):** EBITDA, which stands for earnings before interest, taxes, depreciation, and amortization, is a profitability measure that adheres to generally accepted accounting principles (GAAP). However, it is frequently utilized in financial analysis because it offers a more accurate assessment of a company's continuous production returns, which includes many nonoperational items, as opposed to just using earnings. The EBITDA margin calculates the operational profit as a percentage of revenue. By comparing a company's EBITDA margin to others in the same industry, it becomes possible to evaluate the company's actual performance.

  In the EV sector, for instance, if XYZ Company makes $100 million in revenue, $40 million goes for the battery, motor, and auxiliary components of the EVs, and another $20 million goes on overhead. Operating profit is $30 million after depreciation and amortization costs of $10 million. $5 million is spent on interest, leaving $25 million in revenues before taxes. Table 13.2 shows the data set related to the EBITDA of EV company.

- **EBIT margin (%):** The EBIT margin is a financial metric that assesses a company's profitability by excluding taxes and interest. It is calculated by dividing

Table 13.2 EBITDA of EV company.

Net income	$31,000,000
Depreciation amortization	+$10,000,000
Interest expense	+$6,000,000
Taxes	+$5,000,000
EBITDA	$52,000,000

sales or net income by EBIT, which stands for profits before interest and taxes. The EBIT margin, also known as the operating margin, provides insight into the economic activities of an individual EV company and measures the benefit derived from these activities. It does not consider how the company is financed or any government or national policy interventions.

EBIT = sales − variable cost − fixed cost

- To determine the EBIT, it is crucial to identify the sales of EVs. While EV sales have been increasing globally, the growth pattern varies across regions. Europe experienced only a 10% increase in EV sales in the first half of 2022, compared to the previous two years of fast sales growth, due to low overall auto sales, continued component shortages, and the Ukraine War. In contrast, the United States, Canada, and China saw significant year-over-year increases in EV sales, despite external factors such as a slowdown in the broader light vehicle industry, the real estate crisis, and COVID lockdowns. BYD emerged as the global sales leader by more than doubling their sales to 651,050 units, including PHEV sales, while Tesla continued to dominate the market in terms of BEVs with 566,020 units shipped in H1. Although sales volumes of PEVs are still increasing, the mix of PHEVs is becoming less dominant due to lower incentives and improved BEV options. The growth of EV sales is increasingly influenced by the degree of electrification, and FCEV sales have fallen by 9% annually with only five car models generating the majority of their sales in the United States and South Korea.
- **PBT margin (%):** "Profit before tax" measures how much money a corporation makes before having to pay corporate income tax. In essence, it is the entire amount of a company's profits before any taxes. Profit before taxes is shown on the income statement as operating profit minus interest. Profit before tax is the number used to calculate a company's tax obligation.
- **Net profit margin (%):** The net profit margin, sometimes known as just net margin, is the amount of net income or profit expressed as a percentage of revenue. It is the ratio of net profits to revenues for a company or business segment. In addition to being expressed as a percentage, net profit margin can also be expressed as a decimal. A company's net profit margin reveals how much of its income is turned into profit. Table 13.3 shows the assessment of the net profit margin of the EV industry.
- **Current ratio (X):** The current rate, a liquidity rate, evaluates a business' capability to pay short-term loans or those that are due within a time. It shows to investors and judges how an EV company can make the most use of its current means to pay down its other payables and current arrears.
- **Quick ratio (X):** An EV company's capability to meet its short-term scores using its most liquid means is estimated using the quick rate, which also acts

Table 13.3 Net profit margin of the EV industry.

Parameters	($)
Revenue of XYZ EV industry	14,853,223
Total revenue of XYZ EV industry	**14,853,223**
Cost of revenue, total	8,268,951
Gross profit of XYZ EV industry	**6,584,272**
General expenses of XYZ EV industry	1839,000
Research and development of EV's component	954,862
Depreciation of XYZ EV industry	310,347
Unusual expenses (income)	163,700
Operating income of XYZ EV industry	**3,267,909**
Interest expense of XYZ EV industry	100
Income before tax of XYZ EV industry	**3,267,909**
Income tax, total	809,355
Income after tax	2,458,554
Total extraordinary items	0.0
Net income	2,458,554
Profitability Ratio	
Gross profit ratio = gross profit/total revenue	6,584,272/14,853,223 = 44%
Operating profit ratio = operating income/total revenue	3,267,909/14,853,223 = 22%
Net profit ratio = income after tax/total revenue	2,458,554/14,853,223 = 16.5%

as a measure of the EV company's short-term liquidity position. Because it demonstrates how soon the EV company can use its near-cash means to pay off its present arrears, it is occasionally appertained to as the" acid test rate."

- **Inventory turnover ratio (X):** Inventory turnover is a fiscal statistic that shows how constantly a business rotates its force in comparison to its cost of goods sold (COGS) over a given time frame. The average number of days it takes for an EV company to vend its force can also be calculated by dividing the number of days in the period, which is generally a financial time, by the force development rate. Using the force development rate, businesses can make better choices regarding pricing, product, marketing, and purchasing. One effectiveness rate used to assess how effectively a company uses its coffers is this one.

- **Dividend payout ratio (NP) (%):** The dividend payout rate is the proportion of the EV company's net income to the total quantum of tips paid to shareholders. It is the portion of earnings that is paid out to shareholders as tips. The plutocrat that is not given to shareholders is retained by the pot, which uses it to reduce debt or reinvest in its core businesses. It is also known as the payout rate in some cases.

- **Earning retention ratio (%):** The retention rate is the proportion of earnings that the EV company keeps. The retention rate is the percentage of net gains used to grow the business rather than being dispersed as dividends. It is contrary of the payout rate, which determines the chance of gains delivered as tips to shareholders. The plowback rate is another name for the retention rate.

13.4 Cost Assessment of EVs with Different Parameters

Data assessment is the process of evaluating data and reaching significant conclusions through various logical methods. By employing data analysis, experimenters can organize, manipulate, and summarize data to address important questions. One aspect of data assessment involves considering various parameters, such as acceleration, top speed, range, efficiency, and fast-charging capabilities, to determine the prices (in Euros) of EVs manufactured by different companies. Table 13.4 shows the different parameters of the different types of EVs. Among all the vehicles, Lucid has a minimum acceleration of 2.8 seconds. By the regression analysis, Eq. (13.1) shows the relationship between, Price (Euro) as a dependent parameter and various independent parameters, such as acceleration, speed, range, efficiency, and charging capability. Figures 13.2 and 13.3 show the forecasting of the price (Euro) and charging capability of the EVs.

Table 13.4 Different parameters of different types of EVs.

Company	Acceleration (seconds)	Top speed (km/h)	Range (km)	Efficiency (Wh/km)	Fast charge (km/h)	Price (euro)
Aiways	9	150	335	188	350	36,057
Audi	6.3	180	400	193	540	55,000
Audi	3.5	240	425	197	850	125,000
Audi	6.8	190	280	231	450	67,358
Audi	5.7	200	380	228	610	81,639
Audi	5.7	200	365	237	590	79,445
Audi	6.3	180	410	188	550	57,500
Audi	6.8	190	295	219	470	69,551

(Continued)

Table 13.4   (Continued)

Company	Acceleration (seconds)	Top speed (km/h)	Range (km)	Efficiency (Wh/km)	Fast charge (km/h)	Price (euro)
Audi	4.5	210	320	270	510	93,800
Audi	4.5	210	335	258	540	96,050
BMW	6.8	180	360	206	560	68,040
BMW	4	200	450	178	650	65,000
BMW	7.3	150	235	161	270	38,017
BMW	6.9	160	230	165	260	41,526
Byton	5.5	190	390	244	460	64,000
Byton	7.5	190	325	222	420	53,500
Byton	7.5	190	400	238	480	62,000
Citroen	9.7	150	250	180	380	40,000
CUPRA	6.5	160	425	181	570	45,000
DS	8.7	150	250	180	380	37,422
Fiat	9	150	250	168	330	34,900
Fiat	9	150	250	168	330	37,900
Ford	7	180	450	200	430	54,475
Ford	6	180	430	209	410	62,900
Ford	6	180	340	206	360	54,000
Ford	6.6	180	360	194	380	46,900
Honda	9.5	145	170	168	190	32,997
Honda	8.3	145	170	168	190	35,921
Hyundai	7.9	167	400	160	380	40,795
Hyundai	9.7	165	250	153	210	34,459
Hyundai	9.9	155	255	154	210	33,971
Jaguar	4.8	200	365	232	340	75,351
Kia	7.8	167	370	173	350	38,105
Kia	7.9	167	365	175	340	36,837
Kia	9.8	155	235	167	230	34,400
Kia	7.9	167	365	175	320	36,837
Kia	9.9	157	230	170	220	33,133
Lexus	7.5	160	270	193	190	50,000
Lightyear	10	150	575	104	540	149,000
Lucid	2.8	250	610	180	620	105,000

Table 13.4 (Continued)

Company	Acceleration (seconds)	Top speed (km/h)	Range (km)	Efficiency (Wh/km)	Fast charge (km/h)	Price (euro)
Mazda	9	150	180	178	240	32,646
Mercedes	5.1	180	370	216	440	69,484
Mercedes	5	200	350	171	440	45,000
Mercedes	10	140	330	273	290	70,631
MG	8.2	140	220	193	260	30,000
Mini	7.3	150	185	156	260	31,681
Nissan	7.9	144	220	164	230	29,234
Nissan	7.3	157	325	172	390	37,237
Nissan	7.6	160	440	198	520	50,000
Nissan	14	123	190	200	190	33,246
Nissan	5.7	200	420	207	500	57,500
Nissan	7.5	160	330	191	440	45,000
Nissan	5.9	200	325	194	440	50,000
Nissan	5.1	200	375	232	450	65,000
Opel	8.1	150	275	164	420	29,146
Opel	7.3	150	335	173	210	41,906
Opel	8.5	150	255	176	390	35,000
Peugeot	8.1	150	275	164	420	29,682
Peugeot	8.5	150	250	180	380	34,361
Polestar	4.7	210	400	181	620	56,440
Porsche	2.8	260	375	223	780	180,781
Porsche	4	250	365	195	730	102,945
Porsche	4	250	425	197	890	109,302
Porsche	3.5	250	385	217	770	150,000
Porsche	3.2	260	390	215	810	148,301
Renault	11.4	135	315	165	230	31,184
Renault	9.5	140	310	168	230	33,133
Renault	12.6	135	130	164	220	24,790
Renault	11.4	135	255	161	230	29,234
Renault	22.4	130	160	194	210	38,000
SEAT	12.3	130	195	166	170	20,129

(Continued)

Table 13.4 (Continued)

Company	Acceleration (seconds)	Top speed (km/h)	Range (km)	Efficiency (Wh/km)	Fast charge (km/h)	Price (euro)
Skoda	10	160	290	179	230	35,000
Skoda	12.3	130	195	166	170	24,534
Skoda	8.8	160	420	183	560	40,000
Skoda	7	160	400	193	540	45,000
Skoda	6.2	180	400	193	540	47,500
Skoda	9	160	320	181	440	37,500
Smart	12.7	130	95	176	435	22,030
Smart	11.6	130	100	167	440	21,387
Smart	11.9	130	95	176	270	24,565
Sono	9	140	225	156	270	25,500
Tesla	4.6	233	450	161	940	55,480
Tesla	5.6	225	310	153	650	46,380
Tesla	5.1	217	425	171	930	58,620
Tesla	3.4	261	435	167	910	61,480
Tesla	3	210	750	267	710	75,000
Tesla	3.8	250	515	184	560	79,990
Tesla	2.1	410	970	206	920	215,000
Tesla	4.6	250	450	211	490	85,990
Tesla	2.5	261	505	188	550	96,990
Tesla	3.7	241	410	177	900	65,620
Tesla	5	190	460	261	710	55,000
Tesla	2.8	250	440	216	480	102,990
Tesla	7	180	390	256	740	45,000
Volkswagen	10	160	270	167	250	30,000
Volkswagen	9.6	150	190	168	220	31,900
Volkswagen	7.9	160	440	175	590	40,936
Volkswagen	11.9	130	195	166	170	21,421
Volkswagen	7.5	160	420	183	560	45,000
Volkswagen	9	160	350	166	490	33,000
Volkswagen	7.3	160	340	171	470	38,987
Volkswagen	7.3	160	340	171	470	35,575
Volvo	4.9	180	375	200	470	60,437

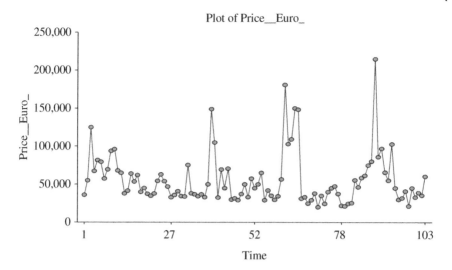

Figure 13.2 Forecasting of the price (euro) of the EVs.

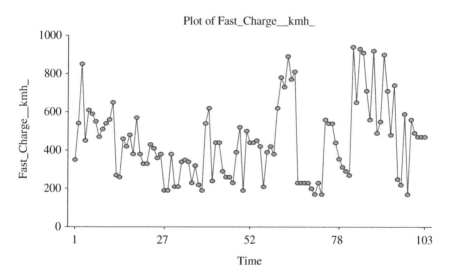

Figure 13.3 Forecasting of the fast charge (kmh) of EVs.

Price (Euro) = − 111,738.870085403 + 1904.36639224141 × acceleration(seconds)
+ 617.648167394431 × top speed (km/h) + 36.4564392881625
× range (km) + 149.297198183294 × efficiency (Wh/km)
+ 5.11001968637254 × fast charge (km/h)

(13.1)

Table 13.5 shows the descriptive analysis (based on the data of Table 13.4 and analysis through the NCSS statistical software) of the EVs with manufacturer

Table 13.5 Descriptive analysis of EVs (Audi).

Variable	Mean	SD	SE	Median
Acceleration (seconds)	5.566667	1.156503	0.3855011	5.7
Top speed (kmh)	200	18.70829	6.236095	200
Range (km)	356.6667	51.8411	17.28037	365
Efficiency (Wh/km)	224.5556	28.53555	9.511851	228
Fast charge (km/h)	567.7778	117.5561	39.18538	540
Price (Euro)	80,593.66	22,003.17	7334.391	79,445
Variable	**Minimum**	**Maximum**	**Range**	**Variance**
Acceleration (seconds)	3.5	6.8	3.3	1.3375
Top speed (kmh)	180	240	60	350
Range (km)	280	425	145	2687.5
Efficiency (Wh/km)	188	270	82	814.2778
Fast charge (km/h)	450	850	400	13,819.44
Price (Euro)	55,000	125,000	70,000	4.841396E+08

company Audi. From the assessment, it is found that the average price of the EVs of this company is €80,593.66 with a mean acceleration of 5.56 seconds, a top speed of 200 km, and a range of 356 km.

Table 13.6 shows the descriptive analysis (based on the data of Table 13.4 and analysis through the NCSS statistical software) of the EVs with manufacturer company BMW. From the assessment, it is found that the average price of the EVs of this company is €53,145.75 with a mean acceleration of 6.25 seconds, a top speed of 172.5 km and a range of 318.75 km.

Table 13.7 shows the descriptive analysis (based on the data of Table 13.4 and analysis through the NCSS statistical software) of the EVs with manufacturer company Kia. From the assessment, it is found that the average price of the EVs of this company is €35,862.4 with a mean acceleration of 8.66 seconds, a top speed of 162.6 km, and a range of 313 km.

Table 13.8 shows the descriptive analysis (based on the data of Table 13.4 and analysis through the NCSS statistical software) of the EVs with manufacturer company Mercedez. From the assessment, it is found out the average price of the EVs of this company is €61,705 with a mean acceleration of 6.7 seconds, a top speed of 173 km, and a range of 350 km.

Table 13.9 shows the descriptive analysis (based on the data of Table 13.4 and analysis through the NCSS statistical software) of the EVs with manufacturer company Nissan. From the assessment, it is found that the average price of the EVs of this company is €45,902.13 with a mean acceleration of 7.625 seconds, a top speed of 168 km and a range of 328 km.

Table 13.6 Descriptive analysis of EVs (BMW).

Variable	Mean	SD	SE	Median
Acceleration (seconds)	6.25	1.515476	0.7577379	6.85
Top speed (kmh)	172.5	22.17356	11.08678	170
Range (km)	318.75	106.174	53.087	297.5
Efficiency (Wh/km)	177.5	20.3388	10.1694	171.5
Fast charge (km/h)	435	199.7498	99.87492	415
Price (Euro)	53,145.75	15,559.13	7779.565	53,263

Variable	Minimum	Maximum	Range	Variance
Acceleration (seconds)	4	7.3	3.3	2.296667
Top speed (kmh)	150	200	50	491.6667
Range (km)	230	450	220	11,272.92
Efficiency (Wh/km)	161	206	45	413.6667
Fast charge (km/h)	260	650	390	39,900
Price (Euro)	38,017	68,040	30,023	2.420865E+08

Table 13.7 Descriptive analysis of EVs (Kia).

Variable	Mean	SD	SE	Median	Minimum	Maximum
Acceleration (seconds)	8.66	1.087658	0.4864154	7.9	7.8	9.9
Top speed (kmh)	162.6	6.0663	2.712932	167	155	167
Range (km)	313	73.53571	32.88617	365	230	370
Efficiency (Wh/km)	172	3.464102	1.549193	173	167	175
Fast charge (km/h)	292	62.20932	27.82086	320	220	350
Price (Euro)	35,862.4	2032.067	908.7679	36,837	33,133	38,105

Variable	Range	Variance
Acceleration (seconds)	2.1	1.183
Top speed (kmh)	12	36.8
Range (km)	140	5407.5
Efficiency (Wh/km)	8	12
Fast charge (km/h)	130	3870
Price (Euro)	4972	4,129,296

Table 13.8 Descriptive analysis of EVs (Mercedes).

Variable	Mean	SD	SE	Median
Acceleration (seconds)	6.7	2.858321	1.650252	5.1
Top speed (kmh)	173.3333	30.5505	17.63834	180
Range (km)	350	20	11.54701	350
Efficiency (Wh/km)	220	51.11751	29.51271	216
Fast charge (km/h)	390	86.60254	50	440
Price (Euro)	61,705	14,478.32	8359.061	69,484

Variable	Minimum	Maximum	Range	Variance
Acceleration (seconds)	5	10	5	8.17
Top speed (kmh)	140	200	60	933.3333
Range (km)	330	370	40	400
Efficiency (Wh/km)	171	273	102	2613
Fast charge (km/h)	290	440	150	7500
Price (Euro)	45,000	70,631	25,631	2.096217E+08

Table 13.9 Descriptive analysis of EVs (Nissan).

Variable	Mean	SD	SE	Median
Acceleration (seconds)	7.625	2.774759	0.9810253	7.4
Top speed (kmh)	168	29.08608	10.28348	160
Range (km)	328.125	87.78779	31.03767	327.5
Efficiency (Wh/km)	194.75	20.88574	7.384225	196
Fast charge (km/h)	395	121.3025	42.88689	440
Price (Euro)	45,902.13	12,228.34	4323.373	47,500

Variable	Minimum	Maximum	Range	Variance
Acceleration (sec)	5.1	14	8.9	7.699286
Top speed (kmh)	123	200	77	846
Range (km)	190	440	250	7706.696
Efficiency (Wh/km)	164	232	68	436.2143
Fast charge (km/h)	190	520	330	14,714.29
Price (Euro)	29,234	65,000	35,766	1.495324E+08

Table 13.10 Descriptive analysis of EVs (Porsche).

Variable	Mean	SD	SE	Median	Minimum
Acceleration (seconds)	3.5	0.5196152	0.232379	3.5	2.8
Top speed (kmh)	254	5.477226	2.44949	250	250
Range (km)	388	22.80351	10.19804	385	365
Efficiency (Wh/km)	209.4	12.60159	5.635601	215	195
Fast charge (km/h)	796	59.8331	26.75818	780	730
Price (Euro)	138,265.8	32,141.87	14,374.28	148,301	102,945

Variable	Maximum	Range	Variance
Acceleration (sec)	4	1.2	0.27
Top speed (kmh)	260	10	30
Range (km)	425	60	520
Efficiency (Wh/km)	223	28	158.8
Fast charge (km/h)	890	160	3580
Price (Euro)	180,781	77,836	1.0331E+09

Table 13.10 shows the descriptive analysis (based on the data of Table 13.4 and analysis through the NCSS statistical software) of the EVs with manufacturer company Porsche. From the assessment, it is found that the average price of the EVs of this company is €138,265.8 with a mean acceleration of 3.5 seconds, a top speed of 254 km, and a range of 388 km.

Table 13.11 shows the descriptive analysis (based on the data of Table 13.4 and analysis through the NCSS statistical software) of the EVs with manufacturer company Renault. From the assessment, it is found that the average price of the EVs of this company is €31,268.2 with a mean acceleration of 13.46 seconds, a top speed of 135 km and a range of 234 km.

Table 13.12 shows the descriptive analysis (based on the data of Table 13.4 and analysis through the NCSS statistical software) of the EVs with manufacturer company Skoda. From the assessment, it is found that the average price of the EVs of this company is €38,255.67 with a mean acceleration of 8.83 seconds, a top speed of 158.33 km and a range of 337.5 km.

Table 13.13 shows the descriptive analysis (based on the data of Table 13.4 and analysis through the NCSS statistical software) of the EVs with manufacturer company Tesla. From the assessment, it is found that the average price of the EVs of this company is €80,272.3 with a mean acceleration of 4.09 seconds, a top speed of 244 km, and a range of 500 km.

Table 13.14 shows the descriptive analysis (based on the data of Table 13.4 and analysis through the NCSS statistical software) of the EVs with manufacturer

Table 13.11 Descriptive analysis of EVs (Renault).

Variable	Mean	SD	SE	Median
Acceleration (seconds)	13.46	5.119375	2.289454	11.4
Top speed (kmh)	135	3.535534	1.581139	135
Range (km)	234	85.24963	38.12479	255
Efficiency (Wh/km)	170.4	13.42758	6.004998	165
Fast charge (km/h)	226	5.477226	2.44949	230
Price (Euro)	31,268.2	4871.005	2178.379	31,184

Variable	Minimum	Maximum	Range	Variance
Acceleration (seconds)	9.5	22.4	12.9	26.208
Top speed (kmh)	130	140	10	12.5
Range (km)	130	315	185	7267.5
Efficiency (Wh/km)	161	194	33	180.3
Fast charge (km/h)	220	230	10	30
Price (Euro)	24,790	38,000	13,210	2.372669E+07

Table 13.12 Descriptive analysis of EVs (Skoda).

Variable	Mean	SD	SE	Median
Acceleration (seconds)	8.883333	2.176618	0.8886006	8.9
Top speed (kmh)	158.3333	16.02082	6.540473	160
Range (km)	337.5	86.5881	35.34945	360
Efficiency (Wh/km)	182.5	10.07472	4.112988	182
Fast charge (km/h)	413.3333	171.542	70.03174	490
Price (Euro)	38,255.67	8166.284	3333.871	38,750

Variable	Minimum	Maximum	Range	Variance
Acceleration (seconds)	6.2	12.3	6.1	4.737667
Top speed (kmh)	130	180	50	256.6667
Range (km)	195	420	225	7497.5
Efficiency (Wh/km)	166	193	27	101.5
Fast charge (km/h)	170	560	390	29,426.67
Price (Euro)	24,534	47,500	22,966	6.668819E+07

Table 13.13 Descriptive analysis of EVs (Tesla).

Variable	Mean	SD	SE	Median
Acceleration (seconds)	4.092308	1.389521	0.3853838	3.8
Top speed (kmh)	244.4615	55.99645	15.53062	241
Range (km)	500.7692	173.2032	48.03793	450
Efficiency (Wh/km)	201.3846	39.07586	10.83769	188
Fast charge (km/h)	730	175.6417	48.71424	710
Price (Euro)	80,272.3	44,432.09	12,323.25	65,620

Variable	Minimum	Maximum	Range	Variance
Acceleration (seconds)	2.1	7	4.9	1.930769
Top speed (kmh)	180	410	230	3135.603
Range (km)	310	970	660	29,999.36
Efficiency (Wh/km)	153	267	114	1526.923
Fast charge (km/h)	480	940	460	30,850
Price (Euro)	45,000	215,000	170,000	1.974211E+09

Table 13.14 Descriptive analysis of EVs (Volkswagen).

Variable	Mean	SD	SE	Median	Minimum
Acceleration (seconds)	8.8125	1.635707	0.5783096	8.45	7.3
Top speed (kmh)	155	10.69045	3.779645	160	130
Range (km)	318.125	93.42444	33.03053	340	190
Efficiency (Wh/km)	170.875	5.792544	2.047974	169.5	166
Fast charge (km/h)	402.5	163.5979	57.84061	470	170
Price (Euro)	34,602.38	7301.637	2581.518	34,287.5	21,421

Variable	Maximum	Range	Variance
Acceleration (seconds)	11.9	4.6	2.675536
Top speed (kmh)	160	30	114.2857
Range (km)	440	250	8728.125
Efficiency (Wh/km)	183	17	33.55357
Fast charge (km/h)	590	420	26,764.29
Price (Euro)	45,000	23,579	5.33139E+07

company Volkswagen. From the assessment, it is found that the average price of the EVs of this company is €34,602.38 with a mean acceleration of 8.81 seconds, a top speed of 155 km, and a range of 318.125 km.

13.5 Financial and Economic Analysis of Green Vehicle Infrastructure by Blockchain

A blockchain refers to a distributed database or ledger that is shared among computer networks. It functions as a digital database for storing data and is most commonly recognized for its use in maintaining secure and decentralized records of cryptocurrency transactions like Bitcoin. Unlike traditional systems that rely on a trusted third party, blockchain technology promotes confidence by ensuring the integrity and security of data records. The arrangement of data in a blockchain is unique, as it is collected in units called blocks, each containing sets of data. Once a block is filled, it is sealed and linked to the previous block to create the blockchain. Blocks have predefined storage capacities, and when a chain is complete, any additional data is combined to create a new block that is added to the chain. This data structure intentionally produces an irreversible record of data when used in a decentralized system. While a database typically organizes its data into tables, blockchain divides its data into blocks that are chained together. As each block is added to the chain, an accurate timestamp is assigned, creating a permanent and immutable record. In a blockchain, the data is gathered in units called blocks, and each block comprises sets of data. Each of the individual blocks that make up the data trail known as the block chain has a specific amount of storage space. A block is sealed when it is full and joined to the block before it. Each piece of data that comes after the just-added block is used to generate a new block, which is then added to the chain after the chain is complete. Figure 13.4 shows the blockchain-based green vehicle transaction infrastructure. Blockchains, as their name suggests, organize their information into units (blocks) that are linked together, in contrast to databases, which typically organize their data into tables. When used decentralized, this data structure purposefully results in an immutable chronology of data. A finished block is sealed forever and added to the timeline. Each block that is submitted to the chain receives a precise timestamp. Digital information can be transmitted and preserved without being modified with the use of a blockchain. Alternatively referred to as distributed ledger technology, unchangeable ledgers or records of interactions that cannot be changed, deleted, or destroyed are constructed on top of a blockchain (DLT).

For EV industry to create a distributed application on "**Ethereum**" platform. In blockchain-based EV transaction use "**Solidity**" an object-oriented language for

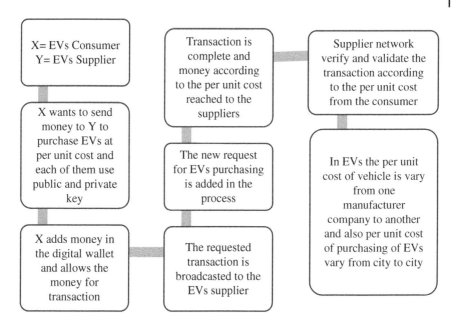

Figure 13.4 Blockchain-based EV transaction.

contract development between EV suppliers and EV consumers. "Remix" based blockchain platform is used for developing and testing EV transaction between supplier and consumer. This transaction changes according to the type of manufacturer company and according to the city and country.

To facilitate the transaction between supplier and consumer Ganache can be used. In blockchain My Ether wallet is used to create a wallet for each EV consumer. The EV supplier will publish the terms and conditions. Any other EV consumer will look at the contract of EVs by using the interface provided by the terms and conditions and send per unit cost to the supplier for executing a part of the transaction [17]. Following are the ways of EV financial transactions through the Ethereum-based blockchain concept.

Decide the name of contract for EV financial transaction:

Terms and conditions of EV financial transaction {<closing flower bracket missing>

We will declare two variables as follows:

```
uint pua;   # pua = per unit amount of EVs
uint voe;   # voe = value of EVs (depend on company and features)
```

The variable "pua" will hold the accumulated money sent by the EV consumer to the supplier. The "voe" shows the value of EVs based on the different features.

In the EV transaction, set the values of two variables

```
Constructor (uint minperunit Amount, uint minvalue of EVs)
public
        {
            pua = 0;
            voe = 10000($);
```

<closing flower bracket missing>

Initially, the amount collected through the EV transaction is zero, so set the "pua" field as zero. Set the amount of electricity required for EVs to some arbitrary number, in this case 10000($).

The supplier decides the value.

In the EV transaction, to examine the collected amount by the consumer at any given point of time, a public contract method called getAmount defined as follows:

```
        function getAmount () public view returns (uint)
{
    return pua;
}
```

To get the balanced EV transaction value at any given point in time, we define **getBalance** EV transaction method as follows:

```
function getBalance EVs transaction () public view
returns(uint) {
    return value;
}
```

Finally, we write an EV transaction method **(Send)**. It enables the EV consumer to send some amount to the supplier for EV transaction:

```
function send (uint newDeposit) public {
    value = value-newDeposit;
    amount = amount + newDeposit;
}
```

The execution of the **Send** method will modify both **"pua"** and **"voe"** fields of the RET.

13.5.1 R3 Corda in EV Transaction

The idea of R3 Corda is that by focusing on enterprise financial applications such as EV transaction, the platform is simplified and eventually mitigates several risks, such as scalability, privacy, and so on, which is also required for efficient EV

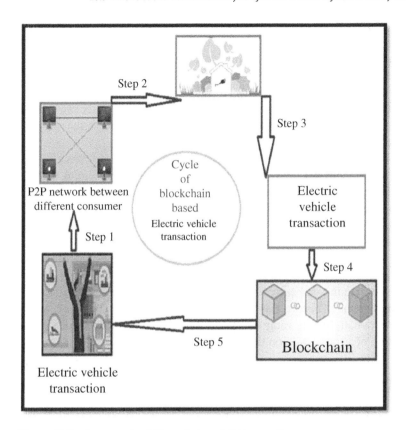

Figure 13.5 Framework of R3 corda-based EV transaction.

transaction. R3 Corda Blockchain is an enterprise blockchain distributed ledger that can be used in EV transaction. A blockchain is a tamper-evident, shared digital ledger that records transactions between EV consumers and EV suppliers through the peer-to-peer network. Distributed to all the electricity consumers into the nodes in the network, the ledger permanently records all the EV transactions in a sequential chain of cryptographic hash-linked blocks, documenting the history of all the EV exchanges that take place between the peers in the network. Framework of R3 Corda-based EV transaction is presented in Figure 13.5.

A Corda network for EV transaction is a fully connected graph, and all nodes can potentially send an information to each EV consumer related to the bidding of the EVs. Nodes are informed about the topology of EVs by a network map service. The Advanced Message Queuing Protocol (AMQP) is used in the R3 Corda system to transmit information between customers and suppliers through Transport Layer Security, enabling resilient queuing and routing even in the event that electric car transaction nodes need to be restarted. Message delivery and persistence guarantees are provided using AMQP, which is asynchronous, performs well under load, and runs without making any assumptions about ongoing communication between manufacturers and buyers of EVs. Messages queue up and retry until delivered, even if the receivers are not online. With this design, a network in Corda, also called as EV Transaction Zone, can process multiple transactions between suppliers and consumers simultaneously, it meets the needs for strict privacy requirements during the payment of per unit cost of EVs.

Corda is a Java virtual machine (JVM)-based application used for transaction of EVs. As a result, many of Corda's features depend on the JVM ecosystem, including platform independence for electricity consumers, portability (the transmission of bytecode), serialization (suspension), and many others. Therefore, a Corda node is a portable JVM run-time process, and by using the "java" command, you may simply operate numerous nodes on a single physical system. Using a configuration known as a network map for transaction system, which is a collection of signed node information that will be kept in a cache on each node, all nodes in an EV transaction system may see each other. It is possible to use a network map server or the file system to restore this cached configuration. The states are saved (updated) internally in the vault relational database on the Corda node upon the recording of a transaction on the ledger.

13.5.2 Blockchain Merkle Tree of EV Transaction

The Merkle tree is the key building component of blockchain technology. It is a sort of mathematical data structure used to describe all the operations of EVs in a block of information. It is made up of hashes of various data blocks between providers and buyers of EVs. Additionally, it enables speedy and secure information verification across a huge volume of EV transaction data. Additionally, it aids in confirming the accuracy and completeness of the data. The Merkle Trees structure is used by both Ethereum and Bitcoin. Hash Tree is another name for Merkle Tree. By creating a digital fingerprint of the collected edition of activities from the consumer, a Merkle tree saves all of the interactions in a block. It enables an EV user to decide whether or not a transaction can be included in a block. Merkle trees are produced by repeatedly hashing node pairs until only one hash remains. The Merkle Root, also known as the Root Hash, is this hash. Merkle Trees are built from the ground up. The non-leaf node is a hash of its previous hashes, whereas each leaf node is a hash of data information. Since Merkle trees are in a binary tree, we need an even number of leaf nodes. The final hash will be duplicated once to produce an even number of leaf nodes if the number of transactions is odd. The most common and fundamental type of Merkle tree, the Binary Merkle Tree, is shown above.

TX1, TX2, TX3, and TX4 are the four transactions that make up a block. As you can see, the top hash, which is the hash of the entire tree, is the root hash, also referred to as the Merkle root. By repeatedly hashing each of these and storing the results in each leaf node, hash 0, 1, 2, and 3 are created. A parent node's subsequent pairings of leaf nodes are summed by hashing Hash0 and Hash1, which results in Hash01, and then separately hashing Hash2 and Hash3, which results in Hash23. The two hashes (Hash01 and Hash23) are hashed once again to get the Root Hash or Merkle Root. The Merkle Root information can be found in block headers. The block header is the area of a Bitcoin block that is hashed during mining. It contains the Root Hash of the most recent block's transactions in a Merkle Tree as well as the Root Hash of the block before that, which is a Nonce. Therefore, by including the Merkle root in the block header, the transaction is rendered impregnable. Because the Root Hash contains the hashes of each transaction in the block, these transactions might save space on the disc. The Merkle Tree preserves the consistency of the data. If a single transactional detail or the order in which it happens changes, the hash of the transaction will change. This modification would affect the value of the Merkle Root, invalidating the block. The Merkle Tree would experience this alteration cascading up to the Merkle Root. Everyone can therefore agree that the Merkle tree allows for a quick and simple test to verify whether a specific transaction is a part of the set or not. Figure 13.6 shows the Merkle tree of different financial parameters of the EVs.

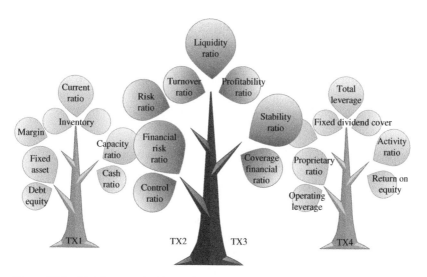

Figure 13.6 Merkle tree of different financial parameters of EVs.

13.6 Applicability of Different Blockchain Cryptocurrencies in EV Transaction

Cryptocurrencies on the blockchain can potentially be used in EVs transactions, as they can provide a secure, decentralized, and transparent method of payment. Here are some different cryptocurrencies and their potential applicability in EV transactions:

- **Bitcoin:** Bitcoin is the most popular and well-known cryptocurrency. Its popularity and widespread acceptance make it a potential payment option for EV transactions. However, due to its high transaction fees and slow processing time, it may not be the most practical option for small transactions.
- **Litecoin:** Litecoin is a cryptocurrency that was created to be a faster and cheaper alternative to Bitcoin. Its faster processing time and lower transaction fees make it a more practical option for smaller EV transactions.
- **Ripple:** Ripple is a cryptocurrency that focuses on cross-border transactions. Its fast processing time and low transaction fees make it a potential payment option for international EV transactions.
- **Bitcoin Cash:** Bitcoin Cash is a fork of Bitcoin that was created to improve its transaction processing time and lower its fees. Its faster processing time and lower fees could make it a more practical option for smaller EV transactions.

13.7 Challenges and Advantages of Using Blockchain for EVs

Blockchain technology offers several potential advantages for the EV industry, including improved security, increased transparency, and reduced fraud. However, there are also challenges associated with using blockchain for EVs.

- **Advantages:**
  - **Secure Transactions:** Blockchain technology can provide an extra layer of security to EV transactions by encrypting them and making them immutable. This reduces the risk of fraud and hacking, ensuring that transactions are transparent and trustworthy.
  - **Smart Contracts:** Blockchain enables the use of smart contracts, which are self-executing contracts with the terms of the agreement written into code. This can automate EV transactions such as charging and billing, reducing the need for intermediaries and improving efficiency.
  - **Decentralization:** Blockchain is a decentralized technology, which means that it operates on a peer-to-peer network without the need for a central

authority. This can reduce transaction costs and increase the efficiency of the overall EV ecosystem.

– **Traceability:** Blockchain can provide a transparent and immutable record of EV transactions, making it easier to track the history of a vehicle and ensuring that the data is trustworthy.

- **Challenges:**
  – **Scalability:** One of the biggest challenges of using blockchain for EVs is scalability. The current infrastructure of blockchain may not be able to handle the large volumes of transactions required for EVs, which can cause delays and increase transaction costs.
  – **Interoperability:** The EV industry is made up of multiple stakeholders, including manufacturers, charging station operators, and utilities. Ensuring that these stakeholders can interact with each other on a blockchain platform requires interoperability, which can be a challenge.
  – **Regulatory Environment:** Blockchain technology is still relatively new, and regulations surrounding its use in the EV industry are still evolving. This can create uncertainty and slow down the adoption of the technology.
  – **Integration:** Integrating blockchain technology into existing EV infrastructure can be a challenge. There are often legacy systems in place that may not be compatible with blockchain technology, which can make it difficult to implement.

At the end, the advantages of blockchain for the EV industry are promising, but the challenges must be addressed to ensure widespread adoption and the realization of its full potential.

13.8 Conclusion

After researching the use of blockchain in the financial sector the blockchain, it becomes evident that a distributed ledger of a closed kind is crucial to the future of finance. Additionally, it is vital to jointly utilize global blockchain consortia, blockchain consortiums of financial institutions, and regional consortiums of local banking organizations. In the near future, it is necessary to identify the basic entity of blockchain in the green vehicle infrastructure. In the future, it is also necessary to identify the financial analysis and application of blockchain technology in the EV charging station. The following are the outcome of this chapter:

- It is found out by the end of 2023 North America will be able to produce 1 million BEVs, with 30 EV types.
- It is also found out Norway is the leading country in the field of EVs with a market share of 74.8%.

- Denmark is trailing behind its neighboring countries like Sweden and Norway in terms of its EVs and EV charging infrastructure. In Denmark, the tax exemption for these vehicles will gradually decrease from 20% in 2020 to 65% in 2021, 90% in 2022, and 100% in 2023.
- Blockchain technology can offer significant benefits to the EV industry, including improved transparency, traceability, data security, efficiency, and decentralization.
- R3 Corda's built-in regulatory and compliance features can help ensure that the financial assessment process complies with relevant laws and regulations, such as anti-money laundering and consumer protection laws. This can help reduce the risk of legal and regulatory issues and improve the reputation of the EV industry.

References

1 Demirguc-Kunt, A., Klapper, L., Singer, D. et al. (2018). The global Findex database 2017. In: *Measuring Financial Inclusion and the FinTech Revolution*. Washington, DC, USA: World Bank.

2 Sahay, R., Allmen, U., Lahreche, A. et al. (2020). *The Promise of Fintech: Financial Inclusion in the Post COVID-19 Era*. Washington, DC, USA: International Monetary Fund.

3 Ziegler, T., Zhang, B., Carvajal, A. et al. (2020). The global COVID-19 finTech market rapid assessment study. CCAF, World Bank and World Economic Forum. https://papers.ssrn.com/sol3/papers.cfm?abstract_id=3770789 (accessed January 2022).

4 Gartner (2022). The CIO's guide to blockchain. https://www.gartner.com/smarterwithgartner/the-cios-guideto-blockchain (accessed September 2022).

5 Khare, V. (2020). Design and assessment of solar-powered EVs by different techniques. *International Transactions on Electrical Energy Systems* 30 (4): e12271.

6 Raghuwanshi, S.S. and Khare, V. (2018). Sizing and modelling of stand-alone photovoltaic water pumping system for irrigation. *Energy & Environment* 29 (4): 473–491.

7 Khare, V., Nema, S., and Baredar, P. (2019). Reliability analysis of hybrid renewable energy system by fault tree analysis. *Energy & Environment* 30 (3): 542–555.

8 Khare, V., Khare, C.J., Nema, S., and Baredar, P. (2021). Modeling, cost optimization and management of grid connected solar powered charging station for EVs. *International Journal of Emerging Electric Power Systems* 23 (4): 587–603.

9 Guo, Y. and Liang, C. (2016). Blockchain application and outlook in the banking industry. *Financial Innovation* 2: 1–12. doi: 10.1186/s40854-016-0034-9.

10 Dorfleitner, D. and Braun, D. (2019). *Fintech, Digitalization, and Blockchain: Possible Applications for Green Finance, in the Rise of Green Finance in Europe*,

207–237. Cham, Switzerland: Palgrave Macmillan https://doi.org/10.1007/978-3-030-22510-0_9.

11 McGhin, T., Choo, K.-K.R., Liu, C.Z., and He, D. (2019). Blockchain in healthcare applications: research challenges and opportunities. *Journal of Network and Computer Applications* 135: 62–75. https://doi.org/10.1016/j.jnca.2019.02.027.

12 Novo, O. (2018). Blockchain meets IoT: an architecture for scalable access management in IoT. *IEEE Internet of Things Journal* 5: 1184–1195. https://doi.org/10.1109/JIOT.2018.2812239.

13 Taylor, P.J., Dargahi, T., Dehghantanha, A. et al. (2020). A systematic literature review of blockchain cyber security. *Digital Communications and Networks* 6: 147–156. https://doi.org/10.1016/j.dcan.2019.01.005.

14 Al-Quraan, M., Mohjazi, L., Bariah, L. et al. Edge-native intelligence for 6G communications driven by federated learning: a survey of trends and challenges. *arXiv 2021* arXiv:2111.07392. https://doi.org/10.48550/arXiv.2111.07392.

15 Xiao, Y., Zhang, N., Li, J. et al. (2020). PrivacyGuard: Enforcing private data usage control with blockchain and at-tested off-chain contract execution. *Proceedings of the European Symposium on Research in Computer Security*, pp. 610–629 (13 September 2020); Springer: Cham, Switzerland. https://doi.org/10.1007/978-3-030-59013-0_30.

16 Gayialis, S.P., Kechagias, E.P., Konstantakopoulos, G.D. et al. (2021). An approach for creating a blockchain platform for labeling and tracing wines and spirits. *Proceedings of the IFIP International Conference on Advances in Production Management Systems*, pp. 81–89, Nantes, France (5–9 September 2021). Springer: Cham, Switzerland. https://doi.org/10.1007/978-3-030-85910-7_9.

17 Okwuibe, G.C. (2020). A blockchain based electric vehicle smart charging system with flexibility. *IFAC-PapersOnLine* 53 (2): 13557–13561.

18 Wang, Q. and Su, M. (2020). Integrating blockchain technology into the energy sector – from theory of blockchain to research and application of energy blockchain. *Computer Science Review* 37: 100275. https://doi.org/10.1016/j.cosrev.2020.100275.

19 Gayialis, S.P., Kechagias, E., Papadopoulos, G.A., and Konstantakopoulos, G.D. (2019). Design of a blockchain-driven system for product counterfeiting restraint in the supply. *Proceedings of the IFIP International Conference on Advances in Production Management Systems*, Austin, TX, USA (1–5 September 2019).

20 Romānova, I. and Kudinska, M. (2016). Banking and Fintech: a challenge or opportunity? In: *Contemporary Issues in Finance: Current Challenges from across Europe*. Bingley, UK: Emerald Group Publishing Limited.

21 Mohammed, A.H., Abdulateef, A.A., and Abdulateef, I.A. (2021). Hyperledger, Ethereum and blockchain technology: A short overview. *Proceedings of the 2021 3rd International Congress on Human-Computer Interaction, Optimization and Robotic*

Applications (HORA), pp. 1–6, Ankara, Turkey (11–13 June 2021). https://doi.org/10.1109/HORA52670.2021.9461294.

22 Wang, G., Zhang, S., Yu, T., and Ning, Y. (2021). A systematic overview of blockchain research. *Journal of Systems Science and Information* 9: 205–238. https://doi.org/10.21078/JSSI-2021-205-34.

23 Moezkarimi, Z., Nourmohammadi, R., Zamani, S. et al. (2019). An overview on technical characteristics of blockchain platforms. *Proceedings of the International Congress on High-Performance Computing and Big Data Analysis*, pp. 265–278, Tehran, Iran (23–25 April 2019). Springer: Cham, Switzerland. https://doi.org/10.1007/978-3-030-33495-6_20.

24 Kuo, T.-T., Rojas, H.Z., and Ohno-Machado, L. (2019). Comparison of blockchain platforms: a systematic review and healthcare examples. *Journal of the American Medical Informatics Association* 26: 462–478. https://doi.org/1093/jamia/ocy185.

14

Unmanned Aerial Vehicles Toward Intelligent Transportation Systems

Fereidoun H. Panahi and Farzad H. Panahi

Department of Electronics and Communication Engineering, University of Kurdistan, Sanandaj, Iran

Abbreviations

EH	Energy Harvesting
ECU	Energy Control Unit
ITS	Intelligent Transportation System
IoT	Internet of Things
LBD	Laser Beam Director
LoS	Line-of-Sight
MUV	Mobile Unmanned Vehicle
PPP	Poisson Point Process
PoI	Points of Interest
RE	Renewable Energy
RF	Radio Frequency
SN	Sensor Node
UAV	Unmanned Aerial Vehicle
UE	User Equipment
VANET	Vehicle Ad hoc Network
WC	Wireless Charging
WET	Wireless Energy Transfer
WPT	Wireless Power Transfer
WSN	Wireless Sensor Network
WRSN	Wireless Rechargeable Sensor Network

Interconnected Modern Multi-Energy Networks and Intelligent Transportation Systems: Towards a Green Economy and Sustainable Development, First Edition. Edited by Mohammadreza Daneshvar, Behnam Mohammadi-Ivatloo, Amjad Anvari-Moghaddam, and Reza Razzaghi.
© 2024 The Institute of Electrical and Electronics Engineers, Inc.
Published 2024 by John Wiley & Sons, Inc.

14.1 Introduction

14.1.1 Context and Motivation

In recent years, interconnected modern multi-energy networks have emerged as a promising solution for achieving a green economy and sustainable development. Intelligent transportation systems (ITSs) play a critical role in this transition. A key aspect of ITSs is the deployment of wireless rechargeable sensor networks (WRSNs) to monitor, control, and optimize transportation systems. This chapter contributes to the broader context of the book by presenting an innovative approach for powering WRSNs through the use of unmanned aerial vehicles (UAVs) supported by locally deployed laser beam directors (LBDs) and renewable energy (RE) sources. The UAV relies on a combination of its onboard battery as well as laser beams and RE to procure sufficient and consistent power for charging sensors in the WRSNs. An energy cost optimization model is proposed to minimize the overall energy procurement costs for the UAV's operation. By integrating these emerging technologies into the interconnected multi-energy networks and leveraging the proposed optimization model, this approach aims to significantly enhance the sustainability, efficiency, and cost-effectiveness of next-generation ITSs. It not only advances the state of the art in WRSN charging schemes but also demonstrates the potential of leveraging modern, multi-energy networks in the development of next-generation ITSs, ultimately contributing to a greener economy and a more sustainable future.

14.1.2 Background

Wireless sensor networks (WSNs) have found widespread use in a variety of fields, including environmental and structural monitoring, smart agriculture, military missions, and intelligent transportation [1–5]. WSNs have the ability to greatly increase the functioning of current transportation systems. For instance, if a WSN is utilized to construct a vehicle ad hoc network (VANET) in an ITS, the data acquired from this system can potentially be leveraged to enhance public safety and provide a variety of useful services, including but not limited to transportation and traffic management [6]. Vehicles employing VANET can communicate with one another directly through their sensors, and a collision warning system can be used to avert traffic accidents. Individual vehicle monitoring systems can be set up for tracking criminal suspects remotely via connecting the vehicle sensors with roadside sensors. This has the ability to end deadly, high-speed chases. Moreover, this new form of ITS has the potential to improve public safety while also automating parking and traffic enforcement services. Another possible advantage is lower traffic enforcement costs for police agencies. Currently, wired sensors are generally used to collect traffic data for traffic planning and control. Existing sensing systems' high equipment and maintenance costs, as well as time-consuming installation, preclude the large-scale deployment of real-time traffic control and monitoring. Tiny, wireless sensors with built-in sensing, processing, and wireless communication functionalities provide enormous cost and installation benefits. WRSNs will be critical in the creation of future smart cities, notably in the transportation sector [5]. These systems employed RE or self-charging capabilities for powering a wide range of distinctive traffic and environmental monitoring applications. The use of wireless charging (WC) with UAVs is an innovative approach for supplying power for WRSNs [7, 8]. This strategy allows the ongoing deployment of WRSNs to previously inaccessible sites while also decreasing the maintenance requirements for constant WC networks. This solution, however, is constrained by present technologies. Since UAVs have a short battery life, they should return to their base stations on a regular basis to recharge. Moreover, battery capacity constraints in these devices restrict their capability to power sensor networks across a vast region. To overcome the UAV's short battery life, previous research concentrated on different aspects of UAVs and proposed possible solutions like employing high-quality devices such as batteries, manufacturing materials, wings, geometry, and motors. A number of works have described optimization algorithms that explore the shortest path for UAVs to their destination. In the majority of cases, the conventional battery swapping approach is employed; a UAV flies to a charging station on a regular basis, where the exhausted battery is withdrawn and an entirely charged one is inserted. This physical battery replacement approach, however, requires human help, hampering UAV operations and generating severe service disruptions in distant places [9].

To address these challenges, UAVs should typically employ a hybrid energy supply architecture, which brings together various energy sources, enabling the use of UAVs for a range of services without hampering UAV operations or causing service disruptions while conserving time and human resources. Therefore, it is essential to choose an appropriate energy source hybridization architecture with an optimal energy procurement system to ensure the efficient operation of modern UAVs. Although all the studies looking into UAV-related problems concentrated on the energy perspective of average or instantaneous power consumed/harvested when flying, hovering, processing data, or communicating, none of them considered the UAV's energy procurement costs, nor did they consider determining the optimal amount of energies to be procured by the UAV from each energy source. In this chapter, we first review the existing energy supply strategies for WSN-based ITSs. Then, to accommodate for the limits of UAVs in the development of the UAV-based WRSNs charging scheme, we propose a consistent and cost-aware energy procurement framework for a UAV powered concurrently by laser beams emitted by locally deployed LBDs and local RE sources. The UAV intends to decrease its overall energy cost for a given operation cycle by optimizing the quantities of energy acquired from its internal battery and laser beams at each time period. Indeed, the objective is to lower total energy costs by making accurate procurement decisions. Furthermore, we estimate the amount of extra RE that can be used to power the sensors in WRSNs using wireless power transfer (WPT) technology.

14.2 WSN for ITSs: The Energy Supply Issue and Existing Solutions

The tremendous development of vehicles as one of the most essential forms of transportation has aided human life, but it has also introduced new sorts of challenges such as traffic congestion, parking issues, and traffic accidents. Researchers are attempting to tackle these issues by developing ITSs. Intelligent transportation should be capable of managing, monitoring, and directing people to a more secure and coordinated mode of transportation. An ITS must, first and foremost, accurately identify, recognize, and track vehicles in real time [10, 11]. It is now unarguable that ITSs provide major benefits to both users and operators of transportation infrastructure. Obtaining information regarding the present state of transportation infrastructure and transportation parameters is critical for successful road-traffic management. A WSN, which can gather relevant information from spatially distributed sources, is one promising means of acquiring the essential information. A WSN is made up of simple, low-cost elements (nodes) that can gather the required data, pre-process it, and send it to a center through wireless transmission channels. The center can process the data and take relevant intervention action, or it can give the data to the transport system's operator or to users. Today, we see that WSNs are becoming an essential component of each ITS [12].

Traditional WSNs are energy-limited and application-specific owing to sensor node size and cost constraints [5, 13]. As a result, these two features provide serious barriers to data collection and supplying energy. The sensors in typical WSNs are primarily supplied with energy-limited batteries, limiting the longevity of sensor networks. In general, sensors are intended to function for years with a restricted-energy supply. As a result, energy consumption becomes the most important design constraint [14].

Extensive research has been undertaken in recent years to extend the lifetime of WSNs, which may be split into two types [5]. One approach is to cut energy usage in all aspects as much as possible. Significant work has been devoted to overcoming the energy-saving issue of WSNs through the development of energy-efficient communication protocols [15–18]. For example, an energy-efficient medium access control (MAC) protocol for mobile WSNs based on IR-UWB technology [19] is proposed in Ref. [18]. In the second category, sensors' stored energy should be replenished, and most recent research has focused on the fundamental challenge of energy replenishment. Wireless energy transfer (WET) is a novel technique, which can be utilized to wirelessly replenish sensor node batteries [20, 21]. Indeed, WET has been identified as a viable approach for increasing the longevity of micro-sensors, and so conventional WSNs can be upgraded to rechargeable WSNs. A number of WC solutions have been developed to offer additional energy supply for WSNs in order to extend the lifetime of WSNs. With latest developments in radio energy harvesting (EH) technology [22–28], sensor nodes can be recharged at a relatively long distance (>10 m). As long-distance charging produces significantly lower harvesting power, mobile charging methods with small distances are widely employed to attain maximum efficiency in charging. Generally speaking, sensor charging is classified into two types [5, 21]. Enabling sensors to harvest energy from their surroundings is one approach. The most significant downside of this strategy is that the EH rates are highly unstable owing to changing environments throughout time. The alternative option is to use mobile charging vehicles to drive to sensor nodes' locations and replenish them via WET. Sensor nodes may, therefore, be charged with very steady charging rates using WET technology [29]. As a result, the challenge of energy replenishment may be turned into a new one of determining an optimal charging route. Thus, current research has focused on determining the optimal or suboptimal recharging trajectory in order to enhance the efficiency of charging.

14.3 UAV-Based WRSN Charging Scheme for ITSs

The wireless rechargeable sensor network (WRSN) [30], which has a self-charging mechanism or harvests energy from environmental energies [31], will play an essential part in the future of smart cities. WRSN offers a broad range of applications, including environmental monitoring [32] and road state and traffic monitoring [33]. The use of WC with UAVs is an innovative approach for supplying power for WRSNs. This strategy allows the ongoing deployment of WRSNs to previously inaccessible sites and decreases the maintenance requirements for constant WC networks. In this context, Detweiler et al. [34] developed a UAV capable of flying to faraway sites to charge sensors utilizing magnetic resonant WPT. Caillouet et al. [35] recommended employing a dedicated charger delivered to the sensor network by a UAV, allowing energy to be transferred to the sensors through WC techniques. This allows for the ongoing deployment of WRSNs to previously inaccessible sites and minimizes the requirement for maintenance for continuous WC networks. However, since the UAV's battery capacity is restricted, it should return to the terrestrial charging station to replenish, reducing charging efficiency. Likewise, the low battery capacity of UAVs makes charging sensors distributed over large regions difficult. According to recent research, UAVs can be powered through WET and used as mobile chargers. Simic et al. [36] describes the use of WET to charge UAVs, which demonstrates that the WC power is adequate to recharge their batteries. WPT can also be used by UAVs to charge Internet of things (IoT) equipment [37]. Despite the fact that the notion of charging UAVs and UAVs charging IoT equipment has been established [38], the development of an effective UAV-specific WC strategy in WRSNs remains a difficulty. Prior research has offered efficient UAV deployment approaches for replenishing energy, increasing coverage of the chosen location, and assuring service continuity [39]. Chen et al. [40] presented a novel charging strategy in WRSNs for increased network-charging capability employing UAVs. They developed a WC pad deployment problem to apply a minimum number of pads for the UAV to establish at least one promising routing path to each sensor node in the WRSN. Due to the widespread penetration and availability of bus networks in metropolitan regions, buses provide ubiquitous charging opportunities for UAVs [7, 41, 42]. Lin et al. [7] presented a bus-network-aided UAV WC system that employs a bus network to supply the UAV's energy while minimizing the UAV's flight energy usage when charging the WRSN. Trotta et al. [41] proposed a new architecture for UAV-based energy-efficient video surveillance of points of interest (PoIs) in a city. Bus rooftops are used to recharge UAVs and carry them to the next PoI through existing bus routes. By combining an online electric vehicle system with a microwave power transfer system, Jin et al. [42] developed a novel, electric-vehicle charging system that uses a bus network. UAVs can recharge and increase the range of charging services by leveraging the bus network. A laser-charged UAV's trajectory optimization for charging WRSNs is identified by Liu et al. [30], which is concerned with optimizing a UAV's flying trajectory and the traveling plans of a mobile unmanned vehicle (MUV). The MUV carries a laser transmitter that transmits laser beams to recharge the UAV's energy. Won [43], proposed an optimization methodology for simultaneously optimizing the UAV trajectories and charging station locations.

14.4 Challenges and Advantages of Using UAVs in WRSN-Based ITSs

As previously stated, wireless rechargeable sensors play an important role in data collection in ITSs; however, charging these devices when they are spread throughout vast and hard-to-reach regions can be time consuming. A system can be constructed to facilitate charging by remotely charging sensors via radio frequency (RF) waves. However, one significant problem is that the RF source must be relatively close to the sensors to adequately charge them. A novel technology can be devised to facilitate charging by employing UAVs to transfer power through RF signals during a flyby. A UAV can also follow an optimized trajectory that increases the delivery of energy to the sensors under consideration. The utilization of WC with UAVs allows the ongoing deployment of WRSNs to previously unreachable places while also reducing the maintenance needs for continual WC networks. However, it is vital to consider the UAVs' restricted capabilities in carrying out long-duration missions owing to the energy limits imposed by battery power. Indeed, UAVs must pause their missions to recharge their batteries, hindering the charging efficiency. As a result, controlling large-scale UAV fleets must take these constraints into account. Similarly, UAVs' low battery capacity makes charging sensors spread across broad areas challenging. The limited battery energy of UAVs can frequently lead to mission failure, potentially damaging these devices [44]. Furthermore, while UAV integration with WRSNs-based ITSs can serve many different purposes, wide-scale deployment of UAV charging infrastructure is required to increase coverage, which can be very costly. Also, UAVs can be supplied with large battery capacities, although this increases the UAV's weight. Another option is to swap out the batteries once the UAV has landed. However, this increases the complexity and cost when swapping the system [9]. As a result, a consistent (no need for the UAV to land and no extra flight) and cost-aware energy procurement framework must be developed to ensure successful, affordable, and safe UAV mission execution.

The UAVs' flexibility and adaptability enable them to undertake difficult and remote tasks such as wireless sensor charging in ITS. However, path planning for UAVs is seen as one of the primary issues that must be solved in order to enhance their navigation, particularly in urban environments. When deploying UAVs in urban locations, it is vital to remember that they are fragile and face a high chance of colliding with environmental impediments. UAV routing and navigation are very critical in order to dramatically enhance UAV operation, reduce hazards, and boost efficiency [44]. Finally, harsh climate conditions provide significant hurdles for the usage of UAVs. Indeed, weather conditions such as cloud, fog, dust, snow, fog, smog, winds, and atmospheric turbulence affect the performance of any UAV-based system.

To accommodate the energy limits of UAVs in the development of the UAV-based WRSNs charging scheme, we propose a consistent and cost-aware energy procurement framework for a UAV powered concurrently by laser beams emitted by locally deployed LBDs and local RE sources. The details are provided in the subsections that follow.

14.4.1 Network Topology

We investigate a network with a UAV, randomly distributed user equipments (UEs), low-power rechargeable sensor nodes (SNs), and LBDs, as illustrated in Figure 14.1. The UAV is intended to power the sensors (at predetermined places)

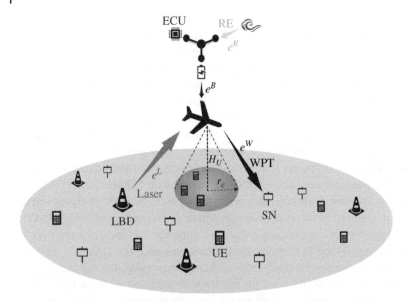

Figure 14.1 System model: a network consisting of a UAV, user equipments (UEs), low-power rechargeable sensor nodes (SNs), and locally deployed laser beam directors (LBDs).

in the WRSN-based ITS via WPT. The LBDs, which are placed on the ground and distributed geographically using a Poisson point process (PPP), Φ_L, with density, λ_L, provide power to the UAV [45]. Each LBD transmits energy with the same, constant level of power, p_L, and the UAV is powered by its closest LBD at any given moment. As seen in Figure 14.1, we assume that the UAV flies at a fixed height, H_U. It is assumed that if the UAV is placed above a predefined altitude, denoted as $H_{U,\min}$, the line-of-sight (LoS) links can be established between the UAV and the LBDs/SNs, i.e., $H_{U,\min} \leq H_U \leq H_{U,\max}$, where $H_{U,\max}$ denotes the maximum altitude the UAV can fly. We will concentrate on a typical UAV that is placed at (X_U, Y_U, H_U), while the serving LBD is positioned at $(X_L, Y_L, 0)$. As a result, $R_L = \sqrt{d_L^2 + H_U^2}$ represents the distance between the UAV and its serving LBD, where $d_L = \sqrt{\left(X_L^2 - X_U^2\right) + \left(Y_L^2 - Y_U^2\right)}$ indicates their horizontal distance. Given that a PPP is used to model the positions of the LBDs, the serving LBD's location and, by extension, d_L, are random variables. Note that a PPP, Φ_G, with density, λ_G, is also used to model the positions of the SNs.

Apart from the nearest LBD, the UAV is also powered by a local RE source, in this chapter wind energy, which can be a sustainable RE source. Note that a combination of different RE sources can be considered to balance out the variability of production of one resource type [46]. Our goal is to minimize the energy procurement cost at the UAV for an operation cycle divided into T time intervals of duration, τ. The quantity of energy obtained from an LBD to power the UAV during the time period $\kappa \in \{1, ..., T\}$ is denoted by e^L_κ (see Figure 14.1). The associated unit price at the κ-th time period is represented by ρ^L_κ. This pricing might change over the operation cycle, depending on the strategy followed by the LBDs. We assume that the available energy from an LBD is not stored in the battery of the UAV but is utilized immediately upon the UAV's energy request. Moreover, as stated earlier, the UAV has its own internal retailer, i.e., the RE generator, which generates a particular quantity of energy at each time period κ, denoted by e^R_κ, which is supposed to be cost-free. The energy obtained from the RE source will be stored in the UAV's internal battery. It is assumed that the UAV's internal battery has a maximum storage capacity, B_U. The quantity of energy obtained by the UAV from its own onboard battery over the κ-th time interval is represented by e^B_κ. As previously stated, the UAV is intended to wirelessly charge a group of rechargeable SNs by the amount of energy e^W_κ at each time period κ, with the assumption that this charging occurs at a given price of ρ^W_κ (where $\rho^L > \rho^W$). In other words, the UAV can offset its energy expenses by selling energy to the sensors via WPT.

The energy usage of the UAV at a time period, κ, is also denoted by e^U_κ, which comprises the propulsion, communication, and on-board processing energies. In fact, the energies e^B_κ and e^L_κ are used together to fulfill the UAV's energy consumption at each time period (e^U_κ), i.e., $e^B_\kappa + e^L_\kappa = e^U_\kappa, \forall \kappa \in \{1, ..., T\}$. We will see later that for minimizing the energy procurement costs while meeting the UAV's energy consumption, e^U, the energies e^L and e^W cannot take values at the same time. In other words, the UAV cannot sell energy to the sensor nodes while still harvesting energy from the nearest LBD. Finally, the energy procurement decisions are centrally managed using an energy control unit (ECU) by gathering all necessary data from the UAV and LBDs, as illustrated in Figure 14.1. More specifically, the ECU determines the quantity of energy to be acquired by the UAV from each energy source throughout the course of the whole operation cycle after obtaining the necessary information from various actors, such as the LBD energy price plan, the RE information, and the UAV energy consumption [47].

14.4.2 Energy Procurement Optimization

Based on the above explanations, the UAV's energy procurement cost at each time period, κ, is formulated as follows:

$$C_\kappa = \sum_{i=1}^{\kappa} \left(\rho^L_i e^L_i - \rho^W_i e^W_i \right), \quad \kappa \in \{1, ..., T\}. \tag{14.1}$$

The minimization of the UAV's energy procurement cost for an operation cycle broken down into T time intervals of duration, τ, is our primary goal. Indeed, C, which reflects the cost of the UAV's energy exchange with the LBDs and sensors, can be either positive or negative, based on the quantity of energy received (from the LBDs) or provided (to the sensors). The minimization of C should be conducted while keeping the following constraints in mind: (i) The ECU should comply with the UAV's energy consumption constraint at each time period, κ (Eq. 14.3). (ii) For the UAV and time period, κ, the sum of the quantities of energy procured from the internal battery (e^B) and transferred to the SNs (e^W) up to the time period, κ, should not surpass the internal retailer's produced RE up to the κ-th time period (the left inequality of Eq. (14.4), with B_L denoting the minimum allowable stored energy in the UAV's battery). (iii) Lastly, for the inner battery, the quantity of stored energy cannot be greater than the capacity B_U (the right inequality of Eq. (14.4)). In summary, the minimization problem can be represented as

$$(P) \quad \min_{\{e_\kappa^W, e_\kappa^B, e_\kappa^L\}} C_\kappa \tag{14.2}$$

$$\text{s.t.} e_\kappa^B + e_\kappa^L = e_\kappa^U, \forall \kappa \tag{14.3}$$

$$B_L \leq \sum_{i=1}^{\kappa} e_i^R - \sum_{i=1}^{\kappa} (e_i^B + e_i^W) \leq B_U, \forall \kappa \tag{14.4}$$

$$e_\kappa^B, e_\kappa^W, e_\kappa^L \geq 0, \forall \kappa \tag{14.5}$$

where the UAV's energy consumption at the time period, κ, of duration, τ, is represented by e_κ^U, and is given as

$$e_\kappa^U = \left(p_c + p_p + p_{\text{const}}\right)\tau, \tag{14.6}$$

where p_c is the UAV's communication-related power consumption when communicating with the UEs. The above-mentioned minimization problem can be addressed using linear programming algorithms [48]. More specifically, we use YALMIP [49], a toolbox for modeling and optimization in MATLAB, to solve the above optimization. Lofberg [49] illustrated how straightforward it is to model complex optimization problems using YALMIP.

14.4.3 Superiorities and Limitations of the Proposed Energy Framework for the UAVs

Concerning the advantages of the proposed UAV energy framework, it should be pointed out that UAVs should typically employ a hybrid energy supply architecture combining various sources of energy to maximize longevity and achieve higher performance. Therefore, it is essential to choose an appropriate energy source hybridization architecture with an optimal energy procurement system to ensure the efficient operation of modern UAVs. Although all the studies looking into UAV-related problems concentrated on the energy perspective of average or instantaneous power consumed/harvested when flying, hovering, processing data, or communicating, none of them considered the UAV's energy procurement costs, nor did they consider determining the optimal amount of energies to be procured by the UAV from each energy source. Regarding the proposed energy framework's shortcomings, it is worth emphasizing that, in practice, RE generation from sources such as wind and solar power is not deterministic at each time period, κ, due to its randomness and intermittent nature [50]. In the presence of uncertainty in e_κ^R, both inequalities in the second constraint (Eq. (14.4)) of the specified minimization problem become random quantities, requiring the proposed optimization to be adjusted, making it complicated and difficult to solve. Thus, the extension of the current work to model the energy supplied by the internal retailer with a stochastic process is interesting and will be considered in future work.

The UAV is powered by laser power transfer in our suggested hybrid energy supply architecture. Laser power transfer has emerged as a potential option for infinite endurance, and it is envisioned that it would power numerous energy-hungry UAV missions over large distances. Although various research works (e.g., Refs. [51, 52]) have provided information about the feasibility of laser power transfer for UAVs, some major issues are blockage, mobility, and performance in long-range missions. This approach is limited to specific regions, such as military bases, airports, and other environments where laser beams are detrimental to human health and living surroundings [9]. As a result, precautions should be taken to eliminate retinal risks. The complexity and costly nature of manufacturing the whole power transfer system is another tangible hurdle in the large-scale deployment of laser-powered UAVs. Extreme weather also poses significant obstacles for laser-powered UAVs. Weather conditions and air turbulence influence the divergence, alignment, and power of the laser beam at the UAV [9, 53]. Therefore, subsequent studies must look at climate impacts and mitigation strategies for UAV laser charging. Table 14.1 compares the proposed energy framework for the UAV with other UAV charging techniques in various applications.

14.5 Simulation Results

A service area of 0.5 km $\times$ 0.5 km including one UAV charging the SNs is considered. We utilize a use-and-store approach, which means that the energy that is obtainable from the internal retailer is utilized before storing it, i.e., $B_L = 0$. As mentioned, the MATLAB toolbox YALMIP is used to solve the minimization

Table 14.1 Proposed energy framework (hybrid charging) compared to other UAV charging methods in different scenarios.

Reference	Name	Type	Advantages	Drawbacks	Energy-cost awareness	Human intervention
[54]	RF aerially charging	Wireless	No landing, On-demand self-recharging	Aerial collision, no autonomous operation	No	No
[55]	RF static charging	Wireless	High charging feasibility, no human intervention	Extra flight	No	No
[56]	RE-based charging	EH from RE sources	No landing, No extra flight	Climate dependent, low harvested energy, extra weight	No	No
[57]	Laser charging	Wireless	No landing, No extra flight	Deployment cost, UAV motion information	No	No
[58]	Battery hot swapping	Swaps a depleted battery with a replenished one	Supporting multi-UAV charging	Round-trip energy cost, autonomous swapping issues	No	Medium
[59]	Charging station	Wired/Wireless	Supporting multi-UAV charging	Round-trip energy cost, low charging feasibility,	No	Medium
Proposed scheme	Hybrid charging	Wireless/EH from RE sources	no landing, no extra flight, consistency, and excellence in the utilization of energy	deployment cost, extra weight	Yes	No

Source: Adapted from Ref. [9].

problem in Eq. (14.2). Indeed, the energy procurement cost at the UAV is minimized for an operation cycle of one hour divided into $T = 60$ time intervals of duration one minute. Unless otherwise specified, we utilize the simulation settings as listed herein. We also focus on wind energy for our RE generation scenario. The UAV altitude is set at $H_U = 5$ m, and it flies at a constant velocity, $v = 10$ m/s. At a particular time period, the wind speed V_w is decided to be uniformly distributed in the range [0 m/s, 20 m/s]. We use the wind energy model explained in [50] to determine the average of e_κ^R, which is formulated as $\frac{1}{2}\rho A V_w^3 C_p \tau$ with air density, $\rho = 1.225$ kg/m^3, blade swept area, $A = 0.04\pi$ m^2, wind speed, V_w^3 that fluctuates with time, and conversion coefficient, $C_p = 0.9$.

In Figure 14.2a, at a given time period κ, we can see how changing the UAV's flight speed (for a fixed-wing UAV [45, 60]) affects the level of the UAV's energy consumption. The figure shows that there is an optimum flying speed at which the UAV consumes the least amount of energy, e^U. For speeds ranging from $v \simeq 10$ m/s to $v \simeq 62$ m/s, the UAV's energy consumption, e^U, is less than the average amount of energy procured from the RE source ($\mu_R = 4.7$Wh) for the given time period; hence, the UAV will be in a position to sell the excess energy to the SNs by charging them through a WPT. This does not apply if the UAV's flight speed goes above $v = 62$ m/s or falls behind $v = 10$ m/s. As can be noticed from Eq. (14.6), e^U is a function of both communication-related energy consumption (i.e., $e^C = p_c \tau$) and propulsion-related energy consumption (i.e., $e^P = p_p \tau$). We can see from the closeness of the e^U and e^P curves that the communication-related energy consumption is much lower than the propulsion energy consumption and hence can be neglected. As seen in Figure 14.2b, for a fixed UAV's flying speed, e^U and e^P will remain constant for an operation cycle of one hour. In other words, if the UAV flies at a constant speed, its energy consumption level stays almost unchanged, assuming that the UE density remains constant throughout the operation cycle. We can see that, e^R, the energy obtained from the RE source (wind energy) fluctuates over an operation cycle as wind speed varies at each time period.

As shown in Figure 14.3, the optimal solution (i.e., $[e^{W^*}, e^{B^*}, e^{L^*}]$) of the minimization of C in Eq. (14.2) has been obtained for each time period, κ, in an operation cycle of one hour. For instance, at the 40-th time period (i.e., $\kappa = 40$), the amount of energy received (at the UAV) from its internal battery is near zero (i.e., $e^{B^*} \simeq 0$) because the procured energy from the wind source has been very small (see Figure 14.2b). In this condition, the UAV should rely on the nearest LBD to receive the required energy to continue its mission; hence, e^{L^*} will be high, and obviously the UAV cannot sell its energy to charge the sensor nodes. When the quantity of energy procured from the nearest LBD (i.e., e^{L^*}) is large, the energy procurement cost is high; however, when the amount of energy procured from the LBD is small, the energy procurement cost is near zero or even negative, based on the quantity of

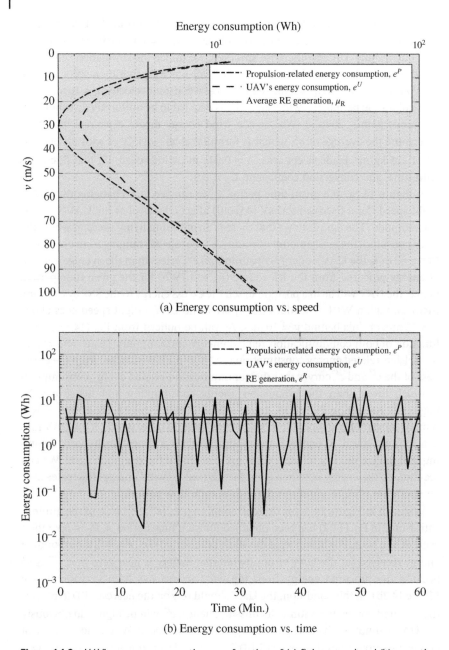

(a) Energy consumption vs. speed

(b) Energy consumption vs. time

Figure 14.2 UAV's energy consumption as a function of (a) flying speed and (b) operation cycle of one hour.

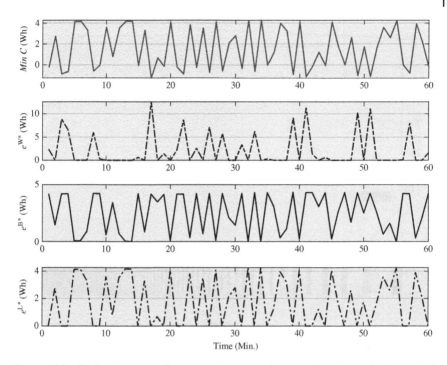

Figure 14.3 Optimal amount of energy to be procured or transferred at each time period (for an operation cycle of one hour) to minimize the energy procurement cost at the UAV.

energy the UAV transfers to the sensor nodes (i.e., e^{W^*}). Figure 14.4 depicts the optimal solution (i.e., $[e^{W^*}, e^{B^*}, e^{L^*}]$) to the minimization of C in Eq. (14.2) for various UAV flight speeds and for a specified time period, κ. For that time period, the average amount of energy procured from the RE source is considered, i.e., $\mu_R = 4.7$ Wh. For speeds ranging from $v \simeq 10$ m/s to $v \simeq 62$ m/s, the UAV will be in a position to sell its excess energy to the sensors (the unfilled bars), while procuring the least amounts of energy from its internal battery (the filled bars). This is because the UAV's energy consumption e^U is minimal for this range of flight speed (see Figure 14.2a). For speeds outside of the specified range, the UAV needs to rely on both the internal battery and the LBDs to fulfill its energy requirements. Furthermore, the higher amount the UAV acquires energy from the LBDs, the greater the cost of energy procurement.

As demonstrated by the simulation results in this section, our proposed hybrid energy supply architecture, which combines various sources of energy to overcome the UAV's battery capacity limitations, is consistent in providing energy to the UAV, allowing the UAV to be used for powering WRSNs-based ITSs without

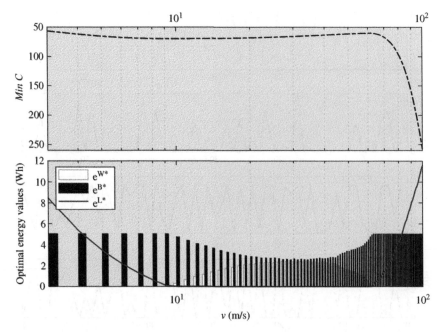

Figure 14.4 Optimal amount of energies to be procured or transferred to minimize the energy procurement cost at the UAV versus the UAV's flying speed.

affecting UAV operation or causing significant service disruptions, and without the need for human assistance. The results indicate that the UAV intends to reduce its total energy cost for a given cycle of operation by optimizing the quantities of energy acquired from its internal battery and laser beams at each time period. Furthermore, the simulations determined the amount of extra energy that can be used to power the sensors in WRSNs-based ITSs utilizing WPT technology by solving the proposed energy procurement optimization.

14.6 Conclusions

In this chapter, we considered the use of WC using UAVs to power WRSNs-based ITSs. To account for UAV battery and energy restrictions in the development of the UAV-based WRSNs charging scheme, we provided a consistent and cost-aware energy procurement framework for a UAV powered concurrently by laser beams emitted by locally deployed LBDs, local RE sources, and the UAV's internal battery. Our proposed hybrid energy supply architecture, which combines various sources of energy to overcome the UAV's battery capacity limitations, enables the UAV to be used for powering WRSNs-based ITSs without affecting UAV operation or causing significant service disruptions, and without the need for human assistance. Indeed, choosing an appropriate energy source hybridization architecture with an ideal energy procurement system is critical in ensuring the efficient functioning of the UAV. Furthermore, unlike previous works that ignored the UAV's energy procurement costs, we looked into the UAV's energy procurement costs to determine the optimal amount of energies to be acquired by the UAV from each energy source. The UAV intends to lower its overall energy cost for a given operation cycle by optimizing the quantities of energy obtained from its own internal battery and laser beams at each time period. Furthermore, by solving the proposed energy procurement optimization, we estimated the amount of extra energy that can be utilized to power the sensors in WRSNs using the WPT technology.

References

1 Panahi, F.H., Panahi, F.H., and Ohtsuki, T. (2023). An intelligent path planning mechanism for firefighting in wireless sensor and actor networks. *IEEE Internet of Things Journal* 10: 1–1.

2 Panahi, F.H. and Panahi, F.H. (2021). Cognitive and energy harvesting-based d2d communication in wireless multimedia sensor networks underlying multi-tier cellular networks. *Energy Harvesting in Wireless Sensor Networks and Internet of Things* 93. https://doi.org/10.1049/PBCE124E_ch.

3 Song, W.-Z., Huang, R., Xu, M. et al. (2010). Design and deployment of sensor network for real-time high-fidelity volcano monitoring. *IEEE Transactions on Parallel and Distributed Systems* 21 (11): 1658–1674.

4 Yick, J., Mukherjee, B., and Ghosal, D. (2008). Wireless sensor network survey. *Computer Networks* 52 (12): 2292–2330.

5 Zhang, M. and Cai, W. (2022). Data collecting and energy charging oriented mobile path design for rechargeable wireless sensor networks. *Journal of Sensors* 2022: 1–14.

6 Hinsen Chan, L.G. and Ho Lin, C. (2020). Intelligent transportation systems with wireless sensor support for law enforcement use. https://www.cse.scu.edu/m1wang/projects/WSN_transportation_13f.pdf (accessed 10 December 2022).

7 Lin, T.-L., Chang, H.-Y., and Wang, Y.-H. (2022). A novel unmanned aerial vehicle charging scheme for wireless rechargeable sensor networks in an urban bus system. *Electronics* 11 (9).

8 Unmanned aerial vehicle-based charging scheme in a bus network. https://www.azom.com/news.aspx?newsID=58993 (accessed 7 December 2022).

9 Mohsan, S.A.H., Othman, N.Q.H., Khan, M.A. et al. (2022). A comprehensive review of micro UAV charging techniques. *Micromachines* 13 (6): 977.

10 Franceschinis, M., Gioanola, L., Messere, M. et al. (2009). Wireless sensor networks for intelligent transportation systems. *VTC Spring 2009 – IEEE 69th Vehicular Technology Conference*, Barcelona, Spain (April 2009), pp. 1–5.

11 Tubaishat, M., Zhuang, P., Qi, Q., and Shang, Y. (2009). Wireless sensor networks in intelligent transportation systems. *Wireless Communications and Mobile Computing* 9 (3): 287–302.

12 Karpis, O. (2013). Wireless sensor networks in intelligent transportation systems. *International Journal of Modern Engineering Research (IJMER)* 3 (2): 611–617.

13 Basagni, C.P.S., Naderi, M.Y., and Spenza, D. (2013). Wireless sensor networks with energy harvesting. In: *Mobile Ad Hoc Networking: Cutting Edge Directions*, 703–736. Wiley.

14 Panahi, F.H., Panahi, F.H., Heshmati, S., and Ohtsuki, T. (2019). Optimal sleep & wakeup mechanism for green internet of things. *2019 27th Iranian Conference on Electrical Engineering (ICEE)*, Yazd, Iran (May 2019), pp. 1659–1663.

15 Lu, R., Lin, X., Zhu, H. et al. (2012). Becan: a bandwidth-efficient cooperative authentication scheme for filtering injected false data in wireless sensor networks. *IEEE Transactions on Parallel and Distributed Systems* 23 (1): 32–43.

16 Wang, Y.-C., Wu, F.-J., and Tseng, Y.-C. (2012). Mobility management algorithms and applications for mobile sensor networks. *Wireless Communications and Mobile Computing* 12 (1): 7–21.

17 Luo, J., Rosenberg, C., and Girard, A. (2010). Engineering wireless mesh networks: joint scheduling, routing, power control, and rate adaptation. *IEEE/ACM Transactions on Networking* 18 (5): 1387–1400.

18 Darif, A., Chaibi, H., and Rachid, S. (2018). An energy-efficient MAC protocol for mobile wireless sensor network based on IR-UWB, *Proceedings of the Ninth International Conference on Soft Computing and Pattern Recognition (SoCPaR 2017)*, Marrakech, Morocco (December 2017), pp. 1–12.

19 Panahi, F. (2012). Spectral efficient impulse radio-ultra-wideband transmission model in presence of pulse attenuation and timing jitter. *IET Communications* 6: 1544–1554.

20 Masotti, D., Shanawani, M., Murtaza, G. et al. (2021). Rf systems design for simultaneous wireless information and power transfer (swipt) in automation and transportation. *IEEE Journal of Microwaves* 1 (1): 164–175.

21 Zheng, K., Liu, X., Wang, B. et al. (2021). Throughput maximization of wireless-powered communication networks: an energy threshold approach. *IEEE Transactions on Vehicular Technology* 70 (2): 1292–1306.

22 Panahi, F.H., Hajimirzaee, P., Erfanpoor, S. et al. (2018) Smart image-processing based energy harvesting for green internet of things. *2018 Smart Grid Conference (SGC)*, Sanandaj, Iran (November 2018), pp. 1–5.

23 Panahi, F.H., Panahi, F.H., and Ohtsuki, T. (2021). Spectrum-aware energy efficiency analysis in k-tier 5g hetnets. *Electronics* 10 (7).

24 Panahi, F.H. and Panahi, F.H. (2020). *Energy Harvesting Technologies and Market Opportunities*, 1–18. Cham: Springer International Publishing.

25 Panahi, F.H., Moshirvaziri, S., Mihemmedi, Y. et al. (2018). Smart energy harvesting for internet of things. *2018 Smart Grid Conference (SGC)*, pp. 1–5.

26 Panahi, F.H., Panahi, F.H., Hattab, G. et al. (2018). Green heterogeneous networks via an intelligent sleep/wake-up mechanism and d2d communications. *IEEE Transactions on Green Communications and Networking* 2 (4): 915–931.

27 Panahi, F.H., Panahi, F.H., and Ohtsuki, T. (2023). A reinforcement learning-based fire warning and suppression system using unmanned aerial vehicles. *IEEE Transactions on Instrumentation and Measurement* 72: 1–16.

28 Panahi, F.H. and Panahi, F.H. (2022, ch. 7). *Cooperative Unmanned Aerial Vehicles for Monitoring and Maintenance of Heat and Electricity Incorporated Networks*, 157–176. Wiley.

29 Ouyang, W., Obaidat, M.S., Liu, X. et al. (2021). Importance-different charging scheduling based on matroid theory for wireless rechargeable sensor networks. *IEEE Transactions on Wireless Communications* 20 (5): 3284–3294.

30 Liu, N., Luo, C., Cao, J. et al. (2022). Trajectory optimization of laser-charged UAVs for charging wireless rechargeable sensor networks. *Sensors* 22 (23).

31 He, S., Chen, J., Jiang, F. et al. (2013). Energy provisioning in wireless rechargeable sensor networks. *IEEE Transactions on Mobile Computing* 12 (10): 1931–1942.

32 Lazarescu, M.T. (2013). Design of a wsn platform for long-term environmental monitoring for iot applications. *IEEE Journal on Emerging and Selected Topics in Circuits and Systems* 3 (1): 45–54.

33 Muñoz-Gea, J.P., Manzanares-Lopez, P., Malgosa-Sanahuja, J., and Garcia-Haro, J. (2013). Design and implementation of a p2p communication infrastructure for wsn-based vehicular traffic control applications. *Journal of Systems Architecture* 59 (10): 923–930, advanced Smart Vehicular Communication System and Applications.

34 Detweiler, C., Eiskamp, M., Griffin, B. et al. (2016). Unmanned aerial vehicle-based wireless charging of sensor. *Networks* 11: 433–464.

35 Caillouet, C., Razafindralambo, T., and Zorbas, D. (2019). Optimal placement of drones for fast sensor energy replenishment using wireless power transfer. *2019 Wireless Days (WD)*, Manchester, UK (April 2019), pp. 1–6.

36 Simic, M., Bil, C., and Vojisavljevic, V. (2015). Investigation in wireless power transmission for UAV charging. *Procedia Computer Science* 60: 1846–1855, knowledge-Based and Intelligent Information & Engineering Systems 19th Annual Conference, KES-2015, Singapore, September 2015 Proceedings.

37 Griffin, B. and Detweiler, C. (2012). Resonant wireless power transfer to ground sensors from a UAV. *2012 IEEE International Conference on Robotics and Automation*, Saint Paul, MN, USA (May 2012), pp. 2660–2665.

38 Jin, Y., Xu, J., Wu, S. et al. (2021). Bus network assisted drone scheduling for sustainable charging of wireless rechargeable sensor network. *Journal of Systems Architecture* 116: 102059.

39 Trotta, A., Felice, M.D., Montori, F. et al. (2018). Joint coverage, connectivity, and charging strategies for distributed UAV networks. *IEEE Transactions on Robotics* 34 (4): 883–900.

40 Chen, J., Yu, C.W., and Ouyang, W. (2020). Efficient wireless charging pad deployment in wireless rechargeable sensor networks. *IEEE Access* 8: 39 056–39 077.

41 Trotta, A., Andreagiovanni, F.D., Di Felice, M. et al. (2018). When UAVs ride a bus: towards energy-efficient city-scale video surveillance. *IEEE INFOCOM 2018 – IEEE Conference on Computer Communications*, Honolulu, HI, USA (April 2018), pp. 1043–1051.

42 Jin, Y., Xu, J., Wu, S. et al. (2021). Enabling the wireless charging via bus network: route scheduling for electric vehicles. *IEEE Transactions on Intelligent Transportation Systems* 22 (3): 1827–1839.

43 Won, M. (2020). Ubat: on jointly optimizing UAV trajectories and placement of battery swap stations. *2020 IEEE International Conference on Robotics and Automation (ICRA)*, Paris, France (August 2020), pp. 427–433.

44 Lucic, M.C., Bouhamed, O., Ghazzai, H. et al. (2023). Leveraging UAVs to enable dynamic and smart aerial infrastructure for its and smart cities: an overview. *Drones* 7 (2).

45 Lahmeri, M.-A., Kishk, M.A., and Alouini, M.-S. (2019). Stochastic geometry-based analysis of airborne base stations with laser-powered UAVs. https://arxiv.org/abs/1910.07794.

46 Boukoberine, M., Zhou, Z., and Benbouzid, M. (2019). A critical review on unmanned aerial vehicles power supply and energy management: solutions, strategies, and prospects. *Applied Energy* 255: 1–22.

47 F. H. Panahi and F. H. Panahi, "Reliable and energy-efficient UAV communications: a cost-aware perspective," *IEEE Transactions on Mobile Computing*. https://doi.org/10.1109/TMC.2023.3284531.

48 Boyd, S. and Vandenberghe, L. (2004). *Convex Optimization*. Cambridge University Press.

49 Lofberg, J. (2004). Yalmip: a toolbox for modeling and optimization in matlab. *2004 IEEE International Conference on Robotics and Automation (IEEE Cat. No.04CH37508)*, Taipei, Taiwan (September 2004), pp. 284–289.

50 Ben Rached, N., Ghazzai, H., Kadri, A., and Alouini, M.-S. (2017). Energy management optimization for cellular networks under renewable energy generation uncertainty. *IEEE Transactions on Green Communications and Networking* 1 (2): 158–166.

51 Chen, Q., Zhang, D., Zhu, D. et al. (2015). Design and experiment for realization of laser wireless power transmission for small unmanned aerial vehicles. In: *AOPC 2015: Advances in Laser Technology and Applications*, vol. 9671 (ed. S. Jiang, L. Wang, C. Tang, and Y. Cheng), 96710N. International Society for Optics and Photonics. SPIE.

52 Lee, S., Lim, N., Choi, W. et al. (2020). Study on battery charging converter for mppt control of laser wireless power transmission system. *Electronics* 9 (10).

53 Lahmeri, M.-A., Kishk, M.A., and Alouini, M.-S. (2022). Charging techniques for UAV-assisted data collection: is laser power beaming the answer? *IEEE Communications Magazine* 60 (5): 50–56.

54 Xu, J., Zhu, K., and Wang, R. (2019). Rf aerially charging scheduling for UAV fleet: A q-learning approach. *2019 15th International Conference on Mobile Ad-Hoc and Sensor Networks (MSN)*, Shenzhen, China (December 2019), pp. 194–199.

55 Li, M., Liu, L., Gu, Y. et al. (2022). Minimizing energy consumption in wireless rechargeable UAV networks. *IEEE Internet of Things Journal* 9 (5): 3522–3532.

56 Sekander, S., Tabassum, H., and Hossain, E. (2021). Statistical performance modeling of solar and wind-powered UAV communications. *IEEE Transactions on Mobile Computing* 20 (8): 2686–2700.

57 Zhao, M.-M., Shi, Q., and Zhao, M.-J. (2020). Efficiency maximization for UAV-enabled mobile relaying systems with laser charging. *IEEE Transactions on Wireless Communications* 19 (5): 3257–3272.

58 Ure, N.K., Chowdhary, G., Toksoz, T. et al. (2015). An automated battery management system to enable persistent missions with multiple aerial vehicles. *IEEE/ASME Transactions on Mechatronics* 20 (1): 275–286.

59 Yin, S., Li, L., and Yu, F.R. (2020). Resource allocation and basestation placement in downlink cellular networks assisted by multiple wireless powered UAVs. *IEEE Transactions on Vehicular Technology* 69 (2): 2171–2184.

60 Zeng, Y. and Zhang, R. (2017). Energy-efficient UAV communication with trajectory optimization. *IEEE Transactions on Wireless Communications* 16 (6): 3747–3760.

15

Autonomous Vehicle Systems in Intelligent Interconnected Transportation Networks

Christos Chronis, Konstantinos Tserpes, and Iraklis Varlamis

Department of Informatics and Telematics, Harokopio University of Athens, Athens, Greece

15.1 Introduction

The field of advanced driver-assistance systems (ADAS) is rapidly evolving as technology advances and the demand for autonomous vehicles increases. Intelligent Transportation Systems (ITS) have gained considerable attention because, as everyone can understand, they can revolutionize the transportation sector. ITS is a convergence of information and communication technology, sensors, controllers, and advanced algorithms aimed at enhancing transportation safety, mobility, environmental sustainability, and productivity [1].

The integration of state-of-the-art technologies such as machine learning and artificial intelligence can help autonomous vehicles to optimize their routes and improve their decision-making abilities in critical tasks. Previous research has demonstrated the benefits of using deep learning, imitation learning, and Reinforcement Learning (RL) algorithms in autonomous driving personalization [2, 3].

Despite the advancements in passive safety technology, many challenges are still open for autonomous driving, including traffic optimization, driver comfort, etc. The development of the Vehicular ad hoc Network (VANET) as part of the ITS worldwide gave rise to many potential solutions to these challenges. The main goal of VANET is to increase the safety of drivers, passengers, and pedestrians. VANET consists of two main components: Vehicle-to-Vehicle (V2V) communication and Vehicle-to-Infrastructure (V2I) communication [4].

Over the past decade, the demand for vehicles equipped with ADAS and autonomous driving capabilities has increased dramatically. AUTOMOTIVE companies have contributed to this trend by promoting their ADAS systems as

Interconnected Modern Multi-Energy Networks and Intelligent Transportation Systems: Towards a Green Economy and Sustainable Development, First Edition. Edited by Mohammadreza Daneshvar, Behnam Mohammadi-Ivatloo, Amjad Anvari-Moghaddam, and Reza Razzaghi.

"autopilots" that provide an extended sense of autonomy during driving. Despite the trend, most cars with ADAS that support autonomous driving have limited personalization options and do not add to the driver's and passengers' comfort and safety. To address these challenges, researchers and companies are working to improve the safety and performance of autonomous driving systems, as well as to personalize the driving experience, by providing various driving modes that better fit the drivers' profiles.

The main objective of this chapter is to study how autonomous vehicles, moving in intelligent interconnected transportation networks, can take advantage of novel machine learning techniques and more specifically to demonstrate how RL and Federated Learning (FL) can be efficiently combined to develop efficient and personalized solutions.

Our previous work [2] proposed the use of an RL model for adapting the driving modes of an autonomous car to the driver's personal preferences, whilst considering the road conditions and the driver's state, including stress and excitement levels. By combining this information by correctly interpreting traffic regulations, it is possible to ensure safe vehicle operation and personalize its operation, to meet the driver's preferences, taking into account the subjectivity of human factors such as stress, excitement, and psychological and physical conditions.

Until now, all the existing autonomous driving solutions lack personalization and do not take advantage of the collective intelligence that can be available from multiple vehicles. FL can be a solution that takes advantage of the collective experiences of multiple autonomous vehicles to enhance the personalization of autonomous driving models. This distributed machine learning technique allows training individual models and aggregating their updates to improve the models across the network. The resulting models learn faster and can be continually refined in real time, taking into consideration the specific driving conditions and behaviors of each driver. By leveraging the collective intelligence of multiple drivers, FL leads to greater accuracy and performance in autonomous driving, thus providing safer and more efficient driving experiences for all users.

The development of intelligent interconnected transportation networks encompasses not just information and communication technologies but also energy management. In the pursuit of sustainable mobility, the significance of energy sources and their efficient utilization cannot be overstated. In this regard, the integration of autonomous vehicle systems with modern multi-energy networks presents a tremendous opportunity. While autonomous vehicles primarily focus on optimizing their routes and decision-making capabilities, it is crucial to also consider the effective utilization of diverse energy sources, including electric power and renewable fuels, which may be accessible within transportation networks. By seamlessly integrating autonomous vehicles into these multi-energy networks, a comprehensive approach to transportation can be realized, fostering energy efficiency, reducing emissions, and advancing sustainability.

The utilization of FL presents compelling advantages in terms of energy efficiency, making it a fitting choice for harnessing the potential of Modern Multi-Energy Networks. By decentralizing data processing and distributing the workload to edge nodes, FL significantly reduces the energy consumption associated with centralized processing in the cloud. This approach optimizes resource allocation and taps into the computational capabilities of edge devices, thereby mitigating the dependence on energy-intensive cloud infrastructure. Moreover, the seamless integration of FL with Modern Multi-Energy Networks allows for dynamic energy management. By leveraging diverse energy sources, including renewable power and smart grid infrastructure, we establish a solid foundation for energy-efficient and sustainable autonomous driving systems. Embracing FL not only enhances energy efficiency but also enables us to leverage the full potential of Modern Multi-Energy Networks, propelling the evolution of intelligent and sustainable transportation forward.

Motivated to provide such experiences for all users, our contribution involves utilizing FL to improve the personalization of autonomous driving models. By leveraging the collective intelligence of multiple autonomous vehicles, this approach allows for faster learning, continual refinement, and ultimately greater accuracy and performance in autonomous driving. It also provides energy efficiency, since the distributed approach of FL leverages the processing load for centralized servers and balances between the edge and the cloud.

The contributions of this chapter can be summarized as follows:

- It performs a comprehensive survey of the various aspects related to the delivery of personalized autonomous vehicle systems in intelligent transportation networks.
- It combines RL with FL to personalize the driving experience while leveraging collective intelligence from multiple vehicles.
- It performs an experimental evaluation of the proposed solution in a simulated environment and proves its superiority.

Our work is structured as follows. In Section 15.2, we present the current advancements in the field of ITS and the usage of machine learning algorithms to strengthen the safety and reliability of ADAS and improve the driving experience for drivers and passengers. Section 15.3 focuses on the importance of RL in the personalization of driving experience, whereas Section 15.4 explains how it can be combined with FL, and how collective knowledge can be beneficial for drivers providing a more personalized and energy-efficient experience that optimally uses the multi-energy network resources. Section 15.5 illustrates the experimental setup and the results we achieved in a simulated environment.

Finally, Section 15.6 provides a comparison of the pros and cons of ADAS and autonomous driving personalization and sets the main concerns around the security and robustness of the proposed approach.

15.2 Related Work

The research work that relates to autonomous vehicular systems in intelligent transportation networks [5] spans various aspects from communications, machine learning, and the use of ADAS in an efficient manner. In this section, we illustrate the various aspects in Figure 15.1 and provide more details on each aspect in what follows.

15.2.1 Vehicle Communications

Various ITS and their applications have made use of wireless technologies such as Wi-Fi, WiMAX, 4G LTE, and 5G to meet their data transfer requirements [6, 7]. In the near future, wireless technology will be utilized to enable swift and secure communication between mobile vehicles and transportation network infrastructure. This will demand fast data acquisition, dependable connections, and stringent data protection measures [4].

VANET is a widely studied method for data delivery in vehicular communication [8]. The mesh network comprises mobile nodes that necessitate routing protocols to steer the network traffic, taking into consideration the vehicle's position and its intended destination. Great examples of VANET-based applications are traffic monitoring and incident detection [9]. The ability of VANET to alert drivers to potential dangers and allow them to adapt their driving mode can help accident prevention and increase road safety [4]. In V2V communication, vehicles exchange information such as position, speed, heading, and brake status, alerting drivers to potential dangers. In contrast to V2V, V2I communication involves the exchange of information between vehicles and the infrastructure. The infrastructure can be streetlights and traffic signals equipped with Roadside Units (RSUs) that communicate with the Onboard Units (OBUs) in vehicles to share real-time environmental conditions, such as weather and road conditions [10].

In the last few years, we have seen great advancements in the field of communications, and VANET to make way for the Internet of Vehicles (IoV), which comprise humans, vehicles, and inroad objects that seamlessly communicate to improve the driving experience [11].

By enhancing the exchange of information between vehicles, infrastructure, and people, Vehicle-to-Everything (V2X) communications equip vehicles with precise environmental information. V2X improves the driving environment by

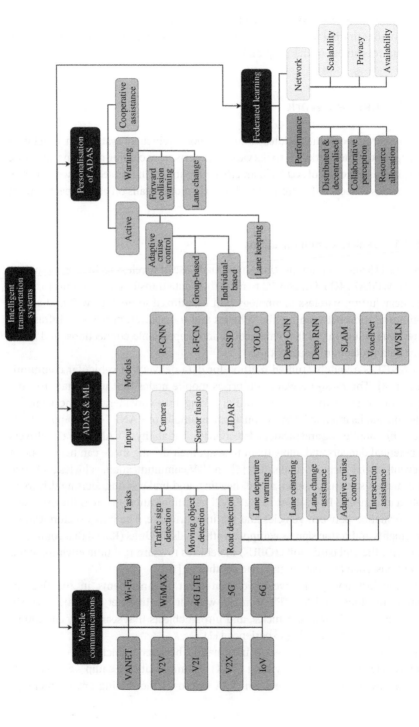

Figure 15.1 Various aspects of autonomous vehicle systems in intelligent transportation networks.

improving traffic flow, reducing pollution and accidents, and providing improved driving experiences with features such as route recommendations and automatic parking [11].

The development of Intelligent Connected Vehicles (ICV) has increased the need for improved wireless network performance in terms of bandwidth, connection stability, and transmission speed. 4G LTE technology has been found to have advantages in meeting these demands and has been applied in several vehicle networking projects with the V2V and V2I. However, the requirement for faster network speeds and greater bandwidth led to the emergence of Fifth-Generation Networks (5G), which use wireless broadband connections to transmit larger data volumes and facilitate the implementation of Vehicle-to-Everything (V2X) communications and applications in autonomous vehicles [7].

The concept behind 6G is to build upon existing 5G technology and to incorporate additional technologies such as satellite communications, AI, and big data. This new technology will enable significant advancements in vehicular network technology and is expected to bring even more advanced applications for the IoV than what is currently possible with 5G. With the integration of these technologies, 6G has the potential to revolutionize the way vehicles communicate and share information, enabling a safer and more efficient driving experience for all [12].

15.2.2 Advanced Driver-Assistance Systems and Machine Learning

A modern ADAS architecture includes four main components – longitudinal control, lateral control, driver's attentiveness monitoring system, and parking assistance [13]. According to Li et al. [14], the functions of ADAS can be categorized into three primary tasks: detecting traffic signs, detecting the road system, and detecting moving objects.

Traffic Sign Detection is an important task for ADAS systems due to its role in ensuring safety on the road. The *camera-based approaches* are the most popular ones, because of their advanced technology for analyzing visual data and processing images effectively. Machine learning-based methods, which are considered fast and simple, have been widely used to recognize traffic signs. A technique for traffic signal recognition using semi-supervised learning was introduced in a study conducted by He et al. [15]. The authors have used a limited number of labeled signs to train models based on color, edge features, and Histogram of Oriented Gradient (HOG) features. They then added high-confidence samples into the labeled dataset for training a Convolutional Neural Network (CNN) model. Arcos-García et al. [16] have compared algorithms such as R-CNN, R-FCN, SSD, and YOLO for object detection in color images. In a related study, Liang et al. [17] employed a ResNet-50 network as a basis for a pyramidal framework for small-size signs detection. Meanwhile, [18] proposed a solution to recognize traffic signals in adverse weather conditions, focusing on small-size and low-resolution images, using a multi-scale cascaded R-CNN.

Sensor fusion has been demonstrated in several studies to improve the accuracy of traffic sign detection tasks. For example, Guan et al. [19] integrated multiple sources of data, such as mobile LiDAR and digital images, to enhance the localization performance of monocular cameras. Similarly, Balado et al. [20] fused input from a LiDAR sensor, a 360° camera, and a GPS-based mobile mapping system to locate traffic signs. Another study by Hirabayashi et al. [21] proposed a method for recognizing traffic lights based on visual analysis and spatial mapping information.

The importance of **Road Detection** in ADAS is well known. Those systems can include a range of automatic assistance systems, such as lane departure warning, lane centering, lane change assistance, adaptive cruise control (ACC), intersection assistance, and more. The *camera-based approach* is the most widely used due to the extensive knowledge of computer vision and the cost-effectiveness of cameras. Among the various deep learning algorithms, CNNs have shown superior performance in image processing, and their application in traffic detection has grown rapidly [22]. In a study, a lane detection technique was proposed by Ye et al. [23] that transforms the lane structure in the image into a waveform. Another system proposed by Zou et al. [24] employs deep CCNs and deep Recurrent Neural Networks (DRNN) to decrease the occurrence of false negatives in lane detection. Li et al. [25] proposed a multi-task deep convolutional network that integrates CNN and RNN detectors for lane structure modeling and lane boundary detection, respectively. Hou et al. [26] introduced a self-train, unsupervised method based on self-attention distillation. Similarly, Xiao et al. [27] in order to detect lane markers presented a module that combines self-attention and channel attention.

To enhance the accuracy and robustness of road detection, researchers have explored the use of *sensor fusion* techniques. For example, Caltagirone et al. [28] introduced a fusion CNN model that integrates camera images and LiDAR point clouds for road detection. This approach employs cross-connections between the two sensor information processing branches in all layers, resulting in superior performance compared to other FCN systems. Another study by Chen et al. [22] proposed a progressive LiDAR adaptation-aided road detection method that aligns 3D LiDAR data with perspective view and applies a cascaded fusion structure for feature adaptation. This approach significantly enhances the reliability of detecting roads in urban environments. In addition, stereo camera systems can provide 3D information about the environment, which may assist in road detection.

The **Moving Object Detection** task refers to the detection and classification of objects in the surrounding environment of the vehicle. For this kind of task, we have three approaches based on camera, LiDAR, and sensor fusion. The *camera-based approaches* are widely used in object detection due to their direct color view capture ability. However, camera-based approaches face limitations like low-resolution capture features during dusky lighting conditions. Kim et al. [29] proposed a modified Faster R-CNN, which added random additive white Gaussian noise to improve the robustness of the system against noise and illumination. To improve the detection performance in bad weather conditions, three deep learning approaches based on YOLO were proposed by Li et al. [30]. Guan et al. [31] presented a pedestrian detection framework that utilized monocular and thermal camera data to achieve high accuracy in various lighting conditions. Moreover, Henein et al. [32] proposed a dynamic Simultaneous Localization and Mapping (SLAM) algorithm that uses an RGB-D camera to extract velocity information.

LiDAR-based approaches are suitable for detecting small objects and for obtaining distance information. Improving pedestrian selection to prevent collisions can be achieved using 3D LiDAR data and a pre-trained support vector machines (SVM) classifier, as proposed by Wang et al. [33]. Various factors are taken into account by this method, including the initial position and velocity of the host vehicle as well as the surrounding information. Real-time object detection and classification using YOLO-v2 was introduced by Ali et al. [34], which uses grid maps generated from a bird's-eye view in order to be used for feature extraction and marks 3D bounding boxes. The VoxelNet architecture was presented by Zhou and Tuzel [35], which employs a trainable deep network to unify feature extraction and 3D bounding boxes. Additionally, Yang et al. [36] developed a multi-view semantic learning network (MVSLN) based on LiDAR data that uses multiple views to preserve features and incorporates a spatial recalibration fusion module for 3D object detection in the RPN module.

Sensor fusion methods combine the advantages of different sensors to provide more accurate and real-time results. Asvadi et al. [37] propose a real-time method for detecting vehicles using multiple modes. The approach employed a fusion of a color camera and 3D LiDAR to achieve its objective. The 3D LiDAR provided dense-depth and reflectance maps, while the camera helped align the 3D LiDAR data to reduce misdetection rates. The data from these two sensors were used as inputs to a Deep ConvNet-based framework to generate bounding boxes of each vehicle in the detection results. A different approach was taken in a study conducted by Gao et al. [38], which proposed a sensor fusion detection method that combines CNN with image up-sampling theory. To overcome the limitations of RGB cameras, this method utilized the point cloud from LiDAR data to extract depth features. In a complementary object identification approach, Zhao et al. [39] fused 3D LiDAR and a vision camera to make better use of the LiDAR sensor. The object region was extracted from the 3D spatial information and used as inputs to a CNN model for feature extraction and object classification. Similarly, Chen et al. [40] developed an object detection system that fused input from a 3D LiDAR and a stereo camera to improve the accuracy of 2D bounding boxes. The utilization of depth and stereo information from the camera enabled the efficient computation of 2D bounding boxes into 3D bounding boxes.

15.2.3 Personalization

Although personalization is a concept that has attracted significant interest in various disciplines, it is a new trend in the area of automotive [41]. ADAS are a critical application area for personalization in the automotive industry [1]. The personalization of driver-assistance systems aims to improve the experience of drivers and passengers and tailor vehicle behavior to individual needs and preferences. This can be accomplished either explicitly by providing drivers with a range of predefined system settings or implicitly by inferring the driver's preferences based on their behavior, according to Fan and Poole [42]. Starting with **ACC**, a driving comfort system that maintains a steady speed as set by the driver while keeping a safe distance from the leading vehicle, personalization can play a critical role. The cruise control system can be a very helpful solution, but still reduces the driver's freedom in choosing custom speeds for varying road conditions (e.g., highway or in the city, day or night, with traffic or not). The personalization of ACC has to consider these conditions and adjust the driving speed accordingly. Two types of approaches have been proposed so far for this task: group-based and individual-based. However, all the approaches focus on deactivating ACC based on a personalized safety distance limit from the leading vehicle or obstacle.

In the research of Rosenfeld et al. [43], a clustering approach has been used to group drivers based on their driving behavior and other demographic information. Then the appropriate safety distance was determined for each of the three driving profile clusters. In the work of Canale et al. [44], three predefined profile classes have been employed and each driver was assigned to one of the driving profile classes. Depending on the class of the driver, different parameters are used by the ACC that controlled the vehicle acceleration.

In order to adapt ACC to the individual driver's style, Bifulco et al. [45] proposed a real-time speed adaptation model that was tested on a linear car following setup. The model used a recursive least squares (RLS) filter that predicted the best speed for keeping a proper safety distance. In a similar car-following scenario, the personalized ADAS of Lefevre et al. [46] combined a hidden Markov-based driver model with a Gaussian mixture regression for predicting the desired acceleration value for each individual driving style. The personalized ACC system of Chen et al. [47] used RL to adjust the vehicle speed to the strategies of different drivers while driving in dynamic traffic conditions. The model combined a PID controller for controlling the brake and throttle with different driving profiles and managed to keep safe distances in varying scenarios, while providing a smooth driving experience. Finally, a longitudinal driver-assistance system that integrates ACC that adapts to the driver preferences was prototyped by Wang et al. [48]. The authors proposed a linear driver model for simulating the use of throttle and brake based on the distance to the leading vehicle and validated their approach in real traffic tests.

The **forward collision warning (FCW)** are increasingly popular safety features in modern vehicles and are highly related to speed control in ADAS. However, false alarms can be frustrating for drivers, so personalized FCW systems have been proposed to reduce false alarm rates and increase early warning times for a longer driver reaction time. One approach is presented by Muehlfeld et al. [49] who use statistical behavior modeling to estimate a driver-specific probability distribution of danger levels for determining the activation threshold of the warning algorithm. Wang et al. [50] proposes a real-time identification algorithm that adjusts warning thresholds online and reduces false warning rates. Govindarajan et al. [51] also proposes a personalized approach for estimating the reaction time for each driver in case of emergency braking and developed an FCW system that better adapts to the driver's preferences.

Lefevre et al. [52] proposes a personalized framework that uses a driver model to predict lane departures and a model predictive controller to keep the vehicle in the lane. The system minimizes false alarm rates by detecting lane departures early. Similarly, Wang et al. [53] employ a personalized technique that involves the use of a hidden Markov model that is based on a Gaussian mixture model to detect early whether a driver will correct a lane deviation without being alerted.

According to Wang et al. [53], a personalized approach for predicting lane departure behavior was developed using a hidden Markov model based on a Gaussian mixture model. The model's aim is to predict whether a driver will correct a lane departure behavior without receiving a warning.

The topic of personalization has been explored in relation to **lane keeping** and departure prediction. To illustrate, Wang et al. [54] put forward a strategy that customizes the steering system and assists drivers to minimize the necessary wheel adjustments for following a path, thus providing a smoother and comfortable change of angle. The steering ratio is modified based on the driver's path-following tendencies. This method was evaluated in a simulation study and was observed to enhance drivers' performance in path following, lessen their cognitive and physical workload, and reduce steering angle variability. Additionally, Schnelle et al. [55] trained personalized models for path following and steering that simulate the driving style of individual drivers for various maneuver types.

Cooperative Assistance is an automation in ADAS that has gained increasing attention as an approach to enhance the safety and convenience of driving. The idea entails offering specific assistance features upon receiving direct requests from the driver, usually through speech commands. Cooperative ADAS, such as the overtaking assistant, aim to work together with drivers by responding to direct requests, usually through spoken commands, regarding adjacent cars during highway overtaking maneuvers. As drivers vary in skills and experience, it is crucial to customize cooperative ADAS to their individual competencies and capacities. In a study using a driving simulator, Schömig et al. [56] showed that most drivers preferred a speech-based on-demand assistance system, which mimics an attentive co-driver, over a visual head-up display of information. The study focused on an intersection scenario in which the driver has to turn left from a smaller to a major road. Moreover, Orth et al. [57] demonstrated that estimating acceptable gaps for each driver could enhance personalized acceptance of the on-demand assistance system. This system allows drivers to control the activation of the assistant system based on the specific situation and automatically adjusts parameters based on observed driving behaviors, showing an active and adaptable approach to personalization.

To provide drivers with personalized assistance during a **Lane change** maneuver based on their driving style, researchers have proposed different approaches. Butakov and Ioannou [58] suggested a methodology for lane change profile modeling, which was based on a parametric lane change model and a Gaussian mixture model that personalizes multiple model parameters, such as gap acceptance and longitudinal fixes that affect the resulting maneuver. This model can operate in real time and continuously update during driving. The approach of Vallon et al. [59] models the driver's behavior in lane-changing scenarios and uses SVM for mapping each driver to the best behavior class. The researchers validated their approach using offline playback and demonstrated its ability to learn individual driving behaviors for various drivers. Both methods aim to assist drivers in performing a lane change while considering safety and comfort and taking into account the surrounding traffic.

In recent years, there has been a significant amount of research devoted to driver modeling. Personalization of ADAS parameters has been a focus, with options ranging from warning thresholds to driving characteristic features for longitudinal and lateral control systems. Although current prototypes typically personalize only one ADAS, a more comprehensive personalization concept is becoming necessary. This concept should consider how personalization information can be shared between different ADAS components. A modular approach that separates ADAS, personalization modules, and human-machine interfaces is recommended for designing interactive, personalized systems. This approach allows for more sophisticated and nuanced system adaptation while still enabling drivers to have direct control and understanding of the system [60].

15.2.4 Federated Learning on ITSs

In 2017, Google introduced FL, a decentralized and collaborative learning approach that enables distributed nodes to train a global model [61]. The FL architecture involves an aggregator agent and multiple federated nodes, each of which trains the model on its locally stored data. The nodes generate model updates, which the aggregator agent combines to create a new global model that is then distributed to all nodes.

Recently, Manias and Shami [62] suggested using RSUs as federated nodes in ITSs to enhance FL applications. Given that RSUs are equipped with sensors and processing capabilities, they are well suited for FL applications. The study focuses on image processing as a key application for FL on RSUs, ranging from pedestrian detection to collision reporting. FL is applied across all RSUs to overcome data imbalance, enabling the collective use of all RSU data and the distribution of a complete model to RSUs that would not otherwise have enough data to train their own model. Furthermore, FL can exploit the differences in conditions, such as lighting, in RSU-collected images, improving object detection and model performance, thereby enhancing safety in ITSs. Overall, the paragraph underscores the crucial role of FL in ITSs and highlights the potential benefits of using RSUs as federated nodes.

Multimodal evaluation of perception technologies is critical in order to overcome the limitations of individual sensors [63]. However, the contradictory conclusions that may arise due to imprecise or inaccurate data highlight the need for collaborative computing and multi-agent **collaborative perception** in vehicular IoT systems [64]. FL can be used to efficiently combine knowledge from multiple vehicles, improving the accuracy of perception systems [65]. To enable decentralized FL in vehicular networks, vehicle clustering approaches can be employed to select central servers for FL [66]. Global model dissemination can be achieved using broadcast protocols, while model updates from clients to the central server can be carried out using unicast protocols [67].

Recent research has shown that **resource allocation** is a significant challenge in vehicular Internet of Things (IoT). The allocation of limited resources such as communication, computing, and storage resources among agents is critical for the overall system performance. However, conventional mathematical optimization approaches have limitations in solving the resource allocation problem in the vehicular environment due to the complexity and imprecise information observed in each vehicle. To address this problem, neural network techniques have been proposed to leverage big data to solve the resource allocation problem. In particular, the FL approach that uses neural networks in local training is a promising way to solve the resource allocation problem for vehicular IoT. FL allows each vehicle to use its own data to train the model and ensures privacy preservation by not sharing the raw data among agents. The sharing of the trained models allows the overall system to benefit from the diversity of data across the vehicles [68].

FL's ability to **preserve privacy** during the model training process makes it an attractive solution for privacy-sensitive applications. The collaboration of multiple agents in FL, each one using its own data for training, leads to the second main advantage of FL, **data availability**. FL is particularly useful in resource-constrained nodes with low processing capabilities, such as those present in MEC IoV systems, as it enables the utilization of global data while processing only local data. FL also exhibits communication efficiency during model training, as there is no data transfer between local nodes and the aggregation agent, and no communication between nodes. Compared to centralized collaborative learning strategies, FL's communication efficiency is a significant advantage. In a fault or failure mitigation scenario, FL demonstrates another key advantage. If a local node goes offline or loses connectivity, the FL training process can continue normally with the remaining nodes. The aggregator agent can quickly push the current global model to the node when communication is restored, and the node can resume operation and restore performance [62].

15.3 Reinforcement Learning for Autonomous Driving Personalization

The personalization of the driving experience can be achieved by adapting the autonomous system behavior to the needs and preferences of the driver and passengers. One way to achieve this is by using machine learning models to adjust the driving mode (output) dynamically to the current driver state and the overall driving conditions (input). The use of RL algorithms is a solution in this direction. The formulation of the problem as an RL task requires a clear definition of the observation space, action space, and reward function. The **observation space** (world state) includes all available sensor data from the car, such as velocity and acceleration, as well as environmental conditions such as weather. These inputs will be utilized by the model to make decisions. The **action space** involves determining the types of actions that can be taken by the car and their forms, such as the selection of a driving profile. The **reward function**, which guides the model by rewarding or punishing actions, is a challenging aspect of RL and requires careful consideration. The components of an RL problem formulation are presented in Figure 15.2.

In our approach, the goal is to create a system that utilizes the vehicle's ADAS components and personalizes the driving experience. Its main objective is to

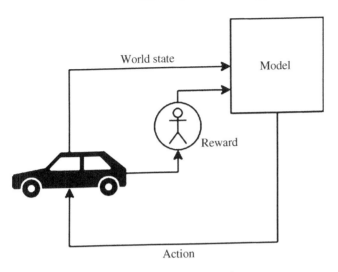

Figure 15.2 Reinforcement Learning mechanism.

minimize the stress levels of the driver. To narrow down the task to a specific implementation, we assume control of the vehicle's advanced cruise control, which has three driving modes (conservative, normal, and sport). Our implementation is based on the **Proximal Policy Optimization (PPO)** algorithm [69], which takes as input information from the world and the driver to change between different driving modes of the cruise control. However, we would like to point out that there is no one-size-fits-all solution when it comes to selecting the best RL algorithm. Each vehicle and its ADAS subsystems have unique characteristics, and there may be algorithms with better performance for specific subsystems. Nevertheless, we aim to demonstrate how RL can be applied to enhance ADAS functionality, rather than determine the best RL algorithm for autonomous driving.

PPO is a model-free, on-policy algorithm equipped with internal mechanisms to fully utilize and collected data. It works by updating its policy using samples collected from the environment and using a surrogate objective function that approximates the true objective of maximizing the expected reward. PPO is designed to be stable and efficient, and it achieves this by using two key techniques: **clipping** and **value function fitting**. The clipping technique is used to prevent the policy from changing too much between updates, which can cause instability. It does this by restricting the change in the policy within a certain range. The value function fitting technique, on the other hand, involves updating the value function separately from the policy and using it to estimate the advantage of a state-action pair. This advantage estimate is used in the surrogate objective function, which helps to improve the stability of the algorithm.

The clipped surrogate objective function that follows is the objective function that is maximized during training to improve the policy:

$$L^{\text{clip}}(\theta) = \hat{\mathbb{E}}_t\left[\min\left(r_t(\theta)\hat{A}_t, \text{clip}(r_t(\theta), 1-\varepsilon, 1+\varepsilon)\hat{A}_t\right)\right] \tag{15.1}$$

It is a clipped version of the policy gradient objective, which includes an additional term that prevents the new policy from deviating too much from the old policy. The clipped surrogate objective function is defined as the minimum of two terms. The first term is the product of the probability ratio $r_t(\theta)$ and the advantage estimate $\hat{A}_t$. The second term is the clipped probability ratio (which is constrained in a range of $1 - \varepsilon$ to $1 + \varepsilon$) multiplied by the advantage estimate.

The advantage estimate is calculated for each time step t using the following equation:

$$\hat{A}t = r_t + \gamma V(s_{t+1}) - V(s_t) \tag{15.2}$$

It is the difference between the discounted sum of rewards from time step t and the estimated value function $V(s_t)$ at time step t. The value function estimate $V(s_{t+1})$ is also included in the calculation to account for future rewards.

The value loss function that is minimized during training is as follows:

$$L^{\text{VF}}(\theta) = \hat{\mathbb{E}}t\left[(V(s_t) - V_t)^2\right] \tag{15.3}$$

It measures the difference between the estimated value function $V(s_t)$ and the empirical estimate of the value function V_t for each time step t. The hat symbol (^) represents an estimate of the expected value.

Finally, the clipped surrogate objective loss for PPO is as follows:

$$L_t(\theta) = \hat{\mathbb{E}}t\left[L_t^{\text{clip}}(\theta) - c_1 L_t^{\text{VF}}(\theta) + c_2 S[\pi_\theta](s_t)\right] \tag{15.4}$$

In these formulas, θ represents the parameters of the policy network, r_t represents the probability ratio between the new and old policies, V represents the value function estimated by the value network, s_t represents the state at time step t, V_t represents the empirical estimate of the value function at time step t, c_1 and c_2 are coefficients, the S denotes an entropy bonus, π_θ is a stochastic policy, and ε is a hyperparameter that controls the magnitude of the clipping.

The architecture of PPO is similar to the Actor-Critic family of algorithms but differs in a few critical points. PPO combines the minimization of a surrogate objective function, which maximizes the expected reward with limited change in the policy, with the clipping mechanism that limits the amount by which the new policy can deviate from the old one. All of them work together and create a robust and sample-efficient algorithm [70].

In our implementation, PPO takes as input the vehicle's velocity and angular velocity, the route's conditions at any moment defined by the expected change in direction in order to stay on route, the speed limit, the weather conditions, and the time of the day (day or night). For the evaluation of every decision of the algorithm, we used the driver's stress level as a reward function. We used a function to simulate the stress level of a driver, based on the vehicle and route conditions at any moment. Of course, in real conditions, we could use a sensor attached to the driver in order to measure the electrodermal activity and the heart rate and determine the stress levels.

15.4 Federated Reinforcement Learning

The utilization of PPO is a promising approach for addressing certain challenges in autonomous driving. However, to fully exploit its benefits, it is necessary to leverage the collective knowledge of multiple agents to overcome certain drawbacks. One such challenge is the very common problem in any personalization task, namely the "cold start" problem [71], which occurs when the system gives bad recommendations to new users with unknown preferences and takes some time to converge. The effect of this problem can be mitigated by utilizing a model from a similar driver with known preferences. Another challenge is the time required for an agent to learn a new task, such as driving in different weather conditions than those it has been trained. This can be achieved by leveraging the experience of multiple agents exposed to different conditions. Finally, it is critical to have generalized models in the vehicles that can handle previously unseen conditions, which can be achieved by utilizing collective knowledge. In the context of ADAS, the combination of PPO and FL can boost the performance of the PPO and overcome the previous challenges.

In this work, we implement a combination of PPO with a common FL technique, the Federated Averaging (FedAvg) [72]. The approach consists of multiple agents with PPO models deployed at the edge. Each edge model is trained with the available states of its corresponding agent, and after a predefined number of episodes, the agents upload the weights of their models to a cloud mechanism. The cloud mechanism then calculates the federated average of all models, generates a new global model, and sends a signal to each agent to download the updated model. Finally, the new global model replaces the current version of the model and the training continues. The total workflow is presented in Figure 15.3.

The objective of the federated approach is to leverage the knowledge learned from similar drivers in different conditions and make the new model applicable to the same drivers in unseen conditions. The choice of the PPO algorithm was

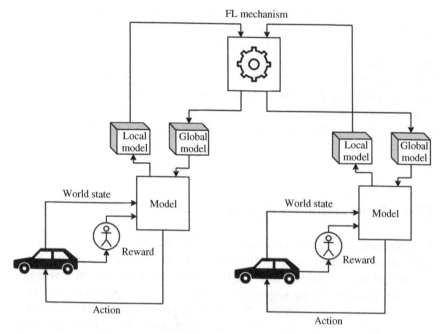

Figure 15.3 Models exchanging workflow.

motivated by its unique policy update mechanisms, which provide stability in the learning process and a stable policy update, even with the use of a nonoptimal federated technique such as FedAvg [73].

15.4.1 Advantages and Challenges from the Combined Use of Reinforcement Learning and Federated Learning

In the field of autonomous driving, the implementation of PPO and FL in combination is proving to be an effective approach to address the various challenges that arise. This approach leverages the collective knowledge of multiple agents to overcome the "cold start" problem, where an agent requires a significant amount of time and data to learn a new task. With the use of PPO and FL, agents can share data and learn from each other, thus reducing the time required for an agent to learn a new task. This, in turn, allows for the quick adaptation of autonomous driving systems to changing driving conditions and situations.

Additionally, this combination allows for the handling of previously unseen conditions, such as unexpected obstacles or road conditions, which can be difficult to anticipate or simulate during the learning process. With PPO and FL, the collective knowledge of multiple agents can be used to identify and adapt to these new conditions, thus improving the overall functionality of ADAS systems. Moreover, FL provides stability in the learning process and a stable policy update, allowing for better coordination and communication between agents, leading to more effective decision-making and smoother operation of autonomous driving systems.

Overall, the combination of PPO and FL offers a promising approach to addressing the challenges of autonomous driving and enhancing the functionality of ADAS systems. This approach leverages the collective knowledge of multiple agents, reduces the time required for an agent to learn a new task, and allows for the handling of previously unseen conditions. By using this approach, autonomous driving systems can quickly adapt to changing driving conditions, making them more reliable, efficient, and safe for all road users.

While the combination of PPO and FL offers many advantages for autonomous driving, there are also several challenges that must be addressed. One of the main challenges is the issue of privacy and security, as sharing data between agents can potentially compromise sensitive information. Additionally, the coordination and communication between multiple agents can be complex, requiring sophisticated algorithms and frameworks to ensure effective collaboration. Another challenge is the potential for overfitting, where the agents may become too specialized in a particular task, leading to a loss of flexibility and adaptability. Finally, there is also the issue of computational complexity, as the use of multiple agents and FL can require significant computational resources and time, making it difficult to implement in real-time applications. Addressing these challenges will be crucial for the successful implementation of the PPO and FL approach in autonomous driving systems.

15.4.2 The Impact of Intelligent Transportation Networks on Interconnected Multi-Energy Networks

In intelligent interconnected energy networks, various multi-energy technologies can be utilized to support the optimal scheduling of energy resources. These technologies encompass a range of energy carriers, including electricity, hydrogen, and natural gas. Each of these energy sources offers distinct advantages and can play a significant role in the efficient operation of transportation systems.

Hydrogen is a pivotal multi-energy technology that holds great promise for fuel cell electric vehicles (FCEVs). FCEVs utilize hydrogen fuel cells to convert stored hydrogen into electricity, emitting nothing more than water vapor as a byproduct. Hydrogen can be produced through various methods, including electrolysis powered by renewable energy sources. The utilization of hydrogen as an energy carrier presents exciting prospects for vehicles with long-range capabilities and rapid refueling, making it a suitable choice for specific transportation applications.

Compressed natural gas (CNG) and liquefied natural gas (LNG) are also employed in multi-energy networks. Natural gas vehicles (NGVs) emit fewer emissions than traditional gasoline or diesel vehicles. They offer a transitional solution towards cleaner energy options while utilizing the existing natural gas infrastructure. Vehicles using natural gas contribute to diversifying the fuel mix, leading to enhanced air quality and reduced greenhouse gas emissions. Finally, electricity is gaining the attention of the automotive industry with electric vehicles (EVs) and plug-in hybrid electric vehicles (PHEVs). The advancements in battery technology and charging infrastructure make electricity a practical and sustainable option for powering autonomous vehicles and allow smart electric (or hybrid) vehicles to optimally use the resource availability of the network. They can easily connect to the grid when the demand is low or when renewable energy is abundant, paving the way for cleaner and more environmentally friendly transportation options.

Intelligent transportation networks, particularly autonomous vehicles that utilize RL and FL, have the potential to influence the optimal scheduling of multi-energy networks. RL-based autonomous vehicles can learn to make informed decisions regarding energy utilization and charging strategies. By considering factors such as traffic conditions, energy availability, and user preferences, RL algorithms enable autonomous vehicles to optimize their energy consumption and charging patterns, thereby reducing the strain on interconnected energy networks. Furthermore, the integration of FL techniques in autonomous vehicles allows for collaborative learning and information sharing among vehicles within the network. Through FL, autonomous vehicles can collectively analyze real-time data related to energy availability, traffic conditions, and user behavior. This collective intelligence can inform the optimal scheduling of multi-energy networks by identifying periods of peak demand, facilitating load balancing, and promoting energy efficiency.

By leveraging RL and FL, autonomous vehicles can adapt their energy consumption patterns to the energy network conditions and user requirements, leading to improved energy utilization and reduced reliance on nonrenewable energy sources. The integration of intelligent transportation networks with multi-energy systems fosters a symbiotic relationship, where the vehicles benefit from optimized energy scheduling, and the networks benefit from reduced energy demand and enhanced sustainability.

In conclusion, the integration of different multi-energy technologies, such as electricity, hydrogen, and natural gas, in interconnected energy networks paves the way for sustainable and efficient transportation systems. Autonomous vehicles using RL and FL techniques have the potential to influence the optimal scheduling of multi-energy networks by optimizing energy consumption patterns, promoting load balancing, and fostering collaborative learning. This integration represents a significant step towards achieving intelligent transportation networks that prioritize energy efficiency and sustainability.

15.5 Experimental Evaluation of Driving Personalization Using Federated RL

In our experiments, we employed the Carla simulator,[1] a highly realistic driving simulator that has been extensively used in driving simulation experiments. In order to mimic the behavior of drivers, we utilized custom functions, drawn from our prior work [2], which are based on a combination of popular curves (sigmoid and skewed normal). These functions simulate the changes in a driver's stress levels based on various conditions, such as the acceleration, velocity, and angular velocity of the vehicle, as well as the surrounding conditions (e.g., day or night driving). By employing these functions, we generate stress profiles for two drivers, who we assume have very similar behaviors but drive in completely different conditions (i.e., day and night). The profiles differ significantly between day and night driving, adding a hard challenge for ADAS models that are not trained in both conditions. More specifically, we assume that the drivers choose the sport driving mode when they drive on a straight road, the normal driving mode when they perform a turn during the day, but at night they switch to the normal driving mode on a straight road and to the conservative profile when they perform a turn. At every step, the RL model takes an action by considering all the conditions, and this action results in a new car state (i.e., new velocities and accelerations), which in turn affects the driver's stress level. This new stress level is used as the reward for the action taken. To examine the impact of FL on training time and final performance, we conducted a series of experiments. In our first experiment, we used the same route in a Carla town and two drivers with similar profiles; we assumed that the first driver operates only during the day and the second only at night, and they implemented different training strategies (Figs. 15.4 and 15.5).

First, we trained two separate models (called **dedicated models** in the following), one for each driver. Second, we trained a single model for both drivers (called **mixed model** in the following), which is fed by episodes from both drivers (from day and night driving). We alternate the training episode types (day and night) every 10 episodes in order to train the mixed model. Finally, we develop an FL mechanism (called the **FL model** in the following) that employs two PPO models, trained separately for each driver, using the specific driving conditions of each. The model aggregation takes place every 10 episodes, and the federated model is sent back to each driver for further training.

---

1 http://carla.org.

Figure 15.4 Day route.

Figure 15.5 Night route.

We train the three alternatives for 200 episodes since after this point no further improvement in the maximum total reward for each episode is noted. Figures 15.6 and 15.7 show the performance (i.e., total reward per training episode) during the training phase in these 200 episodes. As we can see, the mixed model is the most

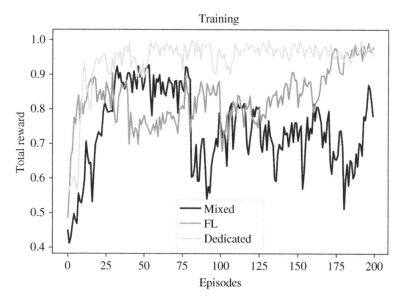

Figure 15.6 Day Task.

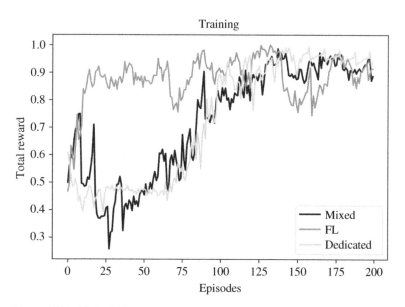

Figure 15.7 Night Task.

unstable one, since when the model is trained for a task (e.g., driving at night) the task changes (i.e., day driving samples arrive). There is still an improvement at the end of the 200 episodes on average. The dedicated models are better in the task they have been trained on, compared to the federated model, which however adapts faster to the conditions of both tasks. Especially in the night task, the federated model quickly achieves a high reward in the first few night episodes.

In the second experiment, we examine the performance of the three training strategies in new unseen routes with varying conditions. For this purpose, we generated 100 random routes of sufficient length, comprising an adequate balance between straights and turns. We tested all three approaches for both day and night driving. We also tested three strategies that do not personalize the autopilot modes but simply use the same mode (sport, normal, or conservative). The results are presented in Figures 15.8 and 15.9. In all cases, we calculate the normalized total reward for each route, assuming that the maximum reward for a route is subject to its length. We report the mean values and standard deviation for the six strategies for 100 test routes in day and night driving conditions.

As expected, the evaluation results indicate the superiority of the dedicated model for the day-driving task. The dedicated model achieved the best performance, which is comparable to that of the mixed model, while the FL model follows. All three models are significantly better than the static autopilot models that always choose a single driving mode. However, in the night task, the FL model has a better mean value, but is comparable in performance with the mixed and

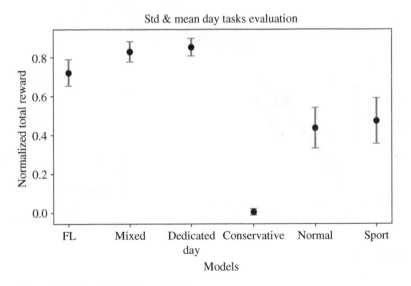

Figure 15.8 Day Task Mean and Std.

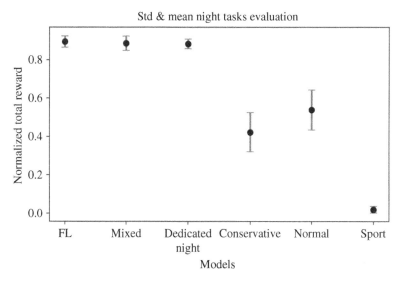

Figure 15.9 Night Task Mean and Std.

dedicated models. The static profiles performed poorly, with the conservative and sport profiles having the lowest performance in the day and night evaluation.

In order to see how the dedicated models for day and night driving are performing in any condition, we tested all models in the 200 test episodes (100 in the day and 100 in night conditions) and display the results in Figure 15.10. The

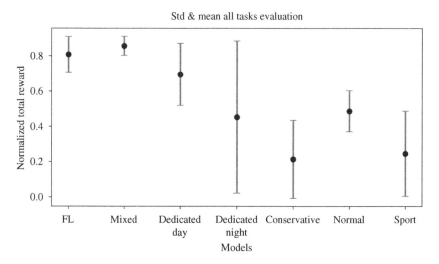

Figure 15.10 Combined Tasks Mean and Std.

performance of the static models is low, as expected, whereas the FL and Mixed models have the best performance. Of course, the mixed model assumes an increased communication overhead for sharing the episode data from each driver with the node that trains the common model. The dedicated day and night models show a medium and very unstable performance, which demonstrates their inability to fit well into unseen tasks.

15.6 Discussion

The improved performance, in combination with the better communication overhead and the privacy it offers, makes the FL technique a perfect candidate for boosting machine learning models used in ADAS. Especially when models are trained on the edge and combine the collective knowledge in global models, they allow the preservation of data privacy, which is particularly important in the context of autonomous vehicles. Another benefit of FL is the scalability and robustness it offers. The number of edge nodes (i.e., vehicles) can be unlimited, leading to a new global model that is exposed to a wider range of conditions and environments.

Despite the benefits, there are also some challenges for FL approaches in autonomous vehicle systems. They relate to the communication overhead that can be high under certain conditions (e.g., in frequent model updates), the heterogeneity of devices, and the quality check of the models shared by the edge nodes. From the communication aspect, we need a stable and reliable connection utilizing the V2I or V2X communications in order to have a continuous connection to the cloud, where an FL aggregator and model validator may reside. The heterogeneity of devices is another major challenge, as different devices may have different hardware and software configurations, which can make it difficult to ensure that the edge models learn and behave in the same way when they receive the global model. Finally, the usage of a cloud-based evaluation mechanism is important in order to ensure that the new models offer better performance in terms of safety. The absence of that mechanism can be fatal because models with bad behavior, either caused intentionally or unintentionally, can infect the global model and consequently the behavior of all edge nodes (in vehicle models).

Except for the concerns and the problems from the implementation or technological aspect, we also consider the impact of vehicle automation on various aspects of human life. This aspect was not a priority in this work, but we believe that it requires further examination. Key areas of future research must include the effects on energy consumption, emissions, safety, economic factors, and social impacts. Long-term effects, potential travel demand changes, and synergistic effects between vehicle automation, sharing, and electrification require exploration. The coexistence of fully automated vehicles and conventional vehicles in mixed traffic situations and the potential impacts of cyberattacks on traffic safety are equally critical research priorities. Additionally, the impact of vehicle automation on public health and its social implications, including the influence on vulnerable social groups and the distribution of benefits, require further examination [74].

The emergence of intelligent interconnected transportation networks presents an extraordinary opportunity for synergistic collaborations among various components of the transportation ecosystem (e.g., the edge nodes in the vehicles and the cloud nodes on the network backbone), encompassing autonomous vehicles and smart energy management systems. The fusion of data-driven intelligence, communication technologies, and advanced energy management systems unlocks a realm of possibilities. It enables dynamic optimization of energy flow, real-time monitoring of energy availability, and adaptive energy allocation tailored to the needs of autonomous vehicles. Furthermore, the integration of renewable energy sources, cutting-edge energy storage solutions, and smart grid infrastructure within these networks lays the groundwork for sustainable and robust autonomous transportation. By harnessing these synergies, autonomous vehicle energy systems can reap the benefits of refined energy management strategies, reduced dependence on fossil fuels, and heightened resilience against disruptions, thus propelling the realization of intelligent and sustainable transportation networks.

Our exploration of utilizing FL for driving personalization carries profound implications in promoting a Green Economy and advancing the sustainability of the transportation network. By harnessing the collective intelligence pooled from multiple autonomous vehicles, our approach facilitates the creation of highly precise and efficient autonomous driving models. This, in turn, cultivates optimized driving behaviors, leading to reduced energy consumption and a diminished environmental footprint. Through the continuous refinement of models in real time, accounting for specific driving conditions and behaviors, we actively contribute to a more sustainable transportation ecosystem. By embracing the power of FL, we not only elevate the personalization of autonomous driving experiences but also foster a future that is greener and more ecologically sound. Our efforts align harmoniously with global aspirations to curtail carbon emissions and embrace eco-friendly mobility solutions.

15.7 Conclusions

Our experiments demonstrate the effectiveness of incorporating FL with RL in the context of personalized ADAS. The main achievements of this work include demonstrating the effectiveness of incorporating FL with RL for personalized

ADAS, reducing training time, and improving performance. Additionally, our approach improved training stability compared to task-alternating training. The results also indicate that the usage of RL algorithms is an effective approach for personalizing ADAS compared to static profiles and can be further optimized when combined with FL techniques. In addition to our future research plans, we also intend to incorporate real driver behavior data into our modeling process to better personalize ADAS. Furthermore, we aim to evaluate the findings of our study using actual stress and excitement labels from car passengers in varying driving conditions. This will allow us to validate our results and provide more accurate and reliable recommendations for improving ADAS. Finally, our future research aims to explore alternative methods for generating global models with evaluation and fail-safe mechanisms and also plans to investigate the feasibility of incorporating FL mechanisms with ICV communication.

Acknowledgment

This work is supported by the TEACHING project funded by the EU Horizon 2020 under GA n. 871385.

References

1 Lin, Y., Wang, P., and Ma, M. (2017). Intelligent transportation system (its): concept, challenge and opportunity. *IEEE International Conference on Big Data Security on Cloud, IEEE International Conference on High Performance and Smart Computing, and IEEE International Conference on intelligent data and security*, Beijing, China (26–28 May 2017), pp. 167–172. IEEE.

2 Chronis, C., Sardianos, C., Varlamis, I. et al. (2021). A driving profile recommender system for autonomous driving using sensor data and reinforcement learning. *25th Pan-Hellenic Conference on Informatics*, Volos, Greece (26–28 November 2021), pp. 33–38.

3 Pan, Y., Cheng, C.-A., Saigol, K. et al. (2020). Imitation learning for agile autonomous driving. *International Journal of Robotics Research* 39 (2–3): 286–302.

4 Khan, A.R., Jamlos, M.F., Osman, N. et al. (2022). Dsrc technology in vehicle-to-vehicle (v2v) and vehicle-to-infrastructure (v2i) iot system for intelligent transportation system (its): a review. *Recent Trends in Mechatronics Towards Industry 4.0: Selected Articles from iM3F 2020*, Pekan, Malaysia (6 August 2020), pp. 97–106.

5 Dimitrakopoulos, G.J., Uden, L., and Varlamis, I. (2020). *The Future of Intelligent Transport Systems*. Elsevier.

6 Chyne, P., Dhilip Kumar, V., and Kandar, D. (2019). Network on wheels: leveraging vehicular communication to newer heights for intelligent transportation system. *International Journal of Electrical Engineering & Education* https://doi.org/10.1177/0020720919891071.

7 Duan, W., Gu, J., Wen, M. et al. (2020). Emerging technologies for 5g-iov networks: applications, trends and opportunities. *IEEE Network* 34 (5): 283–289.

8 Ali, I., Hassan, A., and Li, F. (2019). Authentication and privacy schemes for vehicular ad hoc networks (VANETs): a survey. *Vehicular Communications* 16: 45–61.

9 Darwish, T. and Bakar, K.A. (2015). Traffic density estimation in vehicular ad hoc networks: a review. *Ad Hoc Networks* 24: 337–351.

10 Spyrou, E., Anagnostopoulou, A., and Stylios, C. (2021). Dsrc or lte? Selecting the best medium for v2i communication using game theory. *22nd IFIP WG 5.5 Working Conference on Virtual Enterprises, PRO-VE 2021*, Saint-Étienne, France (22–24 November 2021), pp. 577–587. Springer.

11 Wang, J., Shao, Y., Ge, Y., and Yu, R. (2019). A survey of vehicle to everything (v2x) testing. *Sensors* 19 (2): 334.

12 Lv, Z., Qiao, L., and You, I. (2020). 6g-Enabled network in box for internet of connected vehicles. *IEEE Transactions on Intelligent Transportation Systems* 22 (8): 5275–5282.

13 Moujahid, A., Tantaoui, M.E., Hina, M.D. et al. (2018). Machine learning techniques in adas: a review. *2018 International Conference on Advances in Computing and Communication Engineering (ICACCE)*, Paris, France (22–23 June 2018), pp. 235–242. IEEE.

14 Li, X., Lin, K.-Y., Meng, M. et al. (2022). A survey of adas perceptions with development in China. *IEEE Transactions on Intelligent Transportation Systems* 23 (9): 14188–14203.

15 He, Z., Nan, F., Li, X. et al. (2020). Traffic sign recognition by combining global and local features based on semi-supervised classification. *IET Intelligent Transport Systems* 14 (5): 323–330.

16 Arcos-García, Á., Álvarez-García, J.A., and Soria-Morillo, L.M. (2018). Evaluation of deep neural networks for traffic sign detection systems. *Neurocomputing* 316: 332–344.

17 Liang, Z., Shao, J., Zhang, D., and Gao, L. (2020). Traffic sign detection and recognition based on pyramidal convolutional networks. *Neural Computing and Applications* 32: 6533–6543.

18 Zhang, J., Xie, Z., Sun, J. et al. (2020). A cascaded r-cnn with multiscale attention and imbalanced samples for traffic sign detection. *IEEE Access* 8: 29742–29754.

19 Guan, H., Yan, W., Yu, Y. et al. (2018). Robust traffic-sign detection and classification using mobile lidar data with digital images. *IEEE Journal of Selected Topics in Applied Earth Observations and Remote Sensing* 11 (5): 1715–1724.

20 Balado, J., González, E., Arias, P., and Castro, D. (2020). Novel approach to automatic traffic sign inventory based on mobile mapping system data and deep learning. *Remote Sensing* 12 (3): 442.

21 Hirabayashi, M., Sujiwo, A., Monrroy, A. et al. (2019). Traffic light recognition using high-definition map features. *Robotics and Autonomous Systems* 111: 62–72.

22 Chen, J., Yuan, B., and Tomizuka, M. (2019). Deep imitation learning for autonomous driving in generic urban scenarios with enhanced safety. *2019 IEEE/RSJ International Conference on Intelligent Robots and Systems (IROS)*, The Venetian Macao, Macau (3–8 November 2019), pp. 2884–2890. IEEE.

23 Ye, Y.Y., Hao, X.L., and Chen, H.J. (2018). Lane detection method based on lane structural analysis and cnns. *IET Intelligent Transport Systems* 12 (6): 513–520.

24 Zou, Q., Jiang, H., Dai, Q. et al. (2019). Robust lane detection from continuous driving scenes using deep neural networks. *IEEE Transactions on Vehicular Technology* 69 (1): 41–54.

25 Li, J., Mei, X., Prokhorov, D., and Tao, D. (2016). Deep neural network for structural prediction and lane detection in traffic scene. *IEEE Transactions on Neural Networks and Learning Systems* 28 (3): 690–703.

26 Hou, Y., Ma, Z., Liu, C., and Loy, C.C. (2019). Learning lightweight lane detection cnns by self attention distillation. *Proceedings of the IEEE/CVF international conference on computer vision*, Seoul, Korea (South) (27 October 27–2 November 2019), pp. 1013–1021.

27 Xiao, D., Yang, X., Li, J., and Islam, M. (2020). Attention deep neural network for lane marking detection. *Knowledge-Based Systems* 194: 105584.

28 Caltagirone, L., Bellone, M., Svensson, L., and Wahde, M. (2019). Lidar–camera fusion for road detection using fully convolutional neural networks. *Robotics and Autonomous Systems* 111: 125–131.

29 Kim, J.H., Batchuluun, G., and Park, K.R. (2018). Pedestrian detection based on faster r-cnn in nighttime by fusing deep convolutional features of successive images. *Expert Systems with Applications* 114: 15–33.

30 Li, G., Yang, Y., and Qu, X. (2019). Deep learning approaches on pedestrian detection in hazy weather. *IEEE Transactions on Industrial Electronics* 67 (10): 8889–8899.

31 Guan, D., Cao, Y., Yang, J. et al. (2019). Fusion of multispectral data through illumination-aware deep neural networks for pedestrian detection. *Information Fusion* 50: 148–157.

32 Henein, M., Zhang, J., Mahony, R., and Ila, V. (2020). Dynamic slam: the need for speed. *2020 IEEE International Conference on Robotics and Automation (ICRA)* (31 May–31 August 2020), pp. 2123–2129. IEEE.

33 Wang, H., Wang, B., Liu, B. et al. (2017). Pedestrian recognition and tracking using 3d lidar for autonomous vehicle. *Robotics and Autonomous Systems* 88: 71–78.

34 Ali, W., Abdelkarim, S., Zidan, M. et al. (2018). YOLO3D: end-to-end real-time 3d oriented object bounding box detection from lidar point cloud. *Proceedings of the European Conference on Computer Vision (ECCV) workshops,* Tel Aviv, Israel (23–27 October).

35 Zhou, Y. and Tuzel, O. (2018). Voxelnet: end-to-end learning for point cloud based 3d object detection. *Proceedings of the IEEE conference on Computer Vision and Pattern Recognition,* pp. 4490–4499.

36 Yang, Y., Chen, F., Wu, F. et al. (2020). Multi-view semantic learning network for point cloud based 3d object detection. *Neurocomputing* 397: 477–485.

37 Asvadi, A., Garrote, L., Premebida, C. et al. (2018). Multimodal vehicle detection: fusing 3d-lidar and color camera data. *Pattern Recognition Letters* 115: 20–29.

38 Gao, H., Cheng, B., Wang, J. et al. (2018). Object classification using cnn-based fusion of vision and lidar in autonomous vehicle environment. *IEEE Transactions on Industrial Informatics* 14 (9): 4224–4231.

39 Zhao, X., Sun, P., Xu, Z. et al. (2020). Fusion of 3d lidar and camera data for object detection in autonomous vehicle applications. *IEEE Sensors Journal* 20 (9): 4901–4913.

40 Chen, X., Kundu, K., Zhu, Y. et al. (2017). 3d object proposals using stereo imagery for accurate object class detection. *IEEE Transactions on Pattern Analysis and Machine Intelligence* 40 (5): 1259–1272.

41 Hasenjäger, M., Heckmann, M., and Wersing, H. (2019). A survey of personalization for advanced driver assistance systems. *IEEE Transactions on Intelligent Vehicles* 5 (2): 335–344.

42 Fan, H. and Poole, M.S. (2006). What is personalization? Perspectives on the design and implementation of personalization in information systems. *Journal of Organizational Computing and Electronic Commerce* 16 (3–4): 179–202.

43 Rosenfeld, A., Bareket, Z., Goldman, C.V. et al. (2015). Learning drivers' behavior to improve adaptive cruise control. *Journal of Intelligent Transportation Systems* 19 (1): 18–31.

44 Canale, M., Malan, S., and Murdocco, V. (2002). Personalization of acc stop and go task based on human driver behaviour analysis. *IFAC Proceedings Volumes* 35 (1): 357–362.

45 Bifulco, G.N., Pariota, L., Simonelli, F., and Di Pace, R. (2013). Development and testing of a fully adaptive cruise control system. *Transportation Research Part C: Emerging Technologies* 29: 156–170.

46 Lefevre, S., Carvalho, A., and Borrelli, F. (2015). A learning-based framework for velocity control in autonomous driving. *IEEE Transactions on Automation Science and Engineering* 13 (1): 32–42.

47 Chen, X., Zhai, Y., Lu, C. et al. (2017). A learning model for personalized adaptive cruise control. *2017 IEEE Intelligent Vehicles Symposium (IV)*, Los Angeles, CA (11–14 June 2017), pp. 379–384. IEEE.

48 Wang, J., Zhang, L., Zhang, D., and Li, K. (2012). An adaptive longitudinal driving assistance system based on driver characteristics. *IEEE Transactions on Intelligent Transportation Systems* 14 (1): 1–12.

49 Muehlfeld, F., Doric, I., Ertlmeier, R., and Brandmeier, T. (2013). Statistical behavior modeling for driver-adaptive precrash systems. *IEEE Transactions on Intelligent Transportation Systems* 14 (4): 1764–1772.

50 Wang, J., Yu, C., Li, S.E., and Wang, L. (2015). A forward collision warning algorithm with adaptation to driver behaviors. *IEEE Transactions on Intelligent Transportation Systems* 17 (4): 1157–1167.

51 Govindarajan, V., Driggs-Campbell, K., and Bajcsy, R. (2018). Affective driver state monitoring for personalized, adaptive adas. *21st International Conference on Intelligent Transportation Systems (ITSC)*, Maui, HI (4–7 November 2018), pp. 1017–1022. IEEE.

52 Lefevre, S., Carvalho, A., Gao, Y. et al. (2015). Driver models for personalised driving assistance. *Vehicle System Dynamics* 53 (12): 1705–1720.

53 Wang, W., Zhao, D., Han, W., and Xi, J. (2018). A learning-based approach for lane departure warning systems with a personalized driver model. *IEEE Transactions on Vehicular Technology* 67 (10): 9145–9157.

54 Wang, W., Xi, J., Liu, C., and Li, X. (2016). Human-centered feed-forward control of a vehicle steering system based on a driver's path-following characteristics. *IEEE Transactions on Intelligent Transportation Systems* 18 (6): 1440–1453.

55 Schnelle, S., Wang, J., Su, H., and Jagacinski, R. (2016). A driver steering model with personalized desired path generation. *IEEE Transactions on Systems, Man, and Cybernetics: Systems* 47 (1): 111–120.

56 Schömig, N., Heckmann, M., Wersing, H. et al. (2016). Assistance on-demand: a speech-based assistance system for urban intersections. *Adjunct Proceedings of the 8th International Conference on Automotive User Interfaces and Interactive Vehicular Applications*, Ann Arbor MI (24–26 October 2016), pp. 51–56.

57 Orth, D., Kolossa, D., Paja, M.S. et al. (2017). A maximum likelihood method for driver-specific critical-gap estimation. *IEEE Intelligent Vehicles Symposium (iv)*, Los Angeles, CA (11–14 June 2017), pp. 553–558. IEEE.

58 Butakov, V.A. and Ioannou, P. (2014). Personalized driver/vehicle lane change models for ADAS. *IEEE Transactions on Vehicular Technology* 64 (10): 4422–4431.

59 Vallon, C., Ercan, Z., Carvalho, A., and Borrelli, F. (2017). A machine learning approach for personalized autonomous lane change initiation and control. *2017 IEEE Intelligent Vehicles Symposium (IV)*, Los Angeles, CA (11–14 June 2017), pp. 1590–1595. IEEE.

60 Haseman, R.J., Wasson, J.S., and Bullock, D.M. (2010). Real-time measurement of travel time delay in work zones and evaluation metrics using bluetooth probe tracking. *Transportation Research Record* 2169 (1): 40–53.

61 Konečný, J., McMahan, H.B., Yu, F.X. et al. (2016). Federated learning: strategies for improving communication efficiency. *arXiv preprint arXiv:1610.05492.*

62 Manias, D.M. and Shami, A. (2021). Making a case for federated learning in the internet of vehicles and intelligent transportation systems. *IEEE Network* 35 (3): 88–94. https://doi.org/10.1109/MNET.011.2000552.

63 Ang, F., Chen, L., Zhao, N. et al. (2020). Robust federated learning with noisy communication. *IEEE Transactions on Communications* 68 (6): 3452–3464.

64 Zhu, G., Wang, Y., and Huang, K. (2019). Broadband analog aggregation for low-latency federated edge learning. *IEEE Transactions on Wireless Communications* 19 (1): 491–506.

65 Yang, H.H., Liu, Z., Quek, T.Q., and Poor, H.V. (2019). Scheduling policies for federated learning in wireless networks. *IEEE Transactions on Communications* 68 (1): 317–333.

66 Amiri, M.M. and Gündüz, D. (2020). Federated learning over wireless fading channels. *IEEE Transactions on Wireless Communications* 19 (5): 3546–3557.

67 Mills, J., Hu, J., and Min, G. (2019). Communication-efficient federated learning for wireless edge intelligence in IoT. *IEEE IoT Journal* 7 (7): 5986–5994.

68 Du, Z., Wu, C., Yoshinaga, T. et al. (2020). Federated learning for vehicular internet of things: recent advances and open issues. *IEEE Open Journal of the Computer Society* 1: 45–61.

69 Schulman, J., Wolski, F., Dhariwal, P. et al. (2017). Proximal policy optimization algorithms. *arXiv preprint arXiv:1707.06347.*

70 Wang, Y., He, H., and Tan, X. (2020). Truly proximal policy optimization. In: *Uncertainty in Artificial Intelligence*, 113–122. PMLR.

71 Camacho, L.A.G. and Alves-Souza, S.N. (2018). Social network data to alleviate cold-start in recommender system: a systematic review. *Information Processing & Management* 54 (4): 529–544.

72 McMahan, B., Moore, E., Ramage, D. et al. (2017). Communication-efficient learning of deep networks from decentralized data. In: *Artificial Intelligence and Statistics*, 1273–1282. PMLR.

73 Li, T., Sahu, A.K., Talwalkar, A., and Smith, V. (2020). Federated learning: challenges, methods, and future directions. *IEEE Signal Processing Magazine* 37 (3).

74 Milakis, D., Van Arem, B., and Van Wee, B. (2017). Policy and society related implications of automated driving: a review of literature and directions for future research. *Journal of Intelligent Transportation Systems* 21 (4): 324–348.

Index

Interconnected Modern Multi-Energy Networks and Intelligent Transportation Systems: Towards a Green Economy and Sustainable Development, First Edition. Edited by Mohammadreza Daneshvar, Behnam Mohammadi-Ivatloo, Amjad Anvari-Moghaddam, and Reza Razzaghi.
© 2024 The Institute of Electrical and Electronics Engineers, Inc.
Published 2024 by John Wiley & Sons, Inc.

IEEE Press Series on Power and Energy Systems

Series Editor: Ganesh Kumar Venayagamoorthy, Clemson University, Clemson, South Carolina, USA.

The mission of the IEEE Press Series on Power and Energy Systems is to publish leading-edge books that cover a broad spectrum of current and forward-looking technologies in the fast-moving area of power and energy systems including smart grid, renewable energy systems, electric vehicles and related areas. Our target audi- ence includes power and energy systems professionals from academia, industry and government who are interested in enhancing their knowledge and perspectives in their areas of interest.

Printed and bound by CPI Group (UK) Ltd, Croydon, CR0 4YY

16/04/2025

14658602-0004